CRAP

Crap

A History of Cheap Stuff in America

WENDY A. WOLOSON

THE UNIVERSITY OF CHICAGO PRESS

Chicago and London

The University of Chicago Press, Chicago 60637
The University of Chicago Press, Ltd., London
© 2020 by The University of Chicago
All rights reserved. No part of this book may be used or reproduced in any
manner whatsoever without written permission, except in the case of brief
quotations in critical articles and reviews. For more information, contact the
University of Chicago Press, 1427 E. 60th St., Chicago, IL 60637.
Published 2020
Printed in the United States of America

29 28 27 26 25 24 23 22 21   2 3 4 5

ISBN-13: 978-0-226-66435-4 (cloth)
ISBN-13: 978-0-226-66449-1 (e-book)
DOI: https://doi.org/10.7208/chicago/9780226664491.001.0001

Library of Congress Cataloging-in-Publication Data

Names: Woloson, Wendy A., 1964– author.
Title: Crap : a history of cheap stuff in America / Wendy A. Woloson.
Description: Chicago ; London : The University of Chicago Press, 2020. |
    Includes bibliographical references and index.
Identifiers: LCCN 2019054880 | ISBN 9780226664354 (cloth) |
    ISBN 9780226664491 (ebook)
Subjects: LCSH: Novelties—United States—History. | Novelties—Social
    aspects—United States. | Consumption (Economics)—Social aspects—
    United States. | Material culture—United States. | United States—
    Civilization.
Classification: LCC TS2301.N55 W65 2020 | DDC 688.7/260973—dc23
LC record available at https://lccn.loc.gov/2019054880

♾ This paper meets the requirements of ANSI/NISO Z39.48-1992
(Permanence of Paper).

To David Miller
100% quality

CONTENTS

OUR CRAP, OUR SELVES

Americans have surrounded themselves with crappy things: consumer goods that are typically low priced, poorly made, composed of inferior materials, lacking in meaningful purpose, and not meant to last. Such crap has insinuated itself into just about every aspect of daily life, filling countless kitchen "junk" drawers and clotting garages and basements across the nation. So ubiquitous, crap is nearly invisible, like white noise in material form.

Crappiness is not just a material condition but a cultural one as well: an often exuberant and wholly unapologetic expression of American excess and waste. Crap's creep into daily life might seem like a new thing, but it began centuries ago. Over time, Americans have decided—as individuals, as members of groups, and as a society—to embrace not just materialism itself but materialism with a certain shoddy complexion. Living in a world of crap was not inevitable. But for various reasons, Americans forged consuming habits that are now ingrained in the nation's very DNA. In an age of material surfeit, we continue to spend money on things we do not need, often will not use, and likely do not even want. Why? This book aims to understand Americans' ongoing and often fraught love affair with crap by telling the larger and longer story about what has motivated us to consume, and how it has shaped who we are as a nation of consumers today.

One of my favorite *Twilight Zone* episodes depicts the dynamic of crap better than almost anything else. In "One for the Angels," affable street seller Lou Bookman tries to distract Mr. Death from taking the soul of a beloved neighborhood girl by giving him the sales pitch of a lifetime. Bookman draws Mr. Death's attention to an array of goods that he brings forth from his traveling case, like a magician pulling rabbits out of a hat.

Figure 0.1. Lou Bookman exhorts Mr. Death on the "demonstration in tensile strength" of ordinary sewing thread. Still from the episode "One for the Angels," *The Twilight Zone*, originally aired October 9, 1959.

Thanks to Lou's persuasive skills, Mr. Death, at first an aloof and skeptical customer, becomes utterly entranced. The peddler's neckties are made not of polyester but rather "the most exciting invention since atomic energy," a fabric that would "even mystify the ancient Chinese silk manufacturers." His sewing thread is even more enthralling: "a demonstration of tensile strength . . . as strong as steel yet as fragile and delicate as Shantung silk . . . smuggled in by Oriental birds specially trained for ocean travel each carrying a minute quantity in a small satchel underneath their ruby throats. It takes 832 crossings," Bookman exhorts, "to supply enough thread to go around one spool." Bedazzled, Mr. Death frantically rifles his pockets for cash, shouting, "I'll take all you have!" (fig. 0.1).[1]

Like Mr. Death, Americans have approached the marketplace of goods with a combination of world-weariness and openmouthed credulity. The promise of endless supplies of new things, ever cheap and accessible, has captivated and enchanted. And the risk is low, since any one thing doesn't seem to cost all that much. Yet the result is a material world of ephemeral, disposable, and largely meaningless goods. It is a world of crap, and it has very real costs, ranging from the material to the mental, the environmental to the emotional.

The encrappification of America dates back centuries. While there were, undoubtedly, once village blacksmiths who forged brittle nails, farm women who adulterated their butter, and tailors who cut corners, these were the exceptions. Most things were made by skilled and reputable hands working with good intentions, supplying the needs of people within local communities. The consumer revolution, which began in the mid-1700s, changed all that.[2] Responding to the increasing demands of farther-flung and more anonymous and democratized markets, cabinetmakers contrived faux finishes to simulate exotic woods and intricate inlays, metalsmiths discovered how to make plated and imitation wares, and jewelers began creating glittering gemstones made of "paste" backed with foils. Even then, however, ersatz goods were still only accessible to the elite and fortunate strivers because they still had to be crafted by hand. And these items were often prized *because* they were clever simulations, allowing people to purchase more than they truly needed and to show material excess off to others.

But crappy goods—inartful and deceptive simulations, shoddily made and not meant to last—followed very quickly. Inferior things became desirable for probably as many reasons as there were people to buy them, including sheer accessibility and affordability, the desire to emulate friends and impress neighbors, and a simple thirst for novelty. Crappy goods would not have become popular, however, without the countless slick-tongued persuaders who helped sell them. These early pitchmen, as essential to the rise of the nation as yeoman farmers and independent artisans, descended from a long line of itinerant salesmen who, by the late eighteenth century if not earlier, were pulling beguiling things from their packs with showmen's flourishes. Though they've long vanished from the commercial landscape, their legacies nevertheless remain. The siren songs promising untold treasures at bargain prices call to us from the jumbled stock of dollar store shelves, the seemingly infinite listings on sites like eBay, countless infomercials, and, once, the Lou Bookmans and Willy Lomans pounding the pavement looking for their next opportunity to make a pitch.

As soon as Americans could get their hands on cheap stuff—often aided by all those roving peddlers and pitchmen—they began encrappifying their lives, tentatively at first, and then with gusto. Not long after the American Revolution domestic markets became inundated with goods from overseas. Great Britain dumped the majority of these items on America's shores, and many of them were inferior in some way: remainders; damaged goods and knock-offs; the unfashionable and outmoded; things dyed with fast-fading "fugitive" colors and constructed with less durable materials; myriad items that had little purpose and likely would not last.

None of that mattered. After domestic boycotts and periodic embargos, Americans of all sorts — rich and middling, urban and rural — not only had access to new markets but came to see themselves as consumers as much as producers. By the early decades of the nineteenth century, retailers advertising bargain wares and shops specializing in cheap variety goods began appearing in large cities and small towns alike. There were profits to be made in selling cheap goods. Precursors to our dollar stores, they offered the beguiling combination of great variety and low prices. Aiding these retailers were the countless itinerant peddlers who introduced their Yankee notions — and the cosmopolitan ideas they embodied — to the hinterlands. For the first time, American consumers began to value cheaper, ephemeral objects over more durable ones, enamored by the low cost and pulsing abundance of these new goods and the material and emotional satisfactions they seemed to promise. Americans quickly came to enjoy the cyclical churning of cheap possessions, avoiding long-term commitments to fewer, better-quality, and more expensive things. America's unapologetically disposable culture has its roots in this era and with these goods.

There is something to be said for the embrace of cheap things over time. Such material access has enabled American consumers to fully participate in the marketplace — not simply the world of goods but the ideas and possibilities they represent. Too, the taste for cheap goods boosted the output of manufacturers, thereby helping to raise the general standard of living. Producers were able to employ more workers to make their wares, wholesalers expanded networks to distribute them, retailers could hire more clerks to sell them, and so on. Facilitating access to cheap goods also helped spur the government to invest in infrastructure. Networks of turnpikes, canals, and railroad lines not only connected people to once-distant markets but also made possible new ways of doing business, like the mail order enterprises of Sears, Roebuck and Montgomery Ward. On a more personal level, the vast majority of American consumers could embrace novelty for its own sake and for the pleasures it provided, since they no longer had to make do with just a few things that would have to last a lifetime. This lessened the burdens of ownership itself: now easily discarded and just as easily replaced, possessions no longer had to be painstakingly cared for. The marketplace of crap turned a broken kettle or cracked dish from a crisis into a mere inconvenience effortlessly—and pleasurably—ameliorated by a new purchase.

Cheap goods made people's lives easier in other ways, too. The number of gadgets—from combination corn grinders to miracle fire extinguishers—began increasing in the 1840s and exponentially so after the Civil War, supplementing reliable and familiar tools. Gadgets embodied the seemingly limitless creativity of American ingenuity and drive toward greater efficiency. "New-fangled" devices offered faster, easier, more enjoyable processes for doing everything from washing clothes to peeling apples. But that wasn't all. Gadgets came to seem like personal servants, promising to turn the burdens of work into entertaining leisure activities. People could now, all by themselves, make magic happen, whether instantaneously rejuvenating their skin or transforming ordinary potatoes into perfectly julienned strips with the simple turn of a crank. At relatively low cost, gadgets—from yesterday's all-in-one tools to today's miracle garden hoers—have delivered outsized wonders and spectacles matched only by their extreme functionality.[3]

More alluring still is the crap that isn't just cheap but free. Since the first decades of the nineteenth century, long before Cracker Jack tokens and cereal box prizes, merchandisers were rewarding loyal customers with giveaways and prizes. Even the most pedestrian of things—fly swatters, calendars, ballpoint pens—have helped kindle warm feelings between sellers and buyers, creating loyalty. While today it manifests in items such as t-shirts and tote bags, nineteenth-century commercial goodwill came in the form of things like calendars, embossed rulers, and cheap jewelry. All of it was crap, but it was free crap, which was all that mattered.

Crappy stuff has also enlivened American homes. Early itinerant peddlers selling cheap plaster figurines helped democratize the trade in bric-a-brac, knickknacks, and tchotchkes. Ornamental wares could now be enjoyed by rich and poor alike. Although they lived in "filthy, damp and dismal conditions," nineteenth-century tenement dwellers, for instance, nevertheless were able to "crowd" their mantelpieces with cheap figurines.[4] However crappy, such knickknacks did not simply adorn people's homes but offered them brief mental respite from their straitened circumstances.[5] Sometimes, too, cheap imitations were in some ways superior: artificial plants and fruits, whether plastic or plaster, and even if "laughably clumsy, and daubed over with green and yellow paint," could be more vibrant than the real thing and lasted forever, defying decay and death.[6]

The growing trade in "giftware"—upscale tchotchkes—enabled Americans to expand their decorative horizons even more broadly and boldly. Sold in specialty boutiques, these affordable items—blown-glass art vases, carved wood figurines, hand-dipped candles—allowed their owners to make more nuanced statements about themselves, their tastes, and even their politics. Gift shops began appearing in America at the dawn of the twentieth century, serving the rising number of leisure travelers who took cross-country tours in their new automobiles. These independent shops, often owned by women, offered customers seemingly unique merchandise—Irish linen tea towels, ashtrays crafted in India, hand-painted woodenware napkin holders. Over time, the number of gift shops expanded, enabling ever more consumers to purchase special things that seemed to be imbued with their own personalities, life histories, and individual marks of artistry. But because it has always been mass-produced, giftware can only be derivative and never unique. Its appropriated stylistic glosses, often described as "looks," such as Colonial Revival, rustic, and contemporary, can only embody a faux authenticity.

Another way that Americans have been able to keenly demonstrate their connoisseurship, taste, and status has been through mass-produced and -marketed collectibles. Produced specifically to be collected, "intentional" collectibles first appeared in the late nineteenth century, when cutlery companies began making souvenir spoons. But the market really took off in the mid-1950s, when ceramics manufacturers began aggressively marketing commemorative plates. In due course the world of collectibles expanded to include figurines, historical replicas, dolls, and other items that purported to be investment opportunities for increasingly prosperous Americans.

The manufacturers of these myriad objects democratized collecting, which had been a fairly exclusive activity. People afflicted with the collecting bug but of limited means had had few choices: some collected stamps; others, matchbooks and luggage stickers. Serious collecting of serious things—the high-rolling world of antiques auctions, the fine art market, and museum patronage—was a practice both economically and socially out of bounds to all but the very elite. Intentional collectibles, however, enabled ordinary people to enjoy the thrill of the hunt, the satisfactions of acquisition and curation, the pride of display, the company and camaraderie of like-minded people, and (nominally) the economic benefits of investing. By the 1960s and 1970s, clubs, magazines, and even special market exchanges were serving collectors of everything from Hummel figurines and scale-model replicas of military machines to commemorative coins and limited-edition dolls.

There is no denying that crap has brought different forms of pleasure to people over time. This is probably no truer than when considering novelty goods like Joy Buzzers, Whoopie Cushions, and plastic vomit. These things, too, have long histories: mass merchandisers were selling things like exploding cans of snakes, trick spiders, fake mustaches, Resurrection Plants, and surprise boxes as early as the 1860s. Americans had never before seen many of these queer and curious things, let alone known what to do with them. But no matter. They seemed to offer untold delights, opportunities, and even mysteries, especially among children and childish-at-heart pranksters.

The nascent novelties market continued to expand, thanks in part to technological innovations that made new things pop and whizz and explode more reliably and in part to the expansion of advertising. Pulp magazines, mail order catalogs, and even bubblegum wrappers had become, in the early twentieth century, prime ad space for x-ray spectacles, fake dog poo, and Chinese finger traps. Although the golden age of novelty goods is now long past, for over a century they enabled even young consumers to explore taboo subjects like sex and death by disguising them in frivolous and playful forms.

*   *   *

America would not be its gloriously, obnoxiously materialistic self today without all of this crap, which has made the "goods life" accessible to just about everyone. But that has come at a cost. There is a reason why cheap things are often referred to in the most pejorative of terms, as a synonym for bodily waste. Crappy things are, in various ways, excrescences—quickly used up and happily, even proudly, disposed of. They can be cynical and insincere things that lack integrity. What is more, crap is often sold via deception, on the thin air of so many promises made by slick salesmen and marketers like the silver-tongued Lou Bookmans of the world. Often, however, we are not at all deceived. We buy cheap stuff knowing full well how crappy it is.

Crap is everywhere and transcends clear boundaries of category and genre. Although a lot of crap tends not to cost much, which is one of its many attractions, low price is not a prerequisite. Rich people have crap, too—it's just more expensive. Although lots of crap is shoddily made of inferior materials not meant to last, it is not necessarily defined by poor quality, either. Many gadgets, for instance, use the best materials and employ state-of-the-art technologies, but their utility might be described, charitably, as quite limited.

Crap is not a particular type of object but an existential state of being: a *quality* of a thing rather than the thing itself. What constitutes crap is highly personal and historically contextual. My crap might not be your crap. What to me is an unnecessary gadget might be to you an essential tool. My collection of commemorative plates, while seemingly priceless, might be impossible to sell. A promotional tape measure given away cynically by a faceless company might become, when found in Grandma's sewing basket, a cherished heirloom. Likewise, objects can move in and out of states of crappiness.[7] Labor-saving devices lose their use value when they take more effort to operate and actually create more work. The decorative and monetary value of collectible spoons ebbs and flows according to the tides of taste and the market. The shock value of novelty goods disappears as soon as they are used. An object's relative crappiness lies in the extent to which it offers false hope, was produced to hasten its own obsolescence, has no clear purpose, and/or has no emotional, utilitarian, or market value. We often do not need, have little use for, and might not even really want a lot of crap. Further, crappy things are not equally crappy; their relative crappiness is determined by just how paradoxical, contradictory, insincere, unnecessary, and fundamentally false they are.

Also, crap is not kitsch. Some kitsch is crap, and vice versa, but the two are fundamentally different. Kitsch objects are defined primarily by their aesthetic properties, by their "over-muchness" and embrace of "more is more." Kitschy things do make several cameo appearances in the pages that follow, including donkey-shaped punchbowls and coffee mugs that look like women's breasts, but they are not the central focus. Kitsch does play a more central role when it comes to forms of crap appreciated for their aesthetics, like some intentional collectibles and giftware items. The market for these objects relies in large part on the enthusiasms of "kitsch-men [and -women]," who value easily graspable art because, in the words of the Italian art critic Gillo Dorfles, "they believe art should only produce pleasant, sugary feelings" rather than be "a serious matter, a tiring exercise, an involved and critical activity."[8] This fairly well characterizes the collectors of "art plates," historical replicas, and even things like Precious Moments figurines. Such collectors enjoy the hunt and challenge of amassing an orderly series of objects among a finite universe of them. But the makers of intentional collectibles also speak directly to those for whom artistic things are a decorative kind of "condiment" or aesthetic "background music."

It is important to understand what drives us to possess crap, whether it comes in the form of maudlin ceramic figurines, boob-shaped coffee mugs, patties of plastic vomit, or all-purpose gadgets. Finely made objects are self-explanatory; crap very much needs to explain itself. That is why this book is both a history of crap and a history of consumer psychology. Why did Americans welcome crappy goods into their lives in the first place, and why have we allowed or encouraged them to stay, especially given that crap's cynicism and inferiority is not only not concealed but often a vital selling point? Crap whispers promises and shouts false hopes. It encourages and enables consumers to value abundance for the sake of abundance, excess for the sake of excess, stuff for the sake of stuff. The act of buying crap is often unapologetically irrational, wasteful, excessive. There is nothing more American than crap.

Crap confounds: Low-priced goods aren't necessarily bargains. *Crap beguiles*: Gadgets don't ease our burdens or miraculously cure what ails us. They often create more problems than they purport to solve. *Crap seduces*: Free stuff isn't free, and the people who give it to us want not our friendship but our money. *Crap feigns*: Pieces purchased in gift shops are not unique works of hand-wrought art but are typically made in factories just like any other commodity. Their romantic backstories are fictions, too. *Crap dissembles*: Those mass-produced collectibles are not unique or exclusive. Nor will they appreciate over time. *Crap tricks*: Novelty goods do not help us participate in light-hearted play but implicate us in violence, turning otherwise dispassionate witnesses into perpetrators (those in on the joke) and victims (those who aren't).

Of course, the objects themselves do none of these things, but they seem to: their many seductions, ventriloquized through snappy sales pitches and sexy marketing campaigns, speak to us in the persuasive language of bullshit and humbuggery that erases the presence of the marketeers and seems to emanate from the objects themselves. These various appeals are integral to understanding not only what motivates us to purchase things that are not what they seem—*or are very much what they seem*—but also why we do it time and again.

Thanks to able persuaders, we have come to imbue objects of all sorts, crappy ones in particular, with certain qualities and properties that extend well beyond the things themselves: the "commodity fetishism" described by Karl Marx. An important concept that animates this book as it does the objects themselves, it can help us better understand our crappy goods and our relationship to them. Objects don't really have their own life force, but we often act as if they do, thanks in large part to the effectiveness of their tricked-out advertising and marketing. These promotional gambits, whether the slick pitches of door-to-door salesmen of the past or today's telegenic infomercial hosts, endow objects with mystical properties divorced from the actual circumstances of their production and consumption. When people purchase something at, say, a big-box retail store, they don't usually consider how or where it was made, by whom and under what conditions, or even how it came to arrive on the store shelf. It simply exists as if by magic, seemingly animated by its own life force. Marx thought this curious and remarkable, writing,

> A commodity appears, at first sight, a very trivial thing, and easily understood. Its analysis shows that it is, in reality, a very queer thing, abounding in metaphysical subtleties and theological niceties. . . . So soon as it steps forth as a commodity, it is changed into something transcendent.[9]

Commodity fetishism not only obscures the means of production; it underlies the myriad forms of persuasion that insinuate that magical and transcendent properties inhere in things themselves. Lou Bookman offered Mr. Death neckties made not of ordinary polyester but of the latest space-age fabrics, equally useful and miraculous. The most successful persuasive campaigns are those that get consumers themselves to impute such characteristics to objects, to literally buy into what the crap purveyors are selling. As one collector of ceramic Hummel figurines remarked, "These figurines will collect you."[10]

* * *

Why does crap matter? For one thing, it is pervasive. At once abundant and impoverished, crap deserves to be taken as seriously as fine art and antiques. It is resource intensive, consuming not only the time and effort that go into its conceptualization and manufacture but all the money people spend on it, to say nothing of its environmental impact, from the moment liquefied petrochemicals are poured into product casting molds to when the results are tossed into landfills. Crap also provides a unique insight into the workings of consumer psychology over time. Americans increased their consumption of goods as those goods came flooding into the marketplace, but *why*? Just because there was more stuff didn't mean we had to buy it, but we did, and have continued to ever since. Getting at consumer motivation over time reveals not just how Americans have shopped but, more importantly, how we have thought. Understanding how advertising and marketing strategies have been deployed to sell crap over time reveals much about Americans' emotional selves—not just our deep-seated desires, needs, anxieties, and passions but how these individual impulses and foibles have created a certain kind of national character. Although it seems contradictory, crap, as a rich archive of everyday artifacts and mass consumption, is profoundly revealing. It offers in material form an intellectual history of ordinary people.

What is more, consumers not only believe in various forms of persuasion but also think of possessions as integral to their identities and fundamental parts of who they are.[11] When I have explained to friends and acquaintances that I have been writing about the history of cheap goods in America, they are intrigued. When I tell them I am referring to these things as crap, some are taken aback, and many are offended. They think I am talking about their stuff and therefore impugning them. Our crap has become a fundamental part of who we are, even though we played no role in its conception or production. We simply chose it, paid for it, and brought it into our homes. Yet if our stuff is maligned, by extension we are, too.

This book is intended not to offend but to explain. Because we all have crap in our lives, we should better understand the role it has played over time in shaping us as individuals and as Americans. We should appreciate the extent to which our commodities fail to live up to our expectations. *That* is truly the measure of crap and its crappiness—the degree to which things fail or fall short: the greater the amplitude of the lie, the crappier something is. But it is more complicated than just that, since we are not always fooled by crap. It is one thing to be the dupe of slick marketing campaigns. It is another to buy shoddy things with eyes wide open, simply for the hell of it. We often do not care.

For various reasons, Americans have been buying crap for a long time. And while the term "crap" is of fairly modern origin, shoddy goods are not.[12] As soon as inferior products appeared on the market, in fact, Americans were calling them out for what they were, using their own suggestive terms: thingums, jimcracks, good-for-nothings, cheap jack, trifles, whatnots, gewgaws, and, later, slum. We might think of crap as bullshit in material form.[13]

Bullshit is as expansive and liberating as it is false. The same is true for crap. It should be celebrated for its democratizing power, bringing modern material abundance to people of all classes. But it should also be measured with a jaded eye, one that can calculate the true costs of these seemingly cheap goods more precisely. Two accounts from the early twentieth century illustrate Americans' long, conflicted relationship with crap. In 1911, Theodore Dreiser praised the stock of the five-and-dime store because it enabled just about everyone to buy what they wanted. Dreiser described the "stock of overproduction" at consumers' fingertips as a "truly beautiful, artistic, *humanitarian* thing."[14] Yet others weren't so sure, seeing such abundance as cynical and nihilistic. Writing a decade after Dreiser, economists Stuart Chase and Frederick Schlink decried the "waste of the consumer's dollar." "We are deluged," they observed,

> with things which we do not wear, which we lose, which go out of style, which make unwelcome presents for our friends, which disappear anyhow—fountain pens, cigar lighters, cheap jewelry, patent pencils, mouth washes, key rings, mah jong sets, automobile accessories—endless jiggers and doodads and contrivances.[15]

In other words, so much unnecessary crap. Were all these "endless jiggers and doodads and contrivances" simply a waste of money? Or were they conduits through which Americans could envision better lives, filled as they might be with objects that by turns promised convenience, conferred status, offered novelty, brought value, or evoked the exotic?

For better and worse, crap epitomizes who Americans are as individuals and as a society. Objects are culture in material form; they speak using the language of signs, and they convey meanings. For most Americans—who can afford or are interested in nothing better—it is crap, rather than more venerated objects, that comprises our world. Americans' embrace of crappy stuff is a celebration, acceptance, and internalization of not just crap itself but the ideas it represents. It reflects the social and economic conditions within which we have lived and continue to live, as well as the decisions we have made and continue to make about how to spend our money, which is a direct expression of material and psychic choices and preferences. Crap is, therefore, deeply revealing, laying bare in ways that nicer things cannot some of our most profound desires, drives, and anxieties: our crap, our selves.

PART 1

A Nation of Cheap Jacks

1

FROM THE CHEAPENING MANIA TO UNIVERSAL CHEAPNESS

The market in crappy goods would not have gotten its start without the efforts of countless peddlers who, beginning in the mid-eighteenth century, delivered the first low-priced petty wares to American consumers.[1] Peddlers not only played an essential role in making cheap goods more accessible to more people but were, in fact, responsible for sparking the consumer revolution itself. As much as the actual goods, peddlers promoted the *idea* of material abundance coupled with cheapness. People did not purchase peddlers' wares simply because they were available but because they offered what their customers did not already possess, namely, novelty, variety, and accessibility. The relatively low cost of "Yankee notions" enticed even the most cash-strapped buyers. At the same time, a peddler's disparate goods tantalized, speaking to potential customers on emotional and, arguably, irrational levels as well. It was this particular admixture of low price and variety that drove the early market in crap.

Variety was, of course, a practical expedient that increased a peddler's chances of having something a potential customer might desire. He arranged his stock as economically as possible—ribbons tucked in boxes of pepper, spools of thread in coffeepots, yards of cloth "stowed away beneath tin-cups and iron-spoons"—and set about tramping the remote back roads with a pack or cart, dealing in items that were mobile and fairly inexpensive. Having different kinds of goods to offer with different exchange values meant the shrewd peddler could readily sell his merchandise for cash, or barter it for the tallow, scrap metal, and cloth rags that a household had on hand (by-products he could easily resell or trade again). This was especially important at a time when uniform paper currency did not yet exist and specie (hard money) was in chronically short supply. What was more, by assembling a diverse stock, the peddler created consumer desires, which he could then conveniently satisfy. Before "admiring eyes," he presented new razors, silk handkerchiefs, sausage stuffers, "fancy" neckcloths, "warranted pure steel" knives and forks, and many other "things till then unknown."[2]

The Allure of Heterogeny

By offering a variety of goods, peddlers created and fulfilled longings that were both material and psychological. They promoted the transformative potential of consumption itself, gradually convincing consumers that the market could answer their many wants and needs. Never mind their true worth, variety goods—new and often strange and immediately available—conjured worlds of wonder, intrigue, and seemingly endless surprise: one new thing begat another. Sewing scissors, mother-of-pearl buttons, and pieces of painted tinware, humble as they might be, were also the stuff of fantasy. And by making them affordable—cheap, even—the peddler was able to turn those fantasies into realities. More than skilled artisans and owners of specialty shops, who catered to the elite, it was the peddler's cheap and easy merchandise that democratized consumption. Rather than expensive things, it was crap that enabled Americans of all stations to buy into the world of goods (fig. 1.1).

"Heterogeny," a neologism Nathaniel Hawthorne coined in 1868, captures the spirit and power of the variety in a peddler's pack. The endless "promise" offered by the juxtaposition of random goods alone was enthralling. The peddler's performance only heightened the effect: "I could have stood and listened to him all day long," noted Hawthorne of a man who brought out "bunches of lead pencils, steel-pens, pound-cakes of shaving-soap, gilt finger-rings, bracelets, clasps, and other jewelry, cards of pearl buttons, or steel . . . bundles of wooden combs, boxes of matches, suspenders, and, in short, everything,—dipping his hand down into his wares with the promise of a wonderful lot, and producing, perhaps, a bottle of opodeldoc, and joining it with a lead pencil."[3] Variety played an essential role in attracting the first generation of American consumers to the world of crap.

Peddlers themselves recognized the emotional appeal of their varietal assortments, especially to middling households.[4] Their merchandise was more than just the ill-soldered tinware, lengths of gaudy gold ribbon, and second-rate pieces of porcelain that it was materially. These objects also

Figure 1.1. The peddler presenting a tantalizing chest of trinkets and Yankee notions. F. O. C. Darley, "The Peddler," from John L. McConnel, *Western Characters*, 1853. Library Company of Philadelphia.

provided entree into the larger marketplace—of things *and* ideas. Rather than mere deliverers of new items to remote places, peddlers were important "cultural agents promoting the message of social transformation through the purchase of goods."[5] This is why, over time, Americans' consuming appetites were not satisfied by all of this cheap stuff but instead grew only more voracious. Cheap became an end in itself.

Peddlers made cheap goods come alive, instilling in them a sense of wonder out of all proportion to their true worth—as people recognized. For example, a fanciful verse from the first decades of the nineteenth century depicted various peddlers working for "Hoax, Bore 'em & Co." (plate 1). The traders used their "skill in nibbling" to persuade customers to purchase myriad "flummeries" and "quirks." An accompanying sketch shows these ridiculous men with their ridiculous goods. A man holding a tray of gaudily painted statuettes stands at the center. Another offers up a tray of tiny hats made with fur warranted as "quite equal to real," and "nutria in place of prime Beaver." A German vendor strains under a stack of books, one arm holding a basket of ephemeral songsheets while balancing a tree branch supporting toys and novelties. Together, the goods were immersive and sensuous, engaging sound (cuckoo clocks and a music box), touch (chamber locks), and even taste (brazen cocks to draw beer).

The First Variety Stores

As early as the 1790s the armies of roving peddlers were joined by brick-and-mortar stores selling variety goods. A wholly new form of retail enterprise, variety stores seemed superior to dry goods stores, which carried a rather limited and predictable stock of staple goods, because they sold merchandise that was cheap, novel, often unnecessary, and perhaps even slightly indulgent. Variety stores embraced abundance and miscellany—the very things that had so enchanted consumers about the peddler's pack. And while they did not displace dry goods stores, their expansion into both smaller towns and larger cities suggested that Americans' appetite for new, superfluous, and often disposable things was only continuing to grow.

From the very beginning, proprietors of variety stores appealed simultaneously to both emotion ("variety!") and rationality ("cheap!"): choice and affordability. Selah Norton promoted his Ashfield, Massachusetts, variety store in a 1794 issue of the *Hampshire Gazette* by exclaiming, "New Goods! Cheap Goods! Fashionable Goods!"—a headline supported by a list of over forty sorts of items, ranging from curry combs and frying pans to snuff boxes, indigo, and chocolate. "I am selling, (some say) cheaper than ever," he noted.[6] Townsend & Ward's New Variety Store in Windsor, Vermont, offered a "complete assortment" of goods—everything from "Staple and Fancy Articles" (hardware, walking canes) to "English, French & India Goods" (cloth, slippers, nutmeg). Because variety stores engaged primarily in cash transactions, they were able to realize profits on very short margins.[7] Taking advantage of urban markets and cheap imports after the lifting of the embargo after the War of 1812, proprietors like Edward Vernon of the Cheap Variety Store in Utica, New York, could promise "CHEAP, CHEAP, CHEAP, *FOR CASH*. FRESH GOODS, *From the Auctions in New-York*."[8]

While peddlers continued to champion cheap across the hinterlands throughout the nineteenth century, variety stores offered an increasingly capacious stock that outpaced their itinerant competitors; they often bartered, too, for pork, wheat, dried apples, butter, old pewter, rags, and even pearlash.[9] Because they were able to capitalize on surplus merchandise that wholesalers were forced to unload at a loss, variety stores enjoyed even more popularity and success after the Panic of 1819. Goods bought at "panic prices" could reach even more consumers who were eager to buy yet able to afford only the cheapest of things.

A Variety of Anxieties

But the presence of articles "too numerous to mention" and "too tedious to describe" also created anxiety for both buyers and sellers. When the basic types of consumer goods were quite limited, and especially if they were made locally, people could fairly judge an article's quality and determine its just price, creating a consensus about its true market worth. The creep of cheap goods created new doubts about value, especially because so many of them were of unknown provenance. This promising new world of goods was also filled with uncertainty.

Even seasoned merchants sometimes had difficulty navigating this new terrain. The "queer, humourous recitative" of the peddler's patois that Hawthorne found so entertaining had a cynical side. Drawing on the dark arts of persuasion, peddlers' wheedling attempts were also intended to "catch gulls who don't understand the quirk." Naïve and sophisticated consumers alike might be easily fooled by the peddler's promises, caught in "a sort of scheme-trap by cunning foxes."[10] "Scheme-traps" came in various forms. "Formerly," noted a critic in 1829, "goods had a distinctive character, and were known by their names." In the case of fabrics, for instance, this meant packages were marked with precise widths and yardages. Their particular names indicated quality, composition, and construction. But, he continued, "names and lengths now really mean nothing," as fabrics were no longer cut to standard measurements or put up in consistent quantities. Similarly, pins used to be numbered according to their length and gauge, "but for years past, all has been confusion in this article."[11]

Novice consumers had even fewer guideposts and even less experience to help them judge value, especially of things they had never encountered before. Purveyors of cheap goods chose not to clear up these marketplace vagaries but instead used them to their advantage. Variety itself helped pique consumer interest and at the same time confounded people's ability to determine the true and fair value of any one item within a larger assortment because it was difficult, if not impossible, to establish accurate equivalencies. We know this from people's experiences at "package" auctions, which sold goods in mixed lots. As they did in variety stores, people looking over the merchandise tended to become enamored of the very best things, which made them less able to assess the value of the lesser-quality goods in the lot. One observer noted that it was "not surprising" that some buyers were "thoughtless enough to bid for the whole upon their estimate of one of the most valuable specimens" in the group, "and thus pay far more than the real value." Goods were often displayed "with great art," to encourage browsers to draw false equivalencies and direct their eyes to only the best things.[12]

Because they sold tinware, fabric, pins, ribbon, toys, hardware, and any number of other necessities and petty luxuries, variety stores likewise

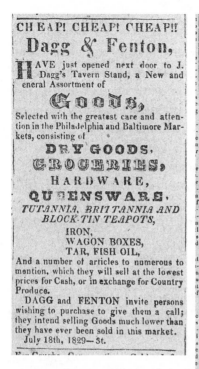

Figure 1.2. Peddlers were not the only ones crying their "cheap" wares. Newspapers did, too, as shown by these advertisements from the early nineteenth century.

obscured relationships between cost and quality. By emphasizing the profusion of goods in stock, advertisements became emotionally charged. Readers responded positively to these appeals, which were in turn coupled with rational ones: by making items superlatively affordable, dealers in variety goods distracted consumers from considering how price might correlate to quality. One merchant pointed out that an auctioneer's insistent cries of "*The cheapest goods ever sold*" "have their effect." The textual version of this hypnotic mantra, "CHEAP! CHEAP! CHEAP!," appeared in newspaper ads all the time and had a similar "effect" on consumers (fig. 1.2).[13] Writing to the editor of the *Connecticut Herald* in 1816, "Solomon Plainly" wryly described his wife as a woman who saved by "purchasing every thing she can get *cheap*"; she recently spent fifty dollars on "trifles we did not want." The "*cheap* advertisements" worked their influence on his entire family; so desirous to consume cheap things, even his children had become "beside themselves." Given all the new cheap goods shops opening up, and their enthralling advertisements, the beleaguered man would, he ventured, "inevitably become a bankrupt."[14]

While relatively unsophisticated compared to today's forms of persuasion, early advertisements for variety stores—with their astonishing lists
of goods—invited this first generation of mass consumers to engage in a
kind of free association that had little do to with practical utility. We might
see these early print advertisements as the textual analogs of a peddler
reaching into his pack and "dipping his hand down into his wares with the
promise of a wonderful lot." As such, variety store ads, and variety stores
themselves, were potent admixtures of emotional and material access that
fueled Americans' desire for more.

Considered through the lens of consumer psychology, these advertising
appeals could also arouse "smart-shopper feelings," making early Americans feel confident about their nascent shopping prowess because they
were able to acquire goods at deep discount. However, smart-shopper
feelings are based in emotion, not reason. Certain kinds of promotions,
like the "splashy 'Save $$!' promises of print ads," generate "drama" that
can cause consumers to desire products intensely but not rationally. Many
readers perusing variety stores' eye-catching advertisements, then, likely
became emotionally invested in the goods on offer—swayed, too, by their
low prices—before even seeing them in real life. It was enough to form
a "vividly imagined" impression, which would have "the same desire-
arousing effect as one which is physically present in the environment." Be-
cause people could imagine these things, they had, in effect, already laid
claim to them. Now all they had to do was to go down to the variety store
and take possession.[15]

Abundant Miscellanies

Early Americans were not, however, simply passive dupes in the face of the
burgeoning market, standing agog and powerless before a new world of
goods. There was something undeniably and inherently pleasurable about
all this crap. Material miscellanies encouraged infinite imaginative asso-
ciations and possibilities. What was more, variety goods' low costs trans-
formed them from the stuff of dreams into material realities. Quasi-exotic
items were now literally within reach.

Such material abundance came from many sources. Certainly, domes-
tic producers furnished some variety store stock (heavy crockery, rough
textiles). Variety goods also came through secondhand channels such as

bankruptcy auctions and liquidation sales. But most cheap things came from overseas—often supplied by British manufacturers and traders who did business as far away as China and India. Satiating increasing American demand for such "lower-class" goods fueled global commerce and established a pattern that continues to this day: most American crap still comes from outside the country, and is often the output of exploited labor.

Abundant choice and low cost came to matter more than anything else, even low quality. A poem from the early 1840s titled "The Scotch Pedlar" derided the ephemerality of the man's glittery wares: "Those muslins fine and showy ginghams / And then that box of gilded *thingums. Those flaunting flowers are born to die / Those colours gay are made—to fly.*"[16] It was no secret that cheap variety goods were cheap for a reason. But, falling under "the mysterious power of influence," people continued to purchase, with gusto.[17]

Yet even shoddy goods could offer wondrous, pregnant possibilities. These contradictions both distressed and amused early Americans. By the antebellum era, people were decrying the "cheapening and cheap-selling spirit" that spread like an infection, degrading everything it touched.[18] "The habit of bargain-hunting, while we laugh at it for its folly, deserves to be denounced for its mischief," wrote a critic in 1845, adding that manufacturers, "in order to gratify the morbid love of cheapness, to produce goods of the most trashy and useless description," had to lower the wages of already poor workers in order to compete. The "cheapening mania" corrupted the quality of goods, destroyed trust in the market, and led to labor exploitation.[19] It degraded everyone and everything.

Describing a wily charismatic peddler and the enticingly cheap goods he sold, the story of "Cheap Jack," published in 1846, illustrates Americans' conflicted relationship with crap. The protagonist travels from town to town selling his wares by puffing them, in the most over-the-top fashion, to "credulous" locals who in an instant become "enchanted" with him and his wares, "the merry mob swallowing down poison as if it were honey." Cheap Jack's engaging personality and humorous exaggerations could make "homely" items like hatchets and salt cellars come to life, as he bent a handsaw into a "bonnet" for an old woman and whirled a hatchet in the air "after a wild Indian fashion." "The appeal," remarked the writer, "is irresistible."

The villagers' gullibility, though, pointed also to the attendant perils of cheap. Often, the fantasies of all-too-trusting customers took cynical material form, as the narrator observed ruefully:

> We must look at the reverse of this picture of rustic whimsicality. A poor car-penter, who purchased one of his planes, lost half a day in setting the tool, and when it was used, it broke in half. Five or six farmers' men discover that their waistcoats . . . are dropping to pieces. The forester finds that his axe-head, instead of being cast-steel, is cast-iron. . . . The instrument breaks after a few blows at the root of a young oak.

The story, about the inevitable loss of people's "confiding simplicity of in-nocence," drove home the reality that as Americans increasingly engaged in consumer society, they also became vulnerable to dubious marketplace temptations delivered by imposturing agents, dressed as they were in re-fined velveteen, "all smiles and insinuations . . . honey and deceit." Risky propositions were a sign of the times, and consumers brought the conse-quences of poor decisions on themselves. The materialist ethos that they increasingly embraced came to define not just inanimate things but society itself. The contemporary "cheap-selling spirit" was pervasive:

> Our cheap shopmen, our cheap tailors, our cheap groce[r]s, bakers, butch-ers, haberdashers, general dealers, and a crowd of others, *are* the Cheap Jacks of the community. The travelling van is replaced by the gaudy, plate-glass-fronted, gilded, and decorated "emporium," "establishment," "mart of commerce"—call it what you will. . . . The bombast comes out in handbills; the jokes in puffs; the long-winded speeches . . . in gigantic advertisements.[20]

This anxiety might explain, then, the tendency of many cheap variety store operators to versify their promotions, tamping down any doubts about the marketplace by making the purchase of shoddy stuff seem like harmless fun. In 1822 John Brown described his store as

> well supplied
> With goods, (worth close attention)
> Of candid minds,) of various kinds, . . .
> Those goods in store, with many more,
> He'll sell for ready money;

> When thus you pay, he's bold to say,
> You need not fear he'll dun ye.[21]

Albany, New York, proprietor R. H. Pease used a similar approach to advertise his Great Variety Store in 1843:

> There you will find a thousand things,
> Of every name and kind,
> And there whatever you may wish,
> You will most surely find.[22]

Even prose endorsements evoked the lyrical and fantastical in an attempt to elevate otherwise pedestrian goods and appeal to consumers' emotional rather than rational selves. "There are a variety of Variety Stores, which by those who fancy such amusement may be variously divided into a variety of classes," read an 1845 advertisement for Van Schaack's Mammoth Variety Store.[23]

The Different Meanings of Cheap

With the rise of variety stores in the mid-nineteenth century came the disambiguation of "cheap." Since the mid-sixteenth century the word had been used to describe both price and quality ("cheap and nasty"), and it was commonly paired with its antonym, "dear." By the early seventeenth century the meaning of cheap had subtly shifted; it could also mean something come by easily, and hence of little intrinsic worth (Samuel Johnson's "The cheap reward of empty praise," for example.) Over time, American variety operators tried to clarify what cheap was. Democratizing the market, as crap proprietors knew quite well, meant putting goods within everyone's reach while rejecting the very notion that these goods were easily procured or of poor quality.

Early store advertisements not only emphasized the more delightful aspects of variety but also illustrated proprietors' attempts to offer the cheap without the nasty. Highlighting variety made it easier to obscure lingering concerns about quality. Several crappy things presented together elevated the entire agglomeration while enabling individual items to transcend their lowly status. This also made it more difficult for consumers to evaluate the quality or value of individual components. What was more, each omnium-gatherum was its own new thing: a novelty made up of other novelties.

No wonder, then, that many purveyors of cheap goods also oversaw entertainment ventures. For example, John Brown ran a saloon next door to his variety store. On the other side sat his bookshop.[24] Beyond the typical store stock, Holden & Cutter's Fancy Goods and Toys in Boston also supplied "The Best Fire Works" in all of New England.[25] Dominicus Hanson not only sold "new and improved" articles "upon their first appearance in the market" at his variety store but also peddled "fresh and pure" patent medicines—magical elixirs promising instant rejuvenation.[26]

Proprietors often euphemistically referred to their stock as "fancy goods," which also helped lift otherwise lowly cuff buttons, five-cent handkerchiefs, and the like into the realm of imagination, since "fancy" was an early modern contraction of "fantasy" and a more refined descriptor than "variety." "Fancy" meant, according to one account, "a great variety of 'good-for-nothing' things which women are so fond of purchasing."[27] Before this new age of cheap and fantastic consumer goods, according to a contemporary novel, "pie-pans had not yet even entered 'the land of dreams,'" and "China dishes and silver plate . . . belonged to the same class of marvellous things, with Aladdin's lamp and Fortunatus's purse."[28] The 1844 *Albany City Guide* described the articles offered at R. H. Pease's Temple of Fancy as "surpassingly rich; exceeding anything in elegance, that we have ever thought, dreamed or read of. . . . His assortment has never been so rich and desirable as at the present time." Crucially, desirability was coupled with accessibility. In the case of Pease, for instance, customers could procure these extraordinary articles "at much less than former prices."[29] It was also fitting that Pease associated himself with the most fantastic peddler of all, Santa Claus, whose overstuffed pack—stocked from the shelves of Pease's Variety Store—contained treats beyond a child's wildest imagination (fig. 1.3).

Over time, appeals to the fantastic, with the occasional nod to affordability, were not enough to convincingly promote cheap goods to ever-more-savvy consumers. The first generation of consumers exposed to cheap variety goods was succeeded by people who had grown up well entrenched in the world of crap. To sell to them required more sophisticated marketing strategies. And so firm prices began appearing in advertisements for variety stores, speaking more to finances than fantasy. Strategies

SANTA-CLAUS

In the act of descending a Chimney to fill the Children's Stockings, after supplying himself with **FANCY ARTICLES, STATIONERY, CUTLERY, PERFUMERY, GAMES, TOYS, &C.**

AT PEASE'S GREAT VARIETY STORE, No. 50 Broadway, Albany,

Where can be found an almost endless variety of "Things to use and things for sport," suitable for CHRIST-MAS AND NEW-YEAR'S PRESENTS, such as

Work boxes, furnish'd & unfurnh'd
Dressing cases, toilet cases
Writing desks, papeteries
Card receivers, portfolios
Card cases, purses
Velvet and silk bags
Rich paper boxes, guard chains
Gold and silver pencils
Infants' rattles of gold, silver, ivory
Gold and silver thimbles
Napkin rings, segar cases
Music boxes, hair brushes
Combs and tooth brushes
Gold and silver toothpicks

Penknives, inkstands
Gloves in a nut shell, watch stands
Bouquet holders of silver and gilt
Breast pins, rings, bracelets
Hair pins, shawl pins, cuff pins
Gilt combs, fans, work baskets
Perfumery in great variety
Games and toys
Splendid and common chess men
Backgammon boards and men
Battledoors, graces, cups and balls
Skipping ropes, dissected maps
Building blocks and alphabets
Mosaic puzzles, panoramas

Magic lanterns, wax dolls
Kid and jointed dolls, tea sets
Nine pins, doll's heads
Rocking horses, whips, swords, guns
Soldiers, pistols, drums and flags
Masks, accordions, sheep and dogs
Printing presses, whistles, rattles
Arks, magnetic fishes, livery stables
Horses and carriages, dominoes, dice
Wheelbarrows and wagons
Cradles, cooking stoves
Kitchens, tea sets
Rocking horses, sofas, tables
&. &c. &c.   de23

Figure 1.3. Santa Claus was perhaps the most beloved peddler of them all. Advertisement for R. H. Pease's Great Variety Store, *Albany Argus*, December 23, 1842. Collection of the Albany Institute of History & Art.

that privileged cost also, presumably, fixed ideas of value in consumers' minds, shifting the terms by which cheap goods were defined. Baltimore proprietor D. Bodge, for instance, claimed, "GOODS AT THE BOSTON CENT STORE ALMOST GIVEN AWAY"—so cheap they were practically free.[30]

Since "cheap" was a vexingly vague modifier, consumers were left to figure out if the descriptor referred to quality or price. Goods that were cheap (low priced) could be a good bargain; those that were cheap (low quality) were the exact opposite. Fixed prices, then, helped shift buyers' focus away from quality toward cost. A representative ad in 1845 read:

> ALWAYS SOMETHING SELLING CHEAP AT THE BOSTON CENT STORE . . . THREAD LACES, 45 cents, 50 cents, 62 ½ cents, 75 cents, worth $1.50 per yd. Come and get some of them before they are all gone. EDGINGS, d[itt]o all prices; Gloves; Mitts; Needles; Pins; Tapes Hosiery, very low; gentlemen's gum Suspenders, very low; Laces, Nets, Edgings, all prices; new style Collars, 25 cents; Net Caps, 3 and 4 cents; Tuck Combs, 2, 3, and 4 cents; fine Tooth d[itt]o, 6 ¼ and 12 ½ cts., worth 25 cents; Frock Bodies, 50 cents each; Almond Soap, 3 cents cake; Hair and Tooth Brushes, and all kinds of PERFUMERY, cheap. Ladies, look once more.[31]

Since much variety store stock *was* of inferior quality, low price became the more viable way to sell it. Before the second industrial revolution, most cheap goods consisted of odd lots of imports that had been broken up dockside or separated from other cargo in transit. These were the easily chipped china cups, the readily fraying cotton handkerchiefs, the fast-fading bolts of fabric, the remaindered books.

The appeal to price rendered quality fairly irrelevant, especially during economic downturns, when people's budgets tightened significantly and the stock of failed businesses presented advantageous buying opportunities for merchants of cheap: one person's failure was another's success. During the Panic of 1857, for instance, New York City auctioneer Thomas Bell announced the sale of "the stock of a fancy dry goods and variety store" that included "damaged goods, toys, &c."[32] Hartford-based Sudgen & Co. bragged that its fabrics, including remnants, were "Bought at 'Panic' Prices."[33] W. W. Palmer, in Salem, boasted that in New York he had been able to purchase miscellaneous items "at ruinously low prices" and was now "enabled to sell at panic prices." He offered variety and economy:

"Whatever kind of goods you want, look at our stock. . . . At present it is prudent to make each dollar go as far as possible."[34]

During the Civil War era, dollars and cents mattered even more to consumers; there was little room for flights of fancy when the nation was at war with itself. One-price stores, which began appearing at this time, presaged the five-and-dime stores of later eras and the dollar stores of today. One of the first was James Kennedy & Bro.'s 10-Cent Store in Pittsburgh, established in 1862.[35] In 1866 George Husted opened his 25-Cent Store in Harrisburg, where he sold such disparate goods as jewelry, ice pitchers, wood, and coal.[36] While Husted's ad listed what he had on offer, many other stores did not, as they tried to further obscure concerns about quality. Boston-based S. S. Houghton & Co.'s Now Then Store, for example, billed itself variously as the One Dollar Store, the Three Shilling Store, the Half Dollar Store, and the Fifty Cent Store. Prices so dominated the business's advertisements that readers had to work doubly hard to discern what Houghton was actually selling, since his hoop skirts, vases, and albums were enumerated in microscopic print.[37] The point was that shoppers would purchase whatever was on offer simply because it was cheap. The advertisement for the One Two Three Dollar Store mentioned no goods at all.[38] A bold figure 5, composed in type from smaller 5s, dominated the ad for the 99-Cent Store in Trenton (fig. 1.4).

As the selection of cheap variety goods expanded due to an increase in domestic manufacturing and the growth of global trade in the second half of the nineteenth century, uniform prices became even more prominent. Low and fixed prices rationalized the chaotic world of material goods, reducing variety to one shared dollar (or cent) figure regardless of the thing, its quality, or its quantity. Ingeniously, one-price stores led consumers to believe they were making smart choices by enabling them to fairly judge and compare goods. But these odd equivalencies only confounded seemingly calculable decisions because they were false. While seeming rational, one-price stores actually muddled people's ability to determine whether they were getting a good deal or not.[39]

An example of this comes from the career of Frank W. Woolworth, who is credited with "inventing" the five-and-dime. (This claim is not quite true; many others preceded his eponymous chain by at least a decade.) Woolworth recognized and popularized the value-obscuring effect of the one-price model. In a memoir, he recalled that while working as a young

Figure 1.4. In variety stores, low prices often overshadowed the goods themselves. *Trenton State Gazette*, May 23, 1879. Special Collections, Rutgers University.

clerk in a Watertown, New York, dry goods store, he heard tell of an ingenious sales method. The proprietor of a Michigan dry goods store had partnered with a traveling salesman intent on selling a line of handkerchiefs for five cents, a price below his cost. The profit would be made by selling the handkerchiefs along with "other notions and stuff." The salesman encouraged the owner to install a five-cent counter that offered the new handkerchiefs mixed with "a lot of old dead goods that had been lying on his counters for years." A prominent sign announced that any of the goods on the counter could be purchased for five cents. According to Woolworth, twelve thousand handkerchiefs "were disposed of promptly, and the customers also bought everything else on the counter *supposing*

them to be of the same value as the handkerchiefs. . . . It was an instantaneous success," he marveled, adding, "People thronged into this store . . .
and it was almost impossible to keep goods on that counter."[40] Others soon
followed suit. Braselman & Co. of New Orleans reported in the mid-1860s
that the store was enjoying "unparalleled success" with its twenty-five-cent
counter. The shopkeeper boasted of his counter, as if it were a theatrical
show, that it "has sustained an uninterrupted run for more than a month,
and is now more attractive than ever, as large accessions of higher cost
goods have been placed upon it."[41] A centerpiece of the advertising campaign for Watson's China Store in Wilmington, North Carolina, was its
counter of crappy stuff and the "Wonderful Bargains on the Five and Ten
Cent Table."[42] Some businesses, like Boston's Heyer Brothers, specialized in
"counter supplies"—that is, lots of assorted merchandise retailing for five
or ten cents for this very purpose.[43]

Convinced by the efficacy of this strategy, Woolworth urged his boss, a
Mr. Moore, to adopt it. Reluctantly, the dry goods proprietor did, placing
a lot of fresh goods recently purchased from a city jobbing house on a
counter along with a bunch "of other old trash." Woolworth observed that
it "seemed almost impossible" to keep the counter well stocked, noting, "It
did not make any difference whether the goods had any value or not. Any
old stuff we could find around the store would be fired on that counter and
would sell immediately."[44] He was able to turn "chestnuts" and "stickers," or
dead stock, into wildly popular merchandise—"plums" and "corkers." According to one account, "The throngs scrapped for the bargains. While in
the store, they also snapped up other surplus goods at higher prices."[45] By
transmuting humble, cheap, shoddy, and unsaleable goods into something
desirable, low price performed the alchemy that variety alone once did,
inciting "excitement and confusion" simultaneously.[46] What was more, the
mere presence of one-price counters encouraged consumers to purchase
full-price "surplus goods" they had not intended to.

Emboldened by the counter's success—sales from the first day alone
brought Moore & Smith back from the brink of bankruptcy—Woolworth
set out on his own to establish an entire store based on the concept. He
opened his first five-cent store in Utica, New York, in 1879, a space he
filled with forty items of low-priced merchandise. A few months later, he
launched his "Great 5¢ Store" in Lancaster, Pennsylvania. By the end of
the first day, over 30 percent of the stock had been sold. The next year he

added ten-cent goods, and although they made the business more profit-able, according to Woolworth, the higher price point took away part of the "charm" of the five-cent gambit. He was onto something, though, opening an outlet in Reading, Pennsylvania, in 1884 and thereafter penetrating towns in upstate New York, Delaware, and New Jersey. By the end of 1889, Woolworth's enterprise consisted of twelve stores with reported sales vol-ume of $246,700, up from $12,000 in less than five years.[47]

By the later decades of the nineteenth century, variety stores, one-price stores, and discount counters could be found around the country, from the Eureka 50 Cent Store in Quincy, Illinois ("The Cheapest Place in the City! To Buy all kinds of Nice Things!"), Mohlenhoff's Cheap Tables in Cin-cinnati ("Gold is Down, and Goods Away [*sic*] Down"), and the 99-Cent store in Omaha, to the New York Fifty Cent Store, located in Chicago ("the best assortment in the city") and Munson's New 99 Cent Store in Boston (fig. 1.5).[48] One-price stores seemed to captivate nearly everyone, not just cash-strapped housewives looking for cut-rate gadgets and dishware. They could, for one, be a boon to broke college students. In the depression year of 1873 the *Yale Courant* called the local ninety-nine-cent store "a favor-ite resort for students," with, according to the *Williams Vidette*, "just the knick knacks a student desires . . . for less than a dollar." The library in Southbridge, Massachusetts, added fifty books to its collection in 1874, all purchased for cheap at the local ninety-nine-cent store. And in 1875 editors of *Publishers' Weekly* suggested that ninety-nine-cent-store mer-chandise might appeal to the striving classes as well: "You can get opera chains, enameled slide and tassels, heavy seal charms for gents, Waverley novels, Mrs. Harriet Beecher Stowe novels, solid gold engraved rings, Dr. Holland's works, sets of dessert-spoons, . . . all of the choicest poetical works, gilt edge, bound beautifully, in diamond and large edition; beautiful engraved lockets, and other articles too numerous to mention."[49]

Yet while the budget-minded and object-mad public welcomed these retail upstarts, others did not. Proprietors' emphasis on low prices could not completely obscure the association of cheapness with dubious quality. As the novelty of cent stores began to wear off, the worrisome valence of cheap that suggested poor quality and bad bargains surfaced once again, putting retailers of low-priced goods on the defensive. Perhaps taking a dig at its competitor the New York 99 Ct. Store, for instance, Guy & Brothers' Union Crockery Company in Lowell, Massachusetts, insisted in an ad for

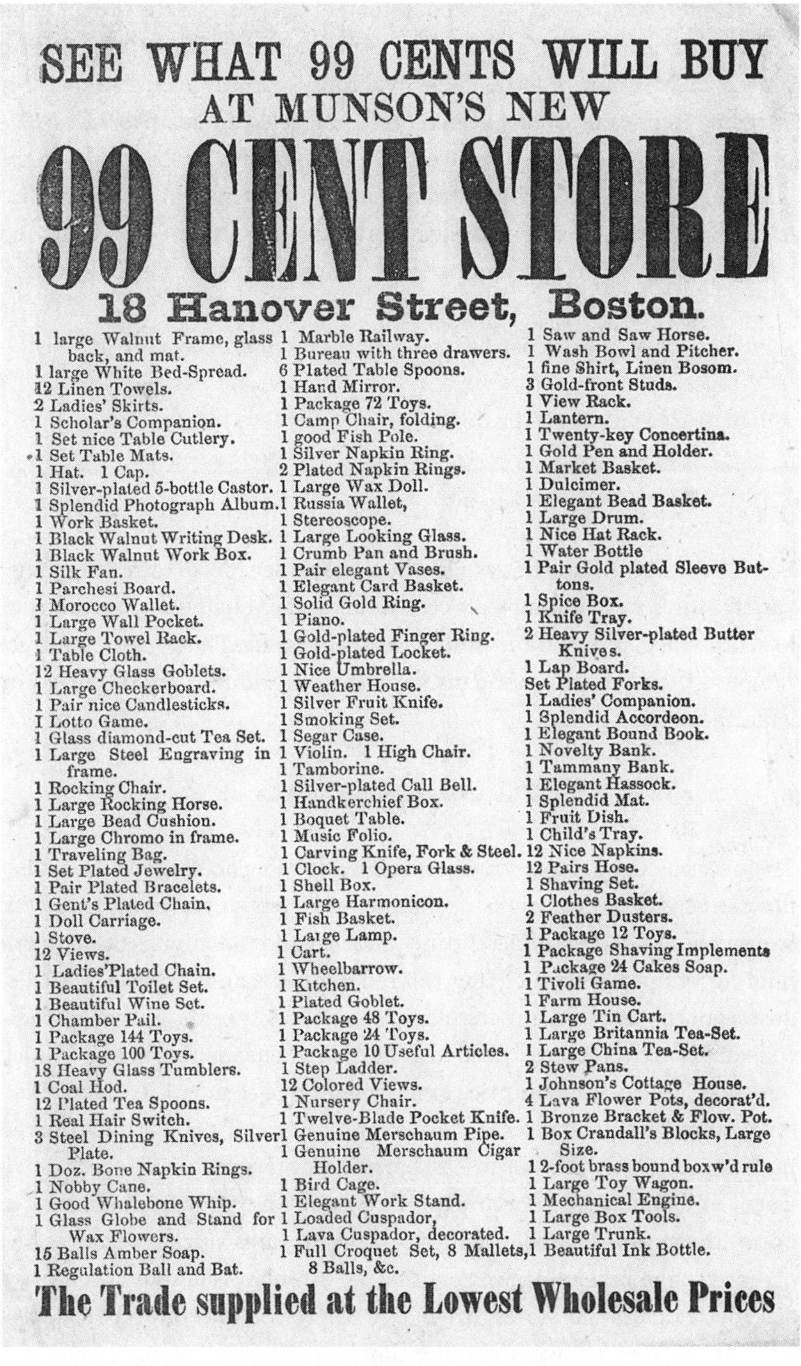

SEE WHAT 99 CENTS WILL BUY
AT MUNSON'S NEW
99 CENT STORE
18 Hanover Street, Boston.

1 large Walnut Frame, glass back, and mat.
1 large White Bed-Spread.
12 Linen Towels.
2 Ladies' Skirts.
1 Scholar's Companion.
1 Set nice Table Cutlery.
1 Set Table Mats.
1 Hat. 1 Cap.
1 Silver-plated 5-bottle Castor.
1 Splendid Photograph Album.
1 Work Basket.
1 Black Walnut Writing Desk.
1 Black Walnut Work Box.
1 Silk Fan.
1 Parchesi Board.
1 Morocco Wallet.
1 Large Wall Pocket.
1 Large Towel Rack.
1 Table Cloth.
12 Heavy Glass Goblets.
1 Large Checkerboard.
1 Pair nice Candlesticks.
1 Lotto Game.
1 Glass diamond-cut Tea Set.
1 Large Steel Engraving in frame.
1 Rocking Chair.
1 Large Rocking Horse.
1 Large Bead Cushion.
1 Large Chromo in frame.
1 Traveling Bag.
1 Set Plated Jewelry.
1 Pair Plated Bracelets.
1 Gent's Plated Chain.
1 Doll Carriage.
1 Stove.
12 Views.
1 Ladies'Plated Chain.
1 Beautiful Toilet Set.
1 Beautiful Wine Set.
1 Chamber Pail.
1 Package 144 Toys.
1 Package 100 Toys.
18 Heavy Glass Tumblers.
1 Coal Hod.
12 Plated Tea Spoons.
1 Real Hair Switch.
3 Steel Dining Knives, Silver Plate.
1 Doz. Bone Napkin Rings.
1 Nobby Cane.
1 Good Whalebone Whip.
1 Glass Globe and Stand for Wax Flowers.
15 Balls Amber Soap.
1 Regulation Ball and Bat.

1 Marble Railway.
1 Bureau with three drawers.
6 Plated Table Spoons.
1 Hand Mirror.
1 Package 72 Toys.
1 Camp Chair, folding.
1 good Fish Pole.
1 Silver Napkin Ring.
2 Plated Napkin Rings.
1 Large Wax Doll.
1 Russia Wallet,
1 Stereoscope.
1 Large Looking Glass.
1 Crumb Pan and Brush.
1 Pair elegant Vases.
1 Elegant Card Basket.
1 Solid Gold Ring.
1 Piano.
1 Gold-plated Finger Ring.
1 Gold-plated Locket.
1 Nice Umbrella.
1 Weather House.
1 Silver Fruit Knife.
1 Smoking Set.
1 Segar Case.
1 Violin. 1 High Chair.
1 Tamborine.
1 Silver-plated Call Bell.
1 Handkerchief Box.
1 Boquet Table.
1 Music Folio.
1 Carving Knife, Fork & Steel.
1 Clock. 1 Opera Glass.
1 Shell Box.
1 Large Harmonicon.
1 Fish Basket.
1 Large Lamp.
1 Cart.
1 Wheelbarrow.
1 Kitchen.
1 Plated Goblet.
1 Package 48 Toys.
1 Package 24 Toys.
1 Package 10 Useful Articles.
1 Step Ladder.
12 Colored Views.
1 Nursery Chair.
1 Twelve-Blade Pocket Knife.
1 Genuine Merschaum Pipe.
1 Genuine Merschaum Cigar Holder.
1 Bird Cage.
1 Elegant Work Stand.
1 Loaded Cuspador,
1 Lava Cuspador, decorated.
1 Full Croquet Set, 8 Mallets, 8 Balls, &c.

1 Saw and Saw Horse.
1 Wash Bowl and Pitcher.
1 fine Shirt, Linen Bosom.
3 Gold-front Studs.
1 View Rack.
1 Lantern.
1 Twenty-key Concertina.
1 Gold Pen and Holder.
1 Market Basket.
1 Dulcimer.
1 Elegant Bead Basket.
1 Large Drum.
1 Nice Hat Rack.
1 Water Bottle
1 Pair Gold plated Sleeve Buttons.
1 Spice Box.
1 Knife Tray.
2 Heavy Silver-plated Butter Knives.
1 Lap Board.
Set Plated Forks.
1 Ladies' Companion.
1 Splendid Accordeon.
1 Elegant Bound Book.
1 Novelty Bank.
1 Tammany Bank.
1 Elegant Hassock.
1 Splendid Mat.
1 Fruit Dish.
1 Child's Tray.
12 Nice Napkins.
12 Pairs Hose.
1 Shaving Set.
1 Clothes Basket.
2 Feather Dusters.
1 Package 12 Toys.
1 Package Shaving Implements
1 Package 24 Cakes Soap.
1 Tivoli Game.
1 Farm House.
1 Large Tin Cart.
1 Large Britannia Tea-Set.
1 Large China Tea-Set.
2 Stew Pans.
1 Johnson's Cottage House.
4 Lava Flower Pots, decorat'd.
1 Bronze Bracket & Flow. Pot.
1 Box Crandall's Blocks, Large Size.
1 2-foot brass bound box w'd rule
1 Large Toy Wagon.
1 Mechanical Engine.
1 Large Box Tools.
1 Large Trunk.
1 Beautiful Ink Bottle.

The Trade supplied at the Lowest Wholesale Prices

Figure 1.5. People could buy all kinds of things for less than a dollar. Advertising circular for Munson's New 99 Cent Store in Boston, ca. 1870. Library Company of Philadelphia.

its 90 Cent Goods store, "We do not intend to sell trash for the sake of making low prices, but GOOD GOODS, such as people are willing to buy a second time."[50] M. M. Cohn, of Little Rock, Arkansas, similarly protested that "My Bargains Represent Goods Which Sell Themselves: I Positively Keep No trash or Auction Goods."[51] They were likely responding to observers like the one who described the "entertaining" stock of the ten-cent store as "of infinite variety, but generally of a cheap sort. There is a little of everything and nothing of value."[52] A popular joke circulating at the time went like this:

> Sol Sodbuster: "Hear about the robbery down t' th' five an' ten cent store last
>   night?"
> Hiram Hayrack: "Nope. D'they git much?"
> Sol Sodbuster: "Yep. They was in there two hours and carried away nearly a
>   dollar's worth of goods."[53]

It had become common knowledge, by the later decades of the century, that variety stores were selling—and consumers were buying—miscellaneous lots, orphaned, ill-used, and otherwise misfit goods. That was their attraction and their curse. Americans were at once seduced and repulsed by what cheap had to offer.

The Cheapening Mania

There were, as well, larger economic issues dogging cheap goods. Even before the Civil War, the increasing number of variety stores prompted critics to decry both retailers' undercutting practices and the throngs of "bargain-hunting" shoppers to which they catered. Rather than turning away in disgust from the increasing materialism enabled by the glut of cheap goods, consumers actively contributed to a prevailing "cheapening mania," which upended principles of fair price and wrought real economic consequences on manufacturers, retailers, and workers. By expecting goods to be offered in the market at ever lower prices, bargain hunters indirectly drove down wages and encouraged labor exploitation by forcing manufacturers "to gratify the morbid love of cheapness." "When the world goes so fast," as one observed, "the passion is for cheapness."[54] Another remarked even more succinctly that dollar stores "are places where you may buy a twenty-five cent article you don't want, for four times its value."[55]

Other businessmen, too, complained about the "chronic" proliferation of variety stores.[56] One-price, cheap variety, and "bazaar stores" competed not only with dry goods emporia and clothing retailers but also with grocers, crockery dealers, and specialty stores. Some magazine and newspaper publishers even refused to run advertisements for dollar stores "on account of character," suspecting that they were merely brick-and-mortar Cheap Jacks. One editor wrote that, "as far as we can see," dollar stores in Chicago "are conducted just as honorably as the dry goods or grocery stores. But they are a novelty, and swindlers will be quick to switch their trains on to that track. We shall never let the latter solicit passengers in our columns if we know it." Ultimately, though, readers were left "to exercise their own judgment" and warned "not to complain if they attempt to get something at half what it is worth, and then find that it was not worth half what it was claimed to be. That," he concluded, "is adding stupidity to dishonesty."[57]

Variety stores even threatened to erode publishers' monopoly control over book prices. In an 1875 letter to *Publishers' Weekly*, "Justice" expressed grave concern about booksellers' undercutting practices, specifically "furnishing the 'dollar stores' and 'ninety-nine cent stores' at a discount which enables them to sell ten, twelve, and fourteen shilling books at one dollar or ninety-nine cents . . . , which they are doing all over the country."[58] As the discards of retail booksellers, variety store books were neither "new and fresh" nor those most in demand.[59] Regardless, many customers seemed to prefer the cheap "everything-you-want establishments" over traditional and presumably more reputable booksellers.[60]

Retailers were not the only ones affected by the "cheapening mania." Consumers themselves—the very people responsible for driving prices down—were beginning to pass off cheap goods to each other as more expensive ones. A bride wrote to *Godey's Lady's Book* in the early 1870s with what was, apparently, becoming a common problem. As a gift for her betrothal, she had received a pair of salt cellars enclosed in a box from one of her city's "leading jewelers." Having already received something similar, she tried to exchange them, only to discover that the items had, in fact, come from a dollar store. "We may add that the bountiful giver of the washed salt cellars was a member of a very wealthy family," the magazine remarked.[61] As an article in 1876 sharply observed, "Most of the stuff purchased at the dollar store is bought to give away."[62] In a *Harper's Bazaar* story, the very same moment a women is imagining what her husband might give her for Christmas—"Something nice, I am sure"—he is at the dollar store, telling the clerk to leave no mark on the package "to indicate that it came from here."[63]

Many questions about variety store goods persisted, including not only their quality and value but also whether consumers should embrace or reject cheap abundance. Equally disquieting was the fear that even the most sophisticated buyers could not tell the difference between high- and low-class goods if they came in boxes and bags from Caldwell's, Bailey's, and Tiffany's. Rumors of misrepresentation and deceit dogged not only variety goods themselves but their sellers and buyers alike. "If the dry goods dealer marks down his goods," remarked one commentator in 1873, "they are either out of season, unsaleable patterns, or inferior goods. Every one knows the chances are largely in favor of being cheated at a dollar store."[64] While cheap removed the barrier of entry—money—that had kept too many from fully participating in the world of goods, its democratization risked leveling everything to the lowest common denominator.

But value was a subjective thing, determined by many criteria. One observer complained, "The miserable gingerbread covers put on the standard books so temptingly displayed in the dollar stores surely add nothing to their value."[65] But was he right? Fancy covers, after all, made for more pleasing objects than the flimsy bright yellow wrappers of cheap paperbacks. In fact, value could be—and often was—determined as much by standards of taste and aesthetics as by price. During his long career, F. W. Woolworth developed a keen sense for customer preferences. Woolworth urged his store managers to pay close attention to patrons' sensibilities rather than their own, saying, "In years gone by, there used to be demand for certain vases, the ugliest ever made, and I was obliged to buy them against my own taste and judgment. And how they did sell!"[66] During one of his buying trips to Germany, Woolworth wrote to his staff that he found thermometers mounted on wood for $7.50 per gross, which was a good price. They "are not so good as the domestic goods," he admitted, but he bought them anyway, because they "make a bigger show."[67] What mattered was that the cheap stuff found in variety stores helped countless consumers attain their material aspirations, critics be damned.

Proprietors and customers alike perpetually balanced low prices against passable quality. A lukewarm contemporary considered dollar stores "legitimate enough," acknowledging the realities of the trade in crap:

Many of the articles sold are good and cheap, others are not so good. Of
course, profit is the object, and on all articles there is doubtless a profit made,
though, of course, less on some than others. . . . Whether better bargains can
be had there than elsewhere is a question for each to decide for himself.[68]

Most critics assumed that people understood they were buying crappy
goods. Since there was purportedly no subterfuge involved, consumers
should be free to buy, a refrain that echoed defenders of auctions in the
late 1810s and early 1820s, who argued that *"in this free and happy republic, every man has a right to be ruined in his own way."*[69] Consumers
could buy "good and cheap" or "not so good"; either way, they had the freedom of choice in the consumers' paradise that Americans envisioned for
themselves (fig. 1.6). As the 98 Cent Store proudly proclaimed succinctly
in 1878, "ECONOMY IS WEALTH."[70]

The siren song of cheap abundance continued its seductive call, especially as world markets opened up, retail sectors expanded, and domestic

Figure 1.6. New goods represented new ideas, and cheap goods only expanded the possibilities for
material and emotional well-being. The New Idea Store, founded by Jacob Shartenberg, in 1886. Rhode
Island Collection, Providence Public Library.

production increased. Cheap not only lured Americans but came to define them as a people who were simultaneously voracious consumers and often-willing market dupes. They felt no need to apologize for either of those things; in fact, Americans lived quite comfortably with this seeming contradiction, for they saw it as no contradiction at all. This mentality was epitomized in the satire "Mrs. Brown Visits the Capital." Published in 1896, it was a traveler's tale well suited to an age transformed by the second industrial revolution, the rise of department stores and mail order catalogs, and, of course, the profusion of crappy goods. Mrs. Brown's trip to the Corcoran Gallery is interrupted because she finds a set of "winder" shades showcased at a nearby forty-nine-cent store more interesting than fine art. Her sojourn through the Capitol is made meaningful only by purchasing a forty-nine-cent pair of carpet slippers to ease her blisters. National markers like the Washington Monument have significance because they are near or remind her of the forty-nine-cent store. Even meeting the president himself only matters because he is wearing a necktie "jes the dead match" to the one she bought for her husband at the forty-nine-cent store. Her travelogue consists of the cheap things she bought at what was to her the most important site of all, a retail outlet for crap.[71]

Mrs. Brown was one of countless Americans swept up in the heady world of cheap variety, which remained a rich source of material satisfaction and satire. A poem from the early twentieth century, written from the perspective of a husband beleaguered by his wife's mindless purchases, included this passage:

> It is not much, this prize she'll bring—
>   'Tis usually something strange,
> Some trifling, useless, little thing
>   To grace the mantel or the range.
> I've even had her say to me;
>   "I don't know what the thing is for,
> But just to think that this could be—
>   I got it at the ten-cent store!"[72]

In "The Fatal Lure of the Whim-Wham," published in 1920, writer Henry Hancmann lampooned his own brush with cheap goods. He tells of stepping into a dime store because he needs some shoelaces, expecting to get "instantaneous service and the guarantee of a fixed price and a probable selection that would prove highly satisfactory." But he is quickly overtaken—by the aroma of scented soap, incense, shoe polish, "wave on wave." He is distracted by the glitter of the jewelry counter, where he buys "a remarkable imitation diamond marquis" for his wife. Forgetting himself, he is transported, "the trade of a varied world exposed before my eyes." Spiriting "up and down the aisles," he considers socks, picture frames, leather shoe soles, coat hangers, pencils. Items are so cheap, he even makes an ironic purchase, buying "a phial of horrible perfume" to laugh about with his wife. He also buys some books, jelly beans, a puzzle, and a map before "debauch[ing]" himself with peanuts, towels, nails, a watch fob, and a tin whistle. Finding yet more stuff in the bargain basement, he finally leaves, with "pockets bulging, the waste [cotton] under my arm, the celluloid fish rattling against the diamond marquis, the hammer thumping upon my hip and the odor of the soap doll trailing behind me." Now Hancmann is running late and has no time to consider the cigarette holders, paper flowers, earrings, fruit syrup, Chinese tassels, and other goods he glimpses on his way out. After all, he still has to go find himself a pair of shoelaces.[73]

Celebrating the thrill of material excess, albeit cynically, these stories captured Americans' aggressively optimistic materialism (fig. 1.7). Others,

Figure 1.7. Americans all over the country enjoyed the cheap and abundant miscellaneous merchandise found in variety stores. "How to Display 5-10-25¢ Goods in the General Store," *The Butler Brothers Way for the General Merchant*, January 1913.

however, were not so sanguine and resisted whatever "progress" cheap goods might have offered. Champions of the Arts and Crafts movement, perhaps the most outspoken, saw the rise of variety and one-price stores as the logical result of the empty status-seeking of the middle classes who were driven to cram their interiors with vulgar bric-a-brac, knickknacks, and what-nots to signal their gentility and refinement. Embracing so much useless stuff, they believed, represented a sickness of spirit brought on and enabled by rampant consumer culture and its handmaiden, encroaching industrialization. According to the cultured elite, this was an era of "cheap goods and inexpensive men" and of "universal cheapness," an age defined by people's preferences for "infinite productivity, abounding things to wear, endless furniture, unlimited chromos" rather than simplicity and minimalism.[74] In contrast, they believed less was more.

What they failed to realize was that in the world of crap, more was always more. Certainly, caught in the "fatal allure" of the expanding market, Americans might not be very smart consumers. But perhaps they just did not care. People might have been satisfied enough to simply be able to imagine and be a part of all-consuming material worlds. Whether they realized it or not, they were continuing a practice that was first made manifest by peddlers carrying their cuckoo clocks and chamber locks, and since realized by retail chain magnates like F. W. Woolworth. Being able to live in an age of "universal cheapness" was entirely the point.

2

CHEAP GOODS IN A
CHAIN STORE AGE

There was no better testament to Americans' embrace of universal cheapness than the rise of chain stores devoted to selling cheap goods. By the late nineteenth century, independent proprietors of variety stores faced greater competition from regional and national chains that brought innovative selling strategies to the retail market. These included more systematized and orderly marketing schemes that continued to draw upon the emotional appeals employed by early peddlers. Rather than becoming obsolete, cheap's ties to fantasy and the carnivalesque were simply updated to suit the modern age.[1]

Bringing Order to Chaos

Variety stores continued to thread the needle between evoking wonder and seeming to promote rational consumption. Woolworth outlets, for example, seeded their staple goods with novelties and seasonal goods that encouraged customers to browse in the hopes of encountering something unusual, thereby enhancing the modern shopping experience by inflecting it with the surprise and randomness of the preindustrial marketplace (fig. 2.1). The most effective variety store encounters were fully immersive, the goods appealing to as many senses as possible. Architectural critic Ada Louise Huxtable had vivid, multisensory memories of these places:

> The stores smelled of sweet candies and cosmetics and burned toast from the luncheonette counters along the wall; they echoed to the sound of feet on hardwood floors, the ringing of old-fashioned cash registers and the clanging

Figure 2.1. Early Woolworth's storefront, featuring a variety of merchandise, including sheet music, bowls, baskets, and figurines. Location unknown, ca. 1900.

of bells for change. And there was the absolute saturation of the eye with every conceivable knickknack, arranged with a geometric precision based on the sales and esthetic theory that more is more. . . . Rows of snap fasteners on cards and stacks of cups and saucers were joined by new colored plastics of infinite uses and rampant kitsch that are now collectors' items. . . . There was a simple cornucopia of counters. "What the bazaar was to the Middle East," the anniversary literature tells us, "Woolworth's was to America."[2]

Department stores might have stoked a more sophisticated and elaborate kind of consumer dreamworld with their dramatic lighting, gleaming display cases, and cathedral-like spaces. But by heightening the immediacy and attainability of their merchandise—turning dreams into realities—variety stores offered something their more sophisticated retail counterparts did not. As a Woolworth's president once remarked, "Each customer who enters a 5-and-10-cent store becomes a rich man—for the moment. He says to himself, 'Anything I see and want I can buy.'"[3]

Retailing trade literature offered "systematic" plans for creating such conducive buying environments, anchoring the chaotic wonder of the bazaar to the firmament of chain store rationality. Strategies included installing effective lighting, building organized display spaces, and marking prices clearly.[4] A professional in the close-out sales business, who traveled from town to town selling discounted stock, tagged everything with prices ending in 6s or 7s, unusual numbers that would draw attention. These were, in his words, "*unbelievable* reductions" that "compelled" people to come and shop, "no matter how great the distance" (fig. 2.2).[5]

Savvy retailers also turned shopping into a treasure hunt. Unlike older dry goods stores, which presumed that patrons knew what they wanted to buy, variety stores made them need things they previously didn't even know existed. Cheap variety merchandise attracted consumers "in spite of themselves," because "they can't resist the temptation to look" at such affordable jumbles.[6] Even the prominent mail order house Sears, Roebuck capitalized on the allure of miscellany and surprise, publishing a monthly *Bargain Counter Bulletin: Low Price Sale of Odds and Ends* to dispose of close-outs, last season's fashions, and merchandise manufactured in "quantities too small to permit of quotation in our regular line." The company created a sense of urgency for discounted watch fobs, shoes, tablecloths, and serving spoons by exhorting readers that to "avoid disappointment," it would be "necessary that you order almost the very day you receive this book."[7]

By combining the random elements of the bazaar with order and rationality, variety store layouts created new kinds of retail spaces.[8] There were more and less advantageous arrangements of goods by which retailers tried to build profitable "associations." "Tooth paste and tooth brushes very obviously suggest each other" and should be displayed close together, "in order that the sale of one will help the sale of the other," advised *The Manual of*

Figure 2.2. The rationality of fixed prices was often offered alongside the sensationalism of variety and cheapness. Page from advertising supplement *"Jim Lane" The Price Wrecking Fool in Charge*, 1920 or 1925.

Variety Storekeeping in 1925.[9] Merchandisers stressed the importance of having well-marked goods placed in an orderly fashion, often illustrating these principles with diagrams showing how to arrange merchandise so as to encourage people to look and touch and make their own determinations about quality and price. Interestingly, putting neatly organized goods next to those heaped in a bin made the jumbled goods seem a better bargain. Earle P. Charlton, who owned an early retail chain, reported that "piling up" things like toothbrushes, combs, and hairbrushes "add[ed] immensely" to their sale. The same was true for other items: "We had rubber heels, wired together. We dumped them in one of these compartments, and showed rubber heels beside them in boxes." Those "dumped" in heaps sold better—"ten to one"—than the same merchandise neatly boxed and stacked. The heaped items seemed cheap by comparison, thus arousing "smart shopper feelings." Jumbled assortments also created a sense of scarcity and urgency, exciting what experts in consumer psychology refer to as the "thrill of the hunt"; even better, shoppers might come upon something they hadn't been looking for—an "unknown object of desire."[10]

Store owners became increasingly aware of the ways that the very arrangement of their stock could influence its desirability and sale. Professional advisors urged proprietors to emphasize a wide assortment of merchandise but cautioned against "grotesque and incongruous combinations" such as sanitary goods next to men's pipes, or hairbrushes "hanging over candy." Merchandisers considered stores as living organisms—containers of commodity fetishism. Charlton couldn't figure out why one of his outlets was failing, since it stocked "goods to make life."[11] The items themselves were thought of as individuals with reputations and agency. Retailing advice books warned proprietors, "Don't embarrass any of your merchandise by irregularly associating it with something that might give rise to ludicrous suggestion."[12] Merchandisers worried equally about "dead stock" that "didn't move" (fig. 2.3).

A Matter of Nickels and Dimes

"Living" variety goods moved rapidly in and out of stores. On average, chain outlets turned over their stock more than four and a half times in a year, and over five times in some years. Working on short margins, dime stores required such efficient circulation. To clear a profit after paying rent, wages, advertising, and other expenses, proprietors had to sell much, sell cheap, and sell often. Despite the benefits of bulk buying, each outlet in the mid-1930s cleared only a few pennies or nickels on each sale, and they made most of their money on items selling for less than twenty cents, such as crockery, glassware, toys, electrical supplies, and stationery.[13]

Figure 2.3. The most enticing variety stores offered a controlled chaos of abundant assortment. Five-and-dime store interior, Skokie, IL, ca. 1930s. Skokie Heritage Museum.

While some dime store goods were sourced from American manufacturers, domestic labor costs proved too high to produce low-priced items selling on short margins. So, as in the past, proprietors turned to overseas producers. By the late nineteenth century, many cheap goods came from Germany, particularly greeting cards, party supplies, novelty goods, "penny toys," and dolls. Soon Japan expanded the cheap goods market, exporting everything from chinaware figurines and lacquerware boxes to novelty animals made of felt and pipe cleaners.[14]

Woolworth's great success derived in large part from his ability to drive down costs by securing low-cost labor. Frequent foreign buying trips enabled him to source cheaper goods directly from factories and gather intelligence about what his competitors were ordering and what they were paying. The mass-production centers in Europe were particularly profitable: the Staffordshire potteries in England, where "some of the finest china in the world is made . . . and some of the poorest"; the German doll-making town of Sonneberg; Lauscha, where marbles and Christmas tree ornaments were produced; Gotha, world's largest manufacturer of tea sets; and glass-making plants in Bohemia. But by focusing so much on what would sell and the bottom line, Woolworth, too, had succumbed to a form of commodity fetishism, since he did not fully understand how the cheap crap supplying *his very own stores* was made until it was explained to him firsthand:

It is no longer a mystery to me how they [the Germans] make dolls and toys so cheap, for most of it is done by women and children at their homes within 20 miles of this place. Some of the women of America think they have got hard work to do, but it is different than the poor women here, who work night and day on toys, and strap them on their backs, and go 10 or 20 miles through the mud with 75 pounds on their back, to sell them. The usual price they get for a good 10¢ doll is about 3¢ here, and they are obliged to buy the hair, shirts and other materials to put them together. . . . They probably get about 1¢ each for the labor they put in them.[15]

By the first decade of the twentieth century, Woolworth was spending millions of dollars a year on goods made by the hands of the young and poor working invisibly day into night. Crap's business model helped build retail empires.[16]

Woolworth was not alone. Thanks to the ready supply of cheap goods flowing to America's shores in the first half of the twentieth century, he faced more and more competition. Over time, the revenue of five-and-dimes steadily increased and companies added more outlets, greatly expanding the presence of cheap chains in the retail landscape. S. S. Kresge started out in 1909 with 42 stores doing $5.1 million in sales; by 1957 the company's 692 stores reported total sales of $377.2 million. The first S. H. Kress store, established in 1896, recorded $31,100 in business and by 1957 had expanded to 261 stores reporting annual sales totaling $158.6 million. Behemoth Woolworth, already with 631 stores in 1912, was doing $60.6 million in annual sales. By 1957 its 2,121 stores were recording annual sales of nearly $824 million, averaging $388,400 each.[17]

Dime stores' continued success was not simply that they carried cheap and crappy goods but that Americans continued to consume them. At first shoppers were attracted to relatively low-risk buying and effortless novelty. By the Depression era, however, low price became a more salient factor influencing the purchasing decisions of the budget-conscious. And because merchandise could not be as easily replaced in straitened times, quality mattered much more, too. One woman in 1932 was frustrated by the "sudden disintegration" of her "bargain" stockings and her "bargain" dresses "made without hems and seam allowances" and her "bargain shoes" that "assumed stranger and stranger shapes" as she wore them.[18] She was especially irked by the drumbeat of the popular press urging women to buy things in order to help lift the country out of its economic decline. Millions of women like her did their part but were victimized by a "destructive system" that used low prices "as an excuse for selling me poor quality goods."[19] Their frugal efforts were thwarted by things that did not last. What did they expect, though? For generations, Americans had been quite aware that products cheap in price might also be cheap in quality.

Dime stores weathered the Depression better than many other retailers, seeing only modest declines in revenue. This they did in part by emphasizing low prices and tamping down the longtime associations of variety with the carnivalesque.[20] Yet even low prices were mere mirages.[21] For years, many dime store items had been sold by their component parts. For instance, a working eggbeater actually cost thirty cents—ten cents each for the bowl, the beater, and the screw-on cover. Curtains were priced at ten cents a yard but made up in sets costing a dollar and up; garters selling at ten cents each had to be purchased in pairs. "It may not be long before it will be selling shirts at 10 cents for each sleeve, 5 cents for each button, 10 cents for the tails and 10 cents for the rest of the garment," joked one writer at the time.[22]

As the Depression deepened, managers increasingly made their stores more orderly, forgoing the heaps of merchandise—better suited to headier times of material abundance—in favor of more rational spaces organized by "department" and neatly arranged. Rather than encouraging serendipitous discoveries, this new approach enabled shoppers to more easily find exactly what they were looking for. Store layouts eventually became so uniform that one adman noted "it was possible to make your purchases with your eyes shut . . . ice cream sandwiches and soda pop on the left, candy in the middle, jewelry over on the right" (plate 2).[23]

While store owners attempted to rationalize the shopping experience by creating orderly spaces and establishing firm prices, more emotional impulses continued to drive consumers, even during hard times, and perhaps especially so.[24] Variety stores regularly changed displays—one week linen curtains, the next electric-cord untanglers. Successful items tended to possess "a strong flash value."[25] And Depression-era marketers claimed that the most successful variety goods would sell "sensationally" when consumers themselves could judge using the "see it-feel it" test.[26]

Balancing the rational and sensational became especially important when dime stores raised their price ceiling to twenty-five cents in the 1930s. Because many people still considered themselves in a "class" too good to shop there, proprietors had to continue to appeal to their core group of regular customers, using both familiar and novel strategies to manage the expectations of cheap.[27] Breaking the ten-cent barrier enabled variety stores to stock more diverse lines of goods. As important, this led shoppers to feel they could be more discerning, directly comparing the quality of higher- and lower-priced goods. Instead of obscuring the cheapness of certain goods, store owners were now, in fact, leveraging poor-quality merchandise to move higher-priced items. Cheap became a foil. One contemporary marketer explained that when shoppers could see different "classes" of goods placed next to each other, "generally they decide in favor of the higher-priced article." This strategy, however, did not mean that those lower-priced picture frames languished on the shelves. They, too, became desirable when offered with a little something for free, like a length of picture wire or "a low-priced print of a popular movie star."[28]

Yet even the most "rationalized" spaces could confound consumers' ability to judge quality and value. Take, for instance, light cookware—a staple of five-and-dime stores. Low-quality cast-aluminum imports were much cheaper than stainless steel pots and pans made domestically, since they were made from melted down aluminum scraps often impregnated with grease, alloys, and other contaminants. They pitted easily and became discolored. Industry insiders referred to these articles as "F&G"—feathers and guts, containing, colloquially, the refuse.[29] But reputable manufacturers commonly made product lines of various grades that were sold alongside one another. Often, better-quality goods carried stamped logos while those at the lower end did not. An executive for the Goods Company, which produced light aluminum ware and the high-quality Mirro line for Kresge, admitted that "we will absolutely not place any brand on goods that we do not consider up to the standard." But the company also produced a generic line "to meet certain conditions"; tellingly, he remarked, "We do not recommend its sale."[30]

Even clearly marked items could not be trusted fully. Imitation silk hosiery was commonly labeled "Art silk," "Novelty silk," "Sylk," and even "Silk." Cotton and other non-wool fibers often went into "wool" fabric. "Irish" lace (actually made in China) was "inferior in quality and value." A fair number of the seventy-three products warranting cease-and-desist orders from the Federal Trade Commission in 1925 alone were commonly found on the shelves of variety stores, such as medical preparations, soap, hosiery, dress snap fasteners, fountain pens, and candy.[31] There was no way for consumers to truly understand each of the tens of thousands of items variety stores were now selling, especially if they were additionally confounded by deceptive goods.[32]

Made in Japan

Japan's long tradition of home manufactures made the country especially well suited to the production of cheap goods for variety stores. As early as the 1870s, soon after the nation opened its ports to global trade, it was making and exporting cheap stuff. Women and children, mostly unpaid, labored in small household units that fell outside of Japanese factory laws. Their hands fashioned everything from paper goods and party supplies to bamboo baskets and wire birdcages. Consisting of many different parts whose assembly could not be mechanized, these items could only be made by hand. Nevertheless, they still had to retail for, at most, a nickel or dime. As they are today, American consumers were the beneficiaries of such exploitive labor practices.[33]

Japanese workers, like their counterparts in Europe, labored in impoverished, unsanitary, and often dangerous conditions for "merchant organizers" who oversaw and coordinated factory production and served as middlemen between producers and wholesalers. Farmers were often recruited with "prizes" to come work in urban factories. And by the 1920s workers were forced to live in dormitories run by companies who monitored their every move, a model maintained by many global producers today.[34]

By the 1920s Japanese producers came to dominate the trade in cheap goods.[35] Because innovations in synthetic fibers led to an erosion of the market for silk, previously one of Japan's chief exports to the United States, Japanese manufacturers began dedicating themselves to producing merchandise destined for the five-and-dimes.[36] By the early 1930s the country was experiencing a veritable "trade boom."[37] American imports of Japanese toys and dolls, to name but one sector, had increased seventy-five-fold in less than four decades and by 1932 had overtaken its signature trade in tea and marine products, despite high protective tariffs.[38]

"Made in Japan" became synonymous with low price and low quality. By the early 1930s variety stores were chock-a-block with all manner of Japanese-made goods besides just toys, such as chinaware, paper goods, wirework, bamboo items, rubber goods, figurines, and tinware. A magazine article from 1933, "Made-in-Japan Christmas in the United States," estimated that some eighty million Japanese-made light bulbs and fifty-four million midget Christmas trees would bedeck American homes that year.[39] The article's author was dismayed that only "price, color, and shape" mattered to consumers, not an item's source country. He continued, "In the long run such goods were more expensive, not cheaper—the bulbs cost more to burn and went out more quickly." This sentiment was echoed in an ad for General Electric lamps urging consumers to "LOOK OUT for inferior 'bargain' lamps that are apt to waste current, die too young, and turn out to be only gay deceivers!"[40] Another observer asserted, not entirely convincingly, that although these goods' low prices found purchase in a depressed economy, "sometimes the quality proved disappointing and Japanese products lost their initial popularity."[41]

Yet until the beginning of World War II, and even after the institution of protective tariffs and quotas, American wholesalers and retailers continued to import cheap Japanese products on such a large scale that consumer advocates and protectionists accused the country of unfairly dumping low-priced merchandice.[42] Japanese exports seemed particularly easy targets for criticism, although cheap variety goods were sourced from around the world, including the no-name inferior lines coming from American manufacturers themselves. One business magazine in 1937 remarked, "Chain store buyers search the world for sources of supply. On the average twenty-five thousand separate items are handled by chain units and sources of supply must be kept open."[43] This "openness" made evaluating cheap goods even more challenging, since countries produced a wide variety of consumer items of various qualities, sometimes emerging from the very same factory. Products were often intentionally mislabeled as well. By the 1930s country of origin no longer provided a trustworthy signal of an item's quality. Some domestic manufacturers placed "imported" labels on their merchandise to

increase its prestige, while others stopped labeling their foreign productions as "Made in U.S.A."[44] In the public imagination, quality sometimes
mapped onto geography, but not always. Consumers thought positively of
Cuban cigars, Chinese tea, Irish linens, and French perfumes, for example,
but were increasingly suspicious of mislabeled and imitation goods that
they suspected or knew came from Japan.

By the late 1930s critics urged American consumers to boycott Japanese
goods for both political and economic reasons, including Japan's invasion
of China, the alleged inferiority of Japanese goods, perceptions of unfair
dumping, accusations of discriminatory trade policies, and a concomitant
rise of pro-American sentiment at home. Despite external pressures, retailers were reluctant to quit selling merchandise from Japan, given that
the country was a key supplier of their store stock.

Consumers themselves had difficulty fully supporting the anti-Japanese
cause, especially since the country provided so many of the cheap choices
they came to desire. "A Shopping Guide for Boycotters," published in 1937,
urged readers to stop buying silk of any kind—silk stockings, silk under-
clothes, silk dresses, and silk neckties. The article's hectoring tone suggested that women likely continued to purchase the offending items anyway.[45] This was due in large part to the "lure of low prices" and the sheer
variety of things coming from Japan.[46] Other cheap variety lines that should
be boycotted, according to the guide, included all items marked "Made in
Japan"; all Christmas tree trimmings, just in case; rag rugs "of the 'hit-and-
miss' type"; pearls, both cultured and imitations "of the cheaper sorts—
the five and dime varieties"; toys, especially inexpensive mechanical toys,
tea sets, celluloid toys, toy musical instruments, and definitely any from
five-and-ten-cent stores; chinaware, both cheap and expensive; bamboo
articles, including pet baskets ("Let your pets sleep in egg crates"); toothbrushes; matches; celluloid combs; sunglasses retailing for less than twenty
cents; umbrellas with wooden handles; brooms; magnifying glasses of the
less expensive variety; small mirrors; lightweight gloves; and, improbably,
mink furs.[47] This was a tall order. One author admitted that overall, "the
sentiment in favor of boycotting is much stronger than the deed itself."[48]
That this burden fell primarily on the solidly lower and middle classes, the
key patrons of dime stores, might have also accounted for its failure.

After the bombing of Pearl Harbor, however, buyers and sellers had
no choice: retailers themselves were removing all Axis-made goods from

their shelves. Some chains, like Woolworth's, "delegat[ed] inconsequential Japanese merchandise to the ashcan" and put better-quality merchandise in storage: after all, it wasn't the items themselves that were problematic, only what they symbolized at that moment. Others removed labels from offending items or obscured their "Made in Japan" imprimatur. Still others donated banished goods to charity.[49]

Japan's manufacturing sector was crippled during the war by the atomic bombs dropped on Hiroshima and Nagasaki and the firebombing of other major cities. While American consumers initially remained disinclined to purchase Japanese goods, by 1947 the two countries were working to rehabilitate Japan's manufacturing sector.[50] Secretary of State John Foster Dulles suggested that exporting cocktail napkins to the United States was one way for the country to rebuild its economy. His statement, however glib, carried a note of truth.[51] Some of the first consumer goods that Japan exported to the United States after the war were, in fact, "inexpensive and often disposable dime store wares," the things that "were exactly what American consumers had come to expect of Japanese manufacturing"— like bamboo fishing poles, Christmas tree ornaments, harmonicas, cotton Easter chicks, and artificial shamrocks for St. Patrick's Day.[52] The revitalization of the country's entire economy relied in large part on Americans buying all of these crappy things.

And demand did continue, even though people associated Japanese goods all the more with cheapness. A 1953 report, "Japanese Industry since the War," noted that subcontracting and the use of low-paid labor had resulted in "some sacrifice of quality and uniformity."[53] Yet this was not only the way small-scale Japanese manufacturing had worked for well over half a century but also the only viable production strategy for the devastated country. "War and occupation have not changed Japan's traditional tendency to dump poor-quality products on the world markets," complained one writer in 1949, who listed several inferior articles, from radios to rubber conveyor belts.[54] A survey in 1958 found that when considering items equal in price, quality, and style, only 1 percent of American consumers would choose Japanese products first.[55] The prejudice lingered. As late as 1967 a study on consumer behavior noted that Japan faced "strong unfavorable attitudes toward its products."[56] In another study of perceived product quality, Japanese items—from candy and shoes to chinaware and television sets—ranked last in every category.[57]

But American consumers had come to depend on cheap variety store goods and the foreigners who supplied the bulk of them. In 1941 the US government had enumerated the impacts of curtailing Japanese goods, almost all of them low in price and quality. Analysts suggested that American-made earthenware and manufactured glassware could substitute for "the cheap grades" from Japan, but they doubted it would, instead foreseeing a "sharp reduction" in domestic consumption.[58] Likewise, supplies of all plastic goods would be drastically reduced without imports from Japan, as would the stock of cheap straw hats, since there were viable substitutes for neither the braiding material nor the low-paid labor to weave it. Consumption of "N.E.S." (nonessential) paper goods, "consist[ing] almost entirely of toys, novelties, decorative knickknacks, and gewgaws of low unit value," and for which "lasting qualities" were not important, would be curtailed by some 70 percent. These products were so "negligible" that American manufacturers would not bother to make substitutes, and likely could not have done so profitably. "Practically all" lower and medium grades of chinaware dishes came from Japan; without them, American consumers would have to pay between 30 and 50 percent more for better grades. People "with only moderate purchasing power" would have to do without. That same demographic would also have to forgo decorative chinaware novelties "of small intrinsic value," like novelty salt and pepper shakers and animal figurines, of which Japan was practically the sole supplier. Nearly 90 percent of pyroxylin plastic articles such as buckles, brooches, and charms, "distributed principally through the 5- and 10-cent novelty stores," would disappear, perhaps to be replaced by higher-priced domestic versions and perhaps not.[59] A lack of access to cheap goods would not only curtail the growth of variety chains but would also foreclose many people's ability to fully participate in the market, especially in ways they had become accustomed to (fig. 2.4).

So once the war was over, Americans took what they could get—namely, crappy articles made in Japan.[60] Individual companies and the government itself tried to elevate Japanese manufactures and dispel stereotypes. The new Japan External Trade Organization (JETRO) aimed to restore the role and status of large wholesale trading companies, which had been essential middlemen before the war, serving as international selling agents and providing capital infusions for small producers.[61] The country also instituted a variety of quality control efforts. One was the Japan Inspection

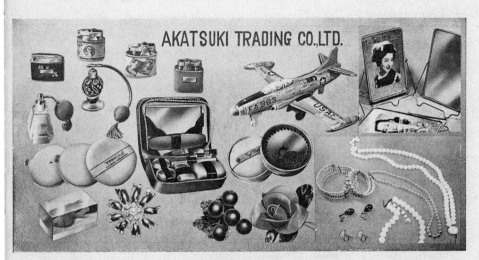

Fancy Goods

Spray, Toilet Set, Brooch, Ear-ring, Bracelet, Powder-Puff, Hair-Pin, Hair Ornament, Clip, False Eyelashes, Comb, Imitation Pearl Jewellery, etc.

Sundries

Tooth Brush, Hair Brush, Shoe Brush, Cloth Brush, Fan, All Net, Nail Clipper, Cutter, Buckle, Button, Fastener, Band, Mirror, Pencil Case, Soap Case, Specta-cles, Roupe, Safety Pin, Needle.

Rubber Goods

Ball, Tennis Ball, Diapper, Elastic String, Bust Pad.

Bags and Purses

Wallet, Suit Case, Hand Bag, Bamboo Basket, Shoulder Bag.

Toys and Games

Celluloid Toy, Mechanical Toy, Cloth Toy, Rubber Toy, Wooden Toy, and Games.

Porcelain Products

Tea Cup, Vase, Dishes, Brooch, Earring and etc.

Hardware & Alminium Wares

All kinds.

Others

Cigarette Lighter, Grass Mat, Glass Cup, Vinyl Sheet, Textile Piece Goods, Gloves, Gold Fish, Paper Weight.

AKATSUKI TRADING CO., LTD.

Trade Mark

| Head Office : | 1 Junkeimachi-dori 2-chome, Minamiku, Osaka | Open Bank : | The Bank of Tokyo, Osaka Branch |

P.O. Box No. 15 Minami, Osaka
Cable Add. : "AKATSUKEI" Osaka

Ref. Bank : The Bank of Tokyo, Shinsai-bashi Branch, Osaka

Figure 2.4. In the postwar era, Japan once again became a key source of cheap goods. These are some of the many "Fancy Goods" exported to the United States. Federation of Foreign Trade Promotion Institutes of Japan, *Merchandise That Japan Offers, 1955.*

Association, which encouraged industry "to strive constantly for better quality."[62] Another was the Japan Pottery Design Center, to ensure that Japanese potters were not pirating foreign designs and to improve quality and "expand sales."[63] Such initiatives were meant, again, to reassure the consuming public that even low-priced Japanese products were not complete crap.

Yet the realities were often much different, not necessarily because Japanese operators were acting in bad faith but because variety store merchandisers regularly practiced "split penny buying"—trading on very small margins—"caus[ing] a cheapening of the product to the point where usefulness is definitely impaired." Many kitchen gadgets, for example, were made of such inferior plastics that if exposed to heat or moisture (!) they would experience "warping and loss of strength." Although plastic toys and novelties came in pleasing colors, they were often made from melted scraps, which compromised durability and made them "barely serviceable."[64] Bamboo goods were susceptible to deterioration, and consumers often made "claims against unsatisfactory articles." Cheap celluloid flowers could catch fire.[65]

In order to keep prices as low as possible, chain store executives continued to squeeze manufacturers. Producers became so dependent on their contracts with variety stores that they were forced to sacrifice product quality, shave labor costs, and at times forgo their own earnings. Seeking to secure the business of a large variety chain, the manufacturer of a toy car, for instance, might be told that his prototype would have to be made in two pieces of two colors rather than one piece of cast metal and be outfitted with rubber instead of metal tires. Producing such a thing that would still retail for five cents was just not possible. And yet it was, at great sacrifice: "Great tears course down the manufacturer's cheeks, but he finally goes home and turns out the two-piece, two-color automobile." Although chain store executives believed that "a merchandiser dictating to a manufacturer produces a better value," it was only because manufacturers were forced to accede to the pressures and dictates of the large chains.[66] The logic of crap trickled up, from consumers expecting low prices to retailers cutting costs by insisting that suppliers scrimp on quality, cut wages, and even pressure *their* suppliers for better deals on raw materials.

American consumers continued to buy cheap Japanese goods out of necessity, ignorance, and apathy. The trade in Japanese ceramics exports, for instance, amounted to $54 million in 1955, almost a thousand times what it had been in the previous decade; almost 50 percent of all ceramics imported to the United States came from Japan.[67] That same year, nearly 50 percent of Japan's rubber toys alone (to say nothing of diaper pants, hot water bottles, fasteners, and other minor rubber goods) were destined for Kresge's, Woolworth's, Kress, McCrory's, Newberry's, and the many other chain stores.[68] Together, the chains were doing over $3 billion in business in 1954, up from $873 million two decades earlier; they could not have realized such gains without cheap Japanese goods (plate 3).[69]

The Persistence of Cheap

In the final decades of the twentieth century, the manufacture of cheap goods moved from Japan to Taiwan and Hong Kong, and finally to mainland China. Large-scale production facilities joined cottage industries to make low-quality goods in even greater quantities. Americans no longer expected that what they were buying might be worth the price, but, according to a trade expert in the 1980s, they had become "less choosy." Cheap's low prices and nastiness had triumphed equally. Therefore, overseas manufacturers "didn't have to worry so much about quality."[70] By the late 1980s Taiwan's producers had become known as "peddlers of cheap rip-offs, plastic toys and disposable junk . . . famous for cheap imitations and dime-store gadgets." Echoing Japan's reputation-burnishing efforts of previous decades, the Taiwanese government launched a $40 million promotional campaign that included a "National Quality Month," in order to improve people's opinions of the country's goods.[71]

Hong Kong joined Taiwan in dominating the late-century cheap goods trade, concentrating on toy production, from Cabbage Patch dolls at the higher end to knock-off plastic Teenage Mutant Ninja Turtle figures at the lower end. By 1989 Hong Kong had become the world's largest toy producer, with annual exports valued at over $2 billion.[72] This achievement came despite the fact that many goods—not only toys—were not just inferior but actually harmful. Consumer watchdogs noted that items coming from Hong Kong often had dangerously sharp edges, might fall apart easily, or were made with noxious plastics. In 1987 alone, the US Consumer Product Safety Commission recorded 113,000 toy-related accidents (from children choking on stray parts to licking toxic paint surfaces), including thirty-five fatalities.[73] Distressingly, many products were convincing imitations of name-brand goods, such as plastic bricks labeled 0937, which when read upside down looked like LEGO. That same year, California toy inspectors identified two hundred kinds of unsafe imported toys, numbering some three hundred thousand items altogether.[74] As in the past, officials recommended that "shoppers adopt a 'consumer beware attitude.'"[75] This proved increasingly difficult. Consumers not only welcomed all this cheap abundance but were unable, just as before, to accurately judge quality, if they even cared.

China's rapid industrialization in the late twentieth century can be largely credited to, or blamed on, Americans' insatiable appetite for crap. By the mid-1990s the country's economy averaged 10 percent growth each year, and in just twenty years the nation became the world's largest exporter of goods.[76] That extraordinary expansion was attributed largely to the production of crap, or what the trade literature refers to as "Miscellaneous Manufactured Articles"—a category comprising clothing, toys, footwear, and "items long identified as characteristic examples of Chinese exports," in other words, cheap stuff.[77]

Servicing the onslaught of cheap stuff also required a new kind of retail space. Mega-discounters, already on the rise, came to dominate the cheap business across the country.[78] They included not just behemoths like Walmart, Target, and Costco but also Big Lots and Deal$, all of which succeeded by offering ever-shifting arrays of cheap products. So-called small-box one-price stores, like Dollar General and Family Dollar, competed in the cheap goods sphere, too. In 1977 Kresge became part of the Kmart Corporation, which sold its remaining stores to McCrory's in 1987. Woolworth's went out of business in 1997. McCrory's and Newberry's followed in 2002.

Like their predecessors, the new chains offered variety as economy, setting up outlets close to what they characterized as "price conscious and budget-consumers," often including Hispanic and elderly populations on low and fixed incomes. Discount chains were able to offer merchandise at such low price points by purchasing "in huge quantities, maintaining strong relationships with vendors and suppliers, and by purchasing manufacturers' over-runs." In addition, companies like Dollar Tree skimped on packaging and put fewer items in a package ("adjusting the quantity") in order to maintain the fixed dollar price—a newer way of confounding value that was very much like the old.[79]

While seeming contemporary, other strategies also borrowed from earlier forms of persuasion. For instance, dollar stores installed freezers and coolers for food products to increase store traffic and thus "provide incremental sales across all categories, including its higher-margin discretionary product," much as early retailers used five-cent counters to increase sales of regularly priced items.[80] A discount on some items did not mean a discount on all, despite bargain hunters' enthusiasms. Staples commonly found in dollar stores, like toothpaste and laundry detergent, were actually more expensive than at larger chains. Shoppers bought them not only because of convenience but also because they *seemed* like bargains in context, stocked as they were amid misfit goods. These included irregulars, discontinued lines, and stale seasonal items. Trade industry reports noted that in this way, stores could sell merchandise that was "*perceived* to offer better value from retailers" but in fact did not.[81]

Other sales strategies established in the distant past also continued to work on newer generations of crap consumers. As dollar stores captured a greater market share among fixed-income consumers by appealing to their price-conscious, rational selves, they also played up the carnivalesque by using the stores' crappy inventories and often chaotic atmosphere as a selling point. Some stores tried to "enhanc[e] the 'shopping experience'" by making it "easy and fun-filled."[82] Others featured "customer-focused assortments."[83] Business analysts seemed to think there was something to this. For example, Dollar General had been consistently performing "well below" its competitors because of "low inventory turnover."[84] As a response, the company worked to "improve its merchandising mix" and "optimize presentation levels"—new-fangled jargon for, basically, offering more dazzling miscellanies of stuff.[85]

Many things had changed since Americans first embraced cheap in the late eighteenth century. Crappy stuff no longer came just from Great Britain, or was funneled through its ports, but was sourced globally, traveling from places like Germany and Japan, Taiwan and China. Independent proprietors of variety goods stores succumbed to competition from farther-flung chain stores and, eventually, big-box retailers, who leveraged their massive buying power to even greater advantage. And American consumers had not just learned to live with items that were routinely inferior and ephemeral but had come to expect nothing more from their stuff.

But in many ways the story had come full circle. Many kinds of cheap goods bound for early American markets, like affordable dishware and textiles, originated in China and India—sources that once again came to supply crap to American consumers. And many newer retailers of crap adopted age-old strategies to lure customers and confound value. While their expectations about cheap goods had grown more realistic, and lower, over time, Americans remained in the thrall of crap. The treasures once hidden in the intrepid peddler's pack—all those petty luxuries, trinkets, and affordable gadgets—had embodied a heady mixture of mystery, surprise, and immediacy. They were cheap, they were accessible, and they were wondrous. Early consumers, many of them market novices, could be excused for being seduced by cheap goods. Cheap not only fueled the nascent consumer revolution but enabled even the middling and lower sorts to participate more fully in the expanding marketplace of goods and ideas. This gave them power, helped them escape, and offered them entree into a world of expansive possibilities. Access to abundance also meant consumers had to determine for themselves how to measure price and quality. Involvement in the market was as precarious as it was liberating.

Over time, increasingly sophisticated marketing strategies only further frustrated consumers' ability to make sound purchasing decisions, and indeed created disincentives for them to do so, sweeping them up in the sensuous romance of cheap abundance. The consumers who inherited the legacy of cheap generations later, one would think, might have been better equipped to approach the world of crap more rationally and realistically. But shoppers remained at the mercy of the market, which, for better and worse, continued to deliver universal cheapness: crap for one and all.

PART 2

Better Living through Gadgetry

3

PERPETUAL IMPROVEMENTS

Americans, so the story goes, have always possessed an innate curiosity, creativity, and impulse to invent things. Their survival as colonists required not just grit and determination but the ability to improvise and make do in an alien world. The nation's expansion was then propelled by forward-thinking visionaries. The never-ending quest to improve sparked innovations that fundamentally and profoundly shaped not only the way people lived their lives but how they came to think about themselves as enterprising individuals and citizens of a nation alive with the spirit of progress. Much of this narrative is true, for better and worse. Undeniably, innovations like the Erie Canal and the cotton gin and the Corliss engine and the Model T helped turn America from a rustic wilderness into the industrialized behemoth it is today, expanding markets and population centers, giving power and light, making work more efficient, and generally raising the standard of living.[1]

Yet a more accurate and nuanced story of American progress can be told through innovations that were more modest and pervasive: everything from the Lightning Butter Churns, Miracle Combination Tools, and Galvanic-Electro Therapy Belts of the past to the Hands Free Hair Rejuvenators, Portable Pet Staircases, Biofeedback Posture Trainers, and LED Pain Relievers of more recent vintage. In their own way, these minor gadgets, too, embody the spirit of Yankee ingenuity—material manifestations of the always creative, often half-baked, and sometimes truly visionary imaginings of geniuses, charlatans, and cranks.

What separates a gadget from a useful tool is often a matter of context. Answering machines, for example, were once just the entertaining novelties of the idle rich; today, voicemail has become an essential part of daily life. But as with other forms of crap, it is only the exceptional gadget that faithfully lives up to its claims. More talk than action, most gadgets are only apparently useful. By embodying the mythic spirit of American enterprise and ingenuity, though, gadgets share something important with more vaunted inventions like Edison's light bulb and Graham's telephone. At the same time, gadgets earnestly embrace and shamelessly promote another vital part of American character, the unapologetic celebration of humbuggery and bullshit. Gadgetization is innovation baldly and boldly encrappified.

Like other forms of crap, useless gadgets have a long history, dating back to the dawn of the nineteenth century, if not earlier. While McCormick was designing his reaper, Singer his sewing machine, Edison his phonographs and light bulbs, countless other gadgeteers were working diligently on their own contrivances. Driven by both inventive impulses and profit motives, they also hoped to capitalize on American consumers' growing appetites for ingenious devices.

Adhering to the logic of crap, most gadgets were (and are) cynical and contradictory. While promising otherwise, they make relatively simple tasks more complicated and create problems where they did not exist before. Sometimes they even generate more work. Yet gadget boosters have convinced consumers that new-fangled devices are better than the outmoded and old-fashioned things they had been using. Because gadgets offer novelty, encourage curiosity, and promise new experiences, it often does not matter, ultimately, if they actually work or not.

Yankee Ingenuity

In ways that more major innovations did not, early gadgets enabled ordinary American citizens to participate in a larger conversation about the nature of progress and improvement. As political economists, manufacturers, and others began touting the benefits of internal improvements and praising American ingenuity in the early decades of the nineteenth century, countless ordinary citizens wondered about the implications of so-called progress, often prompted by the myriad novel things that increasingly entered their households. While there were many enthusiastic early adopters, others believed new devices only undermined their autonomy. The author of an article in 1817 characterized the time as an "age of improvements, when multiplied inventions have rendered useless many acts to which individuals were once called in the common concerns of life." This wasn't a good thing, since machine-made objects were "attained without effort"; too easily crafted, they were "possessed without delight."[2] In other words, people didn't appreciate them as they did hand-crafted items. Further, while labor-saving devices might effectively answer the problem of "self-willed servants," that same "well ordered machinery" risked making employers obsolete, too. Some, who had perhaps witnessed or participated in the "cheapening mania" brought on by variety stores, also worried about the never-ending quest for novelty. Old, reliable devices—durable, long-lasting, and "part of the family"—would be discarded for newer, more exciting, but not necessarily superior models, much like the faithful mate cast aside by a lover with a roving eye.

Over time, people's resistance to innovation, especially in the form of new devices, became even more animated. One self-described "Man Born out of Season" spoke out against gadgets in 1839, arguing that "speed, profit, utility, and convenience are the idols of the age." The "plague of new-fangled inventions," he argued, would result in bodies "dwindled" and minds "dwarfed" through disuse. "In this mechanical age," he declared, "the spiritual portion of man is fast falling into desuetude; and as for the physical part of his apparatus, it will be so nursed and pampered by small and manifold artificialities, that he will gradually lose the use of a great proportion of his faculties."[3] Creators, though, would surely reap mighty profits as machines gained strength and users became weak. As one "obstinate" farmer remarked, "Such jimcracks may suit some people, but they won't suit me; my harrows break the ground quite as well or better than a clod crusher; if some folks ... were half as fond of work as they are of new whims, it would be to their credit."[4] Part of the problem, of course, was the difficulty in distinguishing between truly practical innovations and ridiculous "new whims." What was more, people continued to remain uncertain, and often apprehensive, about the proper roles of machines in daily life.

These issues became more relevant as smaller devices increasingly infiltrated American homes. These were the clever products of "Yankee ingenuity," a term people used equally as a compliment and an epithet, responsible for both significant innovations and impossible flights of fancy.[5] For instance, an advertisement published in 1824 touted a Boston agricultural implement firm whose showroom housed "a display of machines of almost as many forms as there were kinds of animals in Noah's Ark." One might see, perhaps, a plough that would "Set itself to work / And plough an acre in a jerk." Was this a true possibility or hyperbolic nonsense?[6] The "fecundity" of Yankee ingenuity and innovation—giving birth to such things as automatic fans, new ways to saw wood planks from logs, and portable steamer trunks that converted into dentists' chairs—was a subject of marvel, delivering as it could possibilities so miraculous as to verge on the unreal.[7]

A correspondent writing for an 1833 issue of the *Maine Farmer* noted "the criticism and amusement" he faced when introducing a Yankee grinder that he claimed could process corn more quickly and increase its nutritional value simultaneously. Even his so-called friends laughed at him.[8] Non-Yankees were even less bemused by such novel devices, for they exemplified material excess and invention for invention's sake. Not only that, they embodied ideas that threatened local customs and the stability of the social order (fig. 3.1). Midwesterners and southerners resisted what they saw as northerners' encroachment, carrying with them all these new-fangled goods and notions. What was more, these new things were often delivered via sharp-trading itinerant peddlers—a.k.a. "trafficking Jonathans"—who were considered not just unwelcome strangers but foreign commercial

YANKEE INGENUITY.

A company has been formed for the purpose of towing icebergs to every port in the world, where a sale may be anticipated. We wish the project all success.

Figure 3.1. Many equated Yankee ingenuity with ridiculous and impractical ideas, as this cartoon shows. *Illustrated New York News*, June 21, 1851. Courtesy American Antiquarian Society.

agents siphoning local money into northern markets.[9] One agricultural magazine writer who often reviewed new tools admitted his knee-jerk tendency to distrust even "first-rate" devices if they came from "Yankeedom," were "something new," or, like some new religion, were promoted with too much "enthusiasm."[10]

People often expressed their critiques of Yankee ingenuity through comic exaggeration, since the form seemed to best capture gadgets' often excessive foolishness. The 1832 book *Memoirs of a Nullifier*, for instance, lampooned the heroically misguided attempts of an earnest inventor to capitalize on his Hooker's Patent Self-animated Philanthropic Frying-Pan, which would automatically flip pieces of bacon "when exactly half done." What the inventor thought was a perfectly sound business proposition turned out to be a ridiculous, resounding failure that observers relished.[11] An 1834 piece published in the *Virginian and North Carolina Almanack* similarly skewered the superfluous cleverness of Yankee ingenuity—a "fabulous, yet worthless, mechanization of the mundane." It described the fantastical workings of the New-England Sausage and Scrubbing Brush Machine, which ingested live pigs and extruded both *"ready made sausages"* and *"patent scrubbing brushes."*[12] This device was not to be confused with a similarly marvelous confabulation, Jonathan's Patent Labor-Saving, Self-Adjusting Hog Regenerator, which was a farmer's dream:

> One hog will pass through the machine in one second, and come out in the shape of forty hair brushes, one hundred tooth brushes, two hams labeled and smoked, two pig tail candles, (warranted genuine), pork cut in any shape desired, two hundred pounds of sausages, and souse and head-cheese ad libitum.

Even the "mangled skeleton," dropped from the bottom of the machine, could be sold to medical colleges "at the low price of two shillings per carcass" (fig. 3.2).[13]

If only. These whimsicalities worked as jokes because they were so far from the grim, brutal, and labor-intensive realities of hog slaughter. The ordeal entailed harnessing and then stunning often obstinate animals that weighed many hundreds of pounds, slitting their throats, enduring the gush of blood and final death throes, and shouldering the carcasses into and then out of a tub of scalding water. Then they would be debristled and

Jonathan's Patent Labor-Saving, Self-Adjusting Hog Regenerator.

Figure 3.2. Gadgets often differed little from utter confabulations, like Jonathan's Patent Labor-Saving, Self-Adjusting Hog Regenerator, which was able to process everything but the squeal. *Yankee Notions*, May 1, 1853. Courtesy American Antiquarian Society.

disemboweled. Only after that could a hog be fully disassembled and processed into useable parts, which itself was an affair of knives and blood and bone and flesh. Then those parts had to be preserved through salting and curing and turning all that blood into pudding.[14] William Youatt wrote of the pig, "Scarcely an atom of it but is useful." He explained that beyond the meat itself, the feet, head, and "even portions of the intestines" were valued by epicures; the scraps and trimmings were made into sausages and pork pies; the fat was desirable to perfumers, confectioners, and apothecaries; the skin was used for pocketbooks and other purposes; the bristles went into making various consumer goods, including brushes; and bladders, too, were useful (though he didn't explain how).[15] How delusional it was to think that a "labor-saving" machine could efficiently and effortlessly transmute the one living thing, the hog, into all the other things—products to be easily consumed.

In just a few years, though, industrialized meat processing, concentrated in Chicago and Cincinnati, did just that. What were once jokey mechanized confabulations had become all but realities by the mid-1860s. The scale and efficiency of such "labor-saving" devices created such an awesome spectacle that about half a million people each year visited the meat-packing plants. Tours included everything from views of the holding pens to vistas of the killing floors. New-fangled mechanisms made "disassembly lines" possible, on an unprecedented scale and with maximal efficiency.[16] And by alienating them from the processes of livestock slaughter, mechanization fundamentally changed Americans' relationships to their work and to their food.

Criticisms about gadgets, then, did not only center on the fact that many were just dumb ideas made manifest. They also, and more pressingly, addressed the opposite: that these new devices might change people's lives in profound and unintended ways. Whether true improvements or whimsical fancies, gadgets invited intractable conflicts pitting those embracing the new-fangled future against those resolutely stuck in the old-fashioned past. Either position forced people to consider the *true* nature of work and its proper place. Should labor be made less onerous, or would such a shift merely encourage idleness, increase humans' dependence on machines, and ultimately make them irrelevant?

By the antebellum era, vexing questions about the promises and perils of mechanization only intensified, given particular shape and immediacy by a flood of gadgets, particularly those that performed women's work. "Old Lady," writing in 1847, argued that neither old nor new was inherently good and advised carefully evaluating innovations in order to determine their efficacy. She made the case that saving time was as important as making money and that wise and "liberal" investments in new things would provide "a great gain" on the farm. Likening women to beasts of burden, she wrote, "Labor-saving machines in a farm-kitchen are, therefore, of the utmost importance, as they not only save time but strength."[17] "Liberal conveniences," like a new sausage-making machine, that "noiseless friend," offered several benefits, since a woman could now not only take on what had been considered a man's job but could do it more quickly, enabling her to assume additional tasks and accomplish them more efficiently. By simplifying work, gadgets thus created more of it.[18]

Women, perhaps even more than men, appreciated the benefits of technological advances. Printed advertisements, poems, lectures, and essays addressed the relative advantages of household gadgets, often pointing out how the genders had different access to labor-saving technologies. "The Kitchen Song," a popular poem of the late 1840s, for instance, was a proto-feminist call to arms:

Ho, ho Hum, how I wish
That each kettle and dish,
Could be cleansed by some Yankee machine.
It would save much a sight.
Of work, morn and night
To have one that would scour, wash and clean . . .

They're machines to cut glass,
And machines to cut grass.
And machines to fulfill all their wishes.
But they never once think,
While their own healths they drink
Of poor women who have to wash dishes . . .

And when 'tis completed,
The inventor'll be greeted
With praises from all that lack wealth
And ev'ry good lass
Will fill up a glass,
Of bright water to drink to his health![19]

A story in an 1866 issue of the *Prairie Farmer* addressed technological sexism even more directly. Believing in "the old system of hand labor, with no machinery," farmer John Merrill refused to adopt any of the latest agricultural improvements. Life was just fine for him, but his wife's workload was oppressive, and she lacked the authority to demand new tools: "With her children clinging about her, she baked, washed and ironed in the primitive style just as long as she was able to stand." She did laundry not only for her entire family but also for the hired men, without the assistance of a washing machine or wringer, and she daily churned the milk of a dozen cows in a "common wooden churn." Eventually the burdens of work took their toll. A "shadow of her former self," she at last "gave out," like an old horse. Her sister-in-law, a woman "with Yankee go-ahead-ativeness," proved her savior, insisting devices both "useful and ornamental" be purchased for the household, including a patent butter churn and a washing machine with a wringer. Because of her new-fangled devices, "Mary took a new lease on life," and her husband was happy to "keep up with the age."[20] New-fangled as they might be, implements like these could be life-changing tools. Or, as we will see, they could be no more useful than the puffery and promises employed to sell them.[21]

Patents Pending

By the middle decades of the nineteenth century, countless truly useful inventions were being produced for and adopted by the consuming public. Publications such as *Scientific American*, the *Patent Record*, and *Inventive Age* chronicled, publicized, and created a market for inventions while also celebrating the idea of innovation itself. By blurring the lines between the useful and the superfluous, gadget makers traded on the growing cachet and legitimacy of the inventions featured in mainstream publications. In fact, the useless gadgets—"ordinary patented novelties"—far outnumbered the "familiar and important machines known to history" and were even more profitable.[22] As early as 1834, the *Mechanic* magazine criticized the number of designs for "nonsensical machines" that reached the Patent Office. Many proved "of no possible use," "unsound," even "frivolous." "The evil seems to be growing worse," the author complained, especially since the $30 application fee would not "deter a man with more genius than common sense in his head," who had a mind to secure a patent on "every new notion that may enter his cranium."[23] Differing from cranks only in their ability to monetize their harebrained ideas, gadgeteers were a testament, if not to native ingenuity, then at least to the American spirit of trying to make a buck off of anything.

Because of the growing numbers of gadgeteers and their contrivances, consumers found it increasingly difficult to discern the truly useful from the cheap, "quick-sold" things peddled by "some voluble creature more tonguey than truthful."[24] Reviewing various new butter churns in the late 1840s, for instance, the *Ohio Cultivator* remarked on the "great efforts" being made "to palm off upon the farming public various kinds of patent churns, or to induce many mechanics to give large sums of money for the Right to manufacture the same." The author concluded that most examples were "absolutely worthless, or inferior to older kinds that might be procured for less money."[25] Lewis & Johnson's Atmospheric Churn, for one, made such an impressive appearance that even the knowledgeable author was at first "disposed to recommend it." Although it made butter much faster, however, it did not make it better; nor did it make as much. Similarly, Hamilton & Shire's Spring Churn came with "glowing advertisements and pictures to match" that promised women could churn butter "with as much ease and in the same way as rocking a cradle"; they could even knit and read at the same time. Made with a cheap tin can and a faulty spring, however, it proved to be "a humbug." Barlow's Combination Churn was difficult to clean, and Colver's Rotary Concave Butter Churn was "not deserving of the commendations bestowed on it." The recommended model was a simple design, well made of sound materials, and affordable—practical, albeit lacking in the astonishingly transformative capabilities of the most gadgety gadgets.

By adding a layer of legitimacy, patenting only heightened consumer confusion regarding practicality and efficacy. In order to illustrate to its rural readers how cynical profiteers were able to make money from so many bogus "patented improvements," the *Prairie Farmer* presented this scenario:

> Here comes a man with a patent seed sower. All the essential parts are old and of course not patentable. To stamp patent on it, Jones invented (!!) a "double back action ring bolt" and placed it upon the machine, not that it was of any use whatever to the machine, but to enable it to have a name, and it is called "Jones' Double Back Action Seed Sower."[26]

Encouraged by the growing number of patent agents, who stood to gain from trading in intellectual property, "hundreds persist in pushing these useless inventions before people," *Scientific American* grumbled in 1867.[27] The magazine had a vested interest in decrying gadgetization in order to maintain its reputation as a champion of legitimate innovations. Its editors criticized inventors' then-common humbugs of adding superfluous improvement onto superfluous improvement and of designing overly complicated machines that worked less efficiently than doing the job by hand.

Despite the best efforts of the high-minded promoters of "legitimate" innovations, the latest gadgets continued to intrigue both consumers who wanted to purchase the devices and capitalists interested in buying the rights to make and sell them. In the early 1870s one writer estimated that farmers were "swindled" out of tens of thousands of dollars annually "by a set of loafing patent peddlers" who were "palming off worthless patent churns, washing machines, &c., &c." on a credulous public. Professional teamsters observed that families often left this expensive stuff behind when they moved, since "they are found to be as useless for the purpose for which they are recommended as Don Quixote would be for a clergyman."[28]

Gadget puffery itself often unintentionally detailed the many and various ways that inventions might create more problems than they solved. Take the early washing machine. The Home Manufacturing Company's Home Washer claimed it would not harm fabric or injure a user, two key reasons why washing machines were "so universally discarded." And although it claimed to wash the heaviest of articles "perfectly," to wash whiter and cleaner, and to wash articles in one minute, the company added, "We do not claim that our Machine will *wash without* labor, neither will it do a family washing unaided by human hands."[29] (Then what was the point?) The many improvements made to the Vandergrift line of washing machines were also telling. They included removing internal wooden blocks, "which in time caused the bottom to rot out"; using a tighter-fitting lid to keep heat and the "disagreeable" odor of dirty suds from escaping; anchoring the gearing to the machine's cover so that the castings would not break in transit; packing in a newly designed shipping crate that did not require it to be nailed directly to the washer, "as is usually done" and "will invariably cause it to check or split after putting it to use"; incorporating a tub with a wider base to discourage the hoops from coming off; and implementing a new gear system to prevent the cogs from "jumping and slipping."[30] Similarly, Stratton & Terstegge's improved washers were made with galvanized castings to prevent rust, cypress wood that "allows no shrinkage or bad smells," and removable slats for ventilation and mildew prevention.[31] The takeaway was that improved washing machines' gears slipped and rusted. Their basins smelled, split, rotted, and were prone to mildew. They generated foul odors and did not keep the water warm. They also damaged clothing easily. Oh, and they also failed to eliminate hard work and could be dangerous to use.

Despite the myriad problems they might invite, women increasingly incorporated such "labor-saving" machines into their work routines, even though "labor-saving" was relative.[32] Like countless others, the husband who gave his rural Minnesota farmwife a dishwashing machine in the 1890s hoped it would lighten her workload and increase her productivity. The apparatus consisted of "one large sink, two tubs, and cranks and cogs and levers and inside works and a mop"; it required "copious amounts" of water to be heated on the stove and carried to and from the tubs, and the agitating crank had to be turned by hand. Labor-creating rather than labor-saving, this dishwasher was more productive as component parts — "its new owner decided to wash dishes by hand in the larger tub, using the little mop," while her husband used the smaller tub as a footbath.[33]

Puffing Innovation

Potential buyers could do little but rely on the advice of credible sources — their friends, newsmen, and their own judgment — to help discern humbugs from helpers. But this proved challenging. "Advertorial" endorsements in newspapers and magazines could be trusted to an extent, but as the *Plow, Loom and Anvil* acknowledged in 1857, editors, while "doing much to inform the farmer what implements are worthy of his attention, and to warn him against frauds," could be easily bribed to "join in the conspiracy" of puffing a worthless product. Of that, they noted, "we have no doubt."[34]

Predating the appearance of Yelp and Amazon reviews, gadgeteers were publishing user testimonials to convince potential consumers their products were not merely useful but miraculously so. Brooklyn laundress Mrs. Ann Rice, for instance, claimed that the Home Manufacturing Company's Home Washing Machine did not do the work of six women, as the company promised, but the work of *ten*. Not only that, but clothes came out much whiter and underwent less wear than if washed by hand. She also claimed she earned back her initial investment in the machine, $15, *each day* she used it.[35] Likewise, Mrs. Horace A. Seelye, writing from Montgomery, Alabama, vouched for her Leach Roaster and Baker, attesting that, in an age of hearthside cooking over open flames or on poorly regulated coal- and wood-fired stoves, "the mind is relieved of all anxiety about basting meat or poultry. The turkeys we have had cooked in it were very tender and not dried at all, and had a beautiful brown."[36] Easy, hassle-free, perfect results.

Long before the use of product demonstrations and well before the advent of the televised infomercial, these testimonials often came with illustrations that showed, more than words ever could, how gadgets could transform people's lives. Ephraim Brown launched his new-fangled Potato Steamer in 1856 with advertisements featuring illustrations of how

Figure 3.3. Before the infomercial era, gadgeteers like Ephraim Brown with his Potato Steamer relied on glowing testimonials to demonstrate the efficacy of their products. *Life Illustrated*, April 15, 1857. Hartman Center, Rubenstein Library and University Archives, Duke University.

people would benefit from adopting it, and how they might suffer if they didn't. Two scenes show people eating potatoes. One group praises their goodness, exclaiming, "They are steamed!!" The other diners gripe because their nonsteamed potatoes are "bad's poison!" (fig. 3.3).[37]

By envisioning consumers' pre- and post-gadget lives for them, marketers preempted questions about use value: Did one need such a steamer? Did it work? Did it take up too much room? Was it hard to clean? Did it make better potatoes? Instead, they focused on the lifestyle benefits. Overnight, women would become better cooks. Families happy. Dinner guests satisfied. In promotional material for his Steam Washer, "or Woman's Friend," J. C. Tilton showed interior household scenes that illustrated a "New Departure" from the "Old Way" that went well beyond cleaning clothes. Adopting the washer somehow made children behave better and brought gentility to a home through updated furnishings (fig. 3.4). Likewise, the "blessing" of "New Style" clotheslines made order out of chaos,

The Steam Washer, or Woman's Friend.

Figure 3.4. According to their promoters, adopting new gadgets could be life changing. J. C. Tilton, *150,000 Already Sold*, [1873]. Hartman Center, Rubenstein Library and University Archives, Duke University.

clothes now hanging in neat and tidy arrangements. These depictions encapsulated the true essence of the gadget's appeal. People loved these things not because of what they actually could or would do but because of what they *promised* to do. The world of gadgets was (and is) the world of eternal improvements, perpetually dispensing with old ways and buying one's way toward a more hopeful future.

Superlatives best conveyed gadgets' life-altering possibilities. The 1884 sales catalog for J. E. Shepard's kitchen novelties included "unrivaled" stove pipe shelves that were of "unlimited value."[38] In 1890, W. H. Baird & Co., "manufacturers of household necessities," offered an entire kitchen's worth of conveniences. With its Improved Iron City Dishwasher, "the dishes for an ordinary family can be *Washed Perfectly! Dried Perfectly! Polished Perfectly!* in two minutes' time." Iron City's Perfect Scraper "saves Knives, Spoons, Time and Labor." Astonishingly, the company also had "solved the problem of time" with its Lightning Churn ("Now here is your chance to give yourself a little relaxation"). The Beats All Egg Beater and Cream Whipper was "true to its name."[39]

The superlative spuriousness of makers' and marketers' claims was precisely the point. Like the claims made for other kinds of crappy goods, these, too, created desire by offering fantastically unrealistic solutions to real problems. By encouraging consumers to entertain the possibility of a lifestyle graced by convenience, ease, and perfection, the advertising rhetoric actually revealed the burdens under which most Americans—and rural women in particular—labored until well into the twentieth century. Rural Minnesota farmwife Mary Carpenter, for one, experienced her work as "monotony." A neighbor, Britania Livingston, "saw failure everywhere she looked," worrying herself sick. Women were "mere verbs—'to be, to do, and to suffer.'"[40] It was no wonder, then, that so many placed their faith in unbelievably heroic devices like Eureka Broom Holders and Champion Egg Beaters. They needed to, for their mental health if nothing else.

Extravagant Futility

Gadgets more often possessed "extravagant futility" than heroic utility, however. Devices like combination tools and multi-tools embodied not just the spirit of Yankee ingenuity but its ally, the audacious bullshittery of capitalist enterprise. Where innovation was concerned, the market was never far away.[41] The true object of gadgets was not, of course, practical utility but revenue for their producers. Charles Babbage, author of the popular nineteenth-century treatise *On the Economy of Machinery and Manufacture*, explained that for a maker to become a manufacturer, he had to think more about the profitability of his device than how well it actually worked. The entire design, along with the manufacturing process itself, had to be "carefully arranged" so that the product could be "produced at as small a cost as possible." Making something cheap enough to attract the greatest number of consumers necessitated "a saving of expense in some of the process." In other words, profit, rather than inventive genius or the spirit of improvement, drove ingenuity. The successful innovator was the one who figured out how to "undersell his rivals."[42]

And so, propelled by dreams of profits, inventors patented and brought to market the most improbable creations and odd mash-ups that promised to elide, ingeniously, several different processes into a single all-purpose object. An early example of such a gadget was Joseph B. Gilbert's 1812 invention of a tin box that was "so constructed as to combine all the necessary properties of a foot-stove, tea boiler, chafing dish, plate warmer, and butter or liquor cooler, as the season may be."[43] Others included the combination lung tester and bust developer; an all-in-one body-hugging hot water bottle, foot warmer, ice pack, and enema; a gasoline-charged clothes brush; shirt cuffs convertible to note pads; and animal-shaped puppet oven mitts. Thousands of such combination devices were registered at the Patent Office from the late eighteenth century through the late nineteenth, including an eraser and eraser sharpener, flesh-fork and skimmer, clock and fly trap, boot jack and burglar alarm, and spittoon and foot warmer.[44] Gadgeteers could come up with a seemingly infinite number of new confabulations.

Among the most popular, plausible, and long-lived combination gadgets was the multi-tool. (People still buy them today.) P. T. Barnum recalled being enamored of a deluxe pocketknife as a child: it was "a combination of all that was useful and ornamental," holding two blades, a boring tool, and a corkscrew. The young Barnum coveted this "carpenter shop in miniature."[45] More affordable and seemingly more practical than a carpenter's shop, such multi-tools found particular traction in the market, striking the fancy of middling families who hoped to outfit their households with a basic and cheap complement of tools. They were also fitting objects for the rising middle classes, who, performing mind work rather than manual labor, tended to hire outsiders to do home repair and didn't need truly useful tools. Such extravagantly futile devices were tailor-made for this burgeoning group; it was perfectly logical for them to expect the market to provide ready solutions—and often ineffective ones at that—to their various problems. The Household King, for instance, would find use in the kitchen, at the office, and even on board ships. "In fact," its manufacturers insisted, "no place can be mentioned but what needs this Tool" (fig. 3.5).[46] Another was The Washington Hatchet, whose manufacturer described it as a "tool chest full of the finest grade of tools all in compact form." The ten "perfect" and "useful" tools-in-one included a "well-balanced" hammer harrowingly attached, via hinged handle, to a hatchet head "tempered to a degree of perfection" with a "keen, sharp edge." A pipe wrench effected "many repairs without hiring out an expensive plumber." And the nail puller, pinchers, wire cutter, staple puller, and splicer are "all worked out on

Figure 3.5. Combination tools, often sold by traveling sales agents, first became popular in the later decades of the nineteenth century. Like countless similar devices that followed, The Household King claimed to combine many "useful articles" into one. *M. Young's Monthly Publication of New Inventions*, 1875.

scientific principles." The company insisted that unlike inferior examples, "it has not been thrown together in an endeavor to get as many combinations in a bunch as possible, to make a *selling article only*."[47] In other words, it claimed to be a useful tool, not a gadget.[48] That the advertisement was so emphatic on this point suggested otherwise and merely underscored the point Charles Babbage had made about the market logic of spurious innovation decades earlier (fig. 3.6).

Without the humbuggery and puffery that brought them to life and gave them purpose, gadgets were just inert and often inscrutable things. Promotional rhetoric—proffered through advertisements, mail order

TEN SEPARATE, PERFECT, USEFUL TOOLS IN ONE

The illustration on this page shows very plainly the many uses to which this tool may be put. It may be said with truth and honestly, that it is an ever-ready tool, in which is combined ten different, distinct tools. When we say it is a tool that will do the work of ten different tools, we mean just what we say. It was made to perform these duties faithfully. It has not been thrown together in an endeavor to get as many combinations in a bunch as possible, to make a selling article only, but its inventor spent years of hard study and experienced, in order to get a genuine tool of merit and durability. It is a compact, clean cut combination tool, scientifically made and balanced. The Hammer part is well balanced and has just the right weight for heavy driving. The Hatchet is tempered to a degree of perfection, and will carry a keen, sharp edge. The Screw Driver is made to stand the twists and turns to which it will be put, while the Alligator Wrench is fastened after the regular type and will be found a very useful tool. By means of the Pipe Tongs (between the jaws) you will be able to make many repairs without hiring an expensive plumber—a great saving in itself. These together with the Nail Puller, Pinchers, Wire Cutter, Staple Puller and Wire Splicer, are all worked out on scientific principles, and are as perfect as they can be made. The Washington Hatchet is a tool chest full of the finest grade of tools all in compact form. Any tool wished for can be always found without loss of time. These ten tools would cost ordinarily $5.60. The Washington Hatchet costs $1.50, a direct saving of $4.10 to every purchaser of this handy tool. If you were to purchase the ten tools represented in the Washington Hatchet, you would run chances of getting some of the tools imperfectly made. Every combination on this Hatchet is made alike out of the same material and tempered alike. Made in perfect unison. TEN PERFECT TOOLS All in One.

NAIL PULLER

HAMMER

WIRE SPLICER

ALLIGATOR WRENCH

WIRE

PINCHERS

STAPLE PULLER

PIPE TONGS

CUTTER

SC. DRIVER

HATCHET

SAMPLE TO AGENTS FREE---See Particulars on Other Side

Address All Orders to THE THOMAS MANUFACTURING CO., DAYTON, OHIO

PLEASE USE THE ENCLOSED SELF-ADDRESSED ENVELOPE WHEN WRITING US.

Figure 3.6. Promising to do many things but actually doing none of them very well, multi-tools, like The Washington Hatchet, were examples of "extravagant futility." Thomas Manufacturing Co., ca. 1900. Hartman Center, Rubenstein Library and University Archives, Duke University.

catalogs, and the mouths of traveling sales agents—transformed frivolous contraptions into necessary household tools. It did not matter if they didn't really work or last very long, as long as sellers could convince consumers they should have them. This was especially the case with rural customers, for whom cheap gadgets not only helped forge connections to larger markets but also seemed tailored to their particular needs.[49] Makers and marketers of petty goods "were particularly rigorous in working out intricate sales strategies designed to push farmers, or more often farmers' wives, to make a one-time purchase—or 'transactional' sale."[50]

Specialty manufacturers were flourishing by the end of the nineteenth century. Small and agile, they were able to fill orders for small-batch items on very short notice.[51] Working primarily in wood, wire, light steel, and plastic—the staple materials for many gadgets—they made any number of whisks, eggbeaters, fishing tools, card cases, and the like. In 1885 Philadelphia hosted the Novelties Exhibition, where great inventions like the telephone and the elevator competed for attention with "modest, if not trivial" inventions, including but not limited to "culinary novelties" (like countless chopper/slicer/extractor/dipper combinations), all kinds of celluloid corkscrews and letter openers, and myriad new-fangled paper clips and fasteners.[52]

Better Living through Gadgetry

Joining the countless "revolutionary" labor-saving devices like eternal clotheslines and self-cleaning nutmeg graters were gadgets that promised more personal transformations in health, appearance, and well-being. Dubious medical contraptions were simply material manifestations of the snake oil peddled by patent medicine purveyors; like the countless new-fangled devices promising to save labor, these bogus devices became popular because they, too, purported to address the real problems of average Americans. As tempting as it might be to blame credulous consumers for literally buying into cynical gadgeteers' fabulistic claims, the increasing presence of medical contraptions of all sorts suggests the extent to which the public suffered due to the failings of professional medicine. Formally trained doctors were hard to come by and expensive, and their knowledge was limited. Like inventors of other kinds of gadgets, developers of medical apparatuses were addressing real psychic and physical needs while also taking advantage of commercial opportunities. The seductive promise of medical gadgets was not that they made work lighter or faster but that they would be fundamentally life-altering.[53] L. Shaw's Cosmetic Face Glove—a.k.a. Toilet Masque—was a medicated face covering guaranteeing "entire and effectual renovation of the complexion." Like other gadgets, it claimed to be superlatively transformative: it did not just restore a good complexion "to its normal purity and beauty," but would *render recognition impossible* by the most intimate associate after a faithful use of it."[54] Quackery and gadgetry went hand in hand (fig. 3.7).

Deafness, more debilitating than bad skin, was yet another malady for which consumers sought ready cures, and gadgeteers readily obliged. The patented Dentaphone, "a genuine Scientific Invention," was a fan-shaped device "of peculiar composition" that, with its tip resting against one's teeth, was able to collect sound waves and convey them through the teeth and then through the bones of the face to the auditory nerves.[55] Complex diagrams and text explained how the device worked and attested to its efficacy. The testimony of one hundred "living witnesses" lent credibility

22   HOW TO BE BEAUTIFUL.

(Patented Sept. 4, 1877).

Figure 3.7. Medical devices promised instantaneous and magical transformations. The Cosmetic Face Glove so improved users' appearances, their friends no longer recognized them. L. Shaw, *How to Be Beautiful! Ladies' Manual*, ca. 1886. Hartman Center, Rubenstein Library and University Archives, Duke University.

to the company's claims. While truly effective bone-conducting hearing aids would not be invented until almost a century later, companies like American Dentaphone told the deaf what they wanted to hear (ahem). The Dentaphone claimed even to work for people born without ears.

Often, consumers found themselves buying—and buying into—not simply single devices but entire *systems* of innovation. The Andral Broca

Discovery for the cure of consumption was but one example. The company's seventy-page promotional booklet included testimonials from consumptives before and after treatment, anatomical diagrams, and instructions about how to "stay cured." This last could be accomplished only by subscribing to a proprietary treatment system, which entailed the faithful application of not one but several medical gadgets, such as Dr. Graydon's Direct Inhaler, Compound Inhaler, Nasal Douche, and Atomizer and Spraying Apparatus. In his 1906 exposé on quack medicine, Samuel Hopkins Adams condemned Dr. Graydon's system as "a combination of worthless inhalation with worse than worthless medicines."[56] It was just one of many deviously ingenious tuberculosis scams, which snatched the last desperate dollars from the hands of those dying slow deaths.

Makers and marketers of gadgets capitalized on the often blurry line between worthless device and transformative invention. Because electricity *was* responsible for so many life-changing innovations, it also made for especially beguiling gadgets and gadget systems. Much as multi-tools promised to perform any number of tasks, so too did electrical medical devices promise to cure any number of ailments. Late-century improvements in battery technology made it relatively easy to send electrical current pulsing through any number of devices to any number of body parts. Dr. Pierce's quack Joy to Invalids used electricity to generate "vital energy and physical power."[57] German-Electric Belts, patented here and abroad as a "Sanitary Galvanic Appliance," claimed to cure dyspepsia, liver complaints, kidney disease, back pain, sleeplessness, impotency, and constipation, for the low price of $5.[58] Electric finger rings, like The Conquerer, contained an electromagnet that could cure rheumatism.[59] The Twentieth-Century Electrocure fixed "any case of nervous prostration," "ninety-nine cases of lame back out of every hundred," "a large percentage of coughs and lung troubles," "a larger percentage of heart trouble than any other remedy," "indigestion with almost absolute certainty," "chronic headaches," and "weaknesses" in both men and women.[60] And the electricity harnessed by Professor Chrystal's Electric Belts and Appliances not only cured spinal afflictions and lumbago but also improved memory, reduced nervousness, and helped with "the enlargement of diminutive, shrunken, and undeveloped sexual organs" (fig. 3.8). It was a male enhancement device, and so much more.[61]

By the close of the nineteenth century, gadgets had become an integral part of American life. By then, patriotic claims about the country's

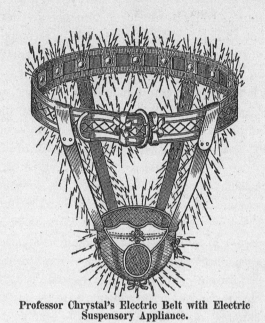

ELECTRIC APPLIANCES.   13

Professor Chrystal's Electric Belt with Electric Suspensory Appliance.

Practical experiments have proven that for all diseases of the sexual organs, *no matter what the disease may be,* and all diseases caused by early indiscretion, excessive sexuality, loss of manhood, nervous debility, etc,, etc., with all their horrible attendants, my electric belt with suspensory appliance, *will give relief, in from three weeks to three months, according to the length of time the disease has been running and the severity of the same.* In no case of this character will it take longer than three months to effect a permanent

Figure 3.8. Electricity, an unseen, mysterious, and often dangerous source of power, inspired countless medical gadgets at the end of the nineteenth century, like this Electric Belt with Electric Suspensory Appliance. Andrew Chrystal, *Catalogue of Professor Chrystal's Electric Belts and Appliances,* [1897].

native Yankee ingenuity had been long proven—ironically, not so much by large and significant innovations of industry but through the countless modest and often useless domestic devices. Forward-looking and innovative, gadgets promised to ease the grind of work and effect significant personal transformations. More important, those countless novel devices and improvements gave consumers a sense of agency—however cynically motivated and futile it might have been. For many, it was a welcome contradiction. Whether they believed in gadgets' outrageous claims or not, Americans remained bedazzled by the transformative actions promised by so many combination tools, miracle face masks, and electro-galvanic belts. By offering a seemingly perpetual stream of affordable improvements and innovations, gadgeteers enabled Americans to buy what they really wanted, which was perpetual optimism.

4

GADGET MANIA

By the early twentieth century, Americans should have realized that most gadgets were a waste of money. After all, by then they had undoubtedly read numerous journalistic exposés, heard public complaints, and cast a fair share of useless gadgets onto the junk heap. But over time, gadgeteers countered with the new technologies and materials they had at their disposal, such as improved plastics and electricity. They also could draw on more sophisticated forms of persuasion. While the new-fangled products of Yankee ingenuity had themselves become old-fashioned, the consumer market continued to be driven by hope and promise, a belief in progress, improvement, and the betterment that democratic materialism supposedly delivered. The market for gadgets flourished not only because of Americans' unflagging optimism but also because of their resolute need to believe, and their willing gullibility.

Works like Magic

As we have seen, early gadgeteers became adept at marshaling print culture to market their devices, leveraging the authoritative claims of "experts," the sincerity of first-hand user testimonials, and the incontrovertible evidence of before-and-after pictures. When those persuasive techniques were no longer enough to seduce increasingly sophisticated consumers, companies began turning to individual sales agents to perform in-person demonstrations showing what their things could accomplish. Sales agents' performances helped turn crappy into credible things. Using techniques borrowed from theater actors and magicians, they were able to harness Americans' innate curiosity, their desire to see and judge things for themselves, their sense of wonder, and their enjoyment of novelty and entertainment—all in order to sell them stuff that typically did not work as claimed and often did not work at all.[1]

Gadgeteers used these techniques to hype any number of products, including devices like the US Chemical Fire Extinguisher, for which the United Manufacturing Company made the kinds of superlative claims that came to define modern gadgetry: it was warranted to "last forever" and "never rust or corrode." The active chemicals would "never cake, lump, absorb moisture, lose their strength, or deteriorate in any way." Oh, and it could even be operated by children.[2] One sales kit enabled agents to perform up to two hundred demonstrations, and United would send "liberal supplies" of additional materials as the orders came in.

Company representatives assured one prospective agent, Henry Jones, that he could make easy money in his spare time selling "the greatest invention of the age." The product practically sold itself, and it would be "your fault if you do not make a fabulous salary."[3] But gadgets, whose value resided in what they purported to do rather than what they actually did, did not sell themselves. Good gadget salesmen well understood that their success depended on an ability to create effective illusions or to conjure reality—and, from the start, Henry Jones wasn't very good at summoning the artifice required to sell United's extinguishers. The devices he was tasked with selling, filled not with "special chemicals" but with sand, according to Jones, could not put out even the smallest fires. When he complained to the company, United's executives accused him of lacking dramatic presence and conducting uninspiring demonstrations: the problem was the inferiority of the show, not the product itself. They wrote:

Dr. Mr. Jones:

You are not a very good fireman.

A good fireman would not think of using chemicals for extinguishing a small paper fire such as you built. He would stamp it out with his feet. Another point, a paper fire is not a fair test for our extinguisher, for the simple reason that there is not sufficient heat created to give the chemicals a chance to act. . . . Now, in making all of your tests be sure that you are a sufficient distance from the fire so that you can throw the chemicals properly. Then, throw them with force so as to form a veritable cloud of chemical dust. In that way you will separate all of the atoms and make each one of them effective.[4]

Setting aside the scientific doublespeak, United made it clear that show-casing the magic of the gadget required effective stagecraft. Hucksters be-came sorcerers mesmerizing gawping customers. A novice seller, Jones had likely pored over United's confidential instruction manual, especially the part headed "How to Give Demonstration," which included lengthy descriptions of indoor and outdoor performances accompanied by an il-lustration of the agent and extinguisher in action.

The performer was as important as the product. As one might expect, the "Gasoline Test" created the "most impressive demonstration." From a small vial, agents were advised to casually decant a thin stream of gasoline onto a nearby surface. They were then to calmly subdue the flames using powder from the extinguisher, thrown at just the right velocity and at just the right part of the flame to extinguish it. This, of course, did not mimic the conditions of a real fire, nor the likely panicked response of the people charged with subduing it. But shock and awe, more than efficacy, was the point. The audience, according to the sales manual, "is at first startled at your seeming carelessness" with the dangerous liquid, and then is "amazed at seeing the flames disappear as if by magic." The successful agent was told to then simply "close your machine, and proceed to take the order" (fig. 4.1).[5] Demonstrators like Henry Jones were thus the human conduits through which gadgets were able to speak and act. The most effective per-formances enabled new-fangled devices to do their work in seemingly ef-fortless fashion and produce their impressive results, as if "by magic." In this way, consumers could see themselves as capable of performing the same feats as well.

The logic of gadgetization, then, rendered tasks accomplished not merely easily and efficiently but magically. It was the magical processes that were sold, rather than the things themselves. Hence the need for a "live, action agent" to sell things like mundane tire repair kits to a nation just beginning to embrace automobile transportation. Repairing automo-bile tires the "old way" took at least an hour and might not work. The Niagara Merchandise Company's new system reduced that time to only a few minutes, necessitated neither cement nor heat nor even jacking up the car, and was permanent. Even a ten-year-old-boy could do it. Real-ity came from the performance, enabling people to glean product truths from the magic: "All you have to do," the company claimed, "is to *show* an automobile owner these facts and he will order from you."[6] The company

This cut shows how to throw a handful of chemicals into the fire. Throw **forcefully** at the **base** of the flames.

Figure 4.1. People were drawn to gadgets as much for their theatrical possibilities as their actual uses. "This cut shows how to throw a handful of chemicals into the fire. Throw **forcefully** at the **base** of the flames." United Manufacturing Company, *Instructions to Salesmen: A Confidential, Man-to-Man Talk with our Representatives by the General Manager*, ca. 1910. Hartman Center, Rubenstein Library and University Archives, Duke University.

assured sales agents that once they successfully walked potential custom-ers through the repair process, "the trick is done."[7]

To be sure, "tricks" were simply convenient fictions. The Cinch Tire Re-pair Kit, like other gadgets, might have been of some use but fell short of the puffed benefits and promises. Plugs used in early Cinch Kits, for example, were made only of brass with no rubber sheathing. Rather than restoring the tire, "in some cases they would chafe," causing even more damage. The kit was also incompatible with newer tires.[8] Finally, because it was part of a "system" rather than a single device, buyers were obligated to use the company's ancillary products, like their proprietary rubber cut-ters, pressure clamps, and plugs, all of which could be purchased—at extra expense, naturally—only from authorized sales agents.[9]

Puffing Scientific Principles

While keeping one foot firmly planted in the Barnum-esque tradition of bombast, half-truths, and showmanship, gadget demonstrators increas-ingly appropriated the rhetoric of science and rationality to create magic as well. The new imperative to maximize personal and technological effi-ciency, promoted by Frederick Winslow Taylor, created new principles and standards by which gadgets were to be marketed, used, and conceived of as labor-saving devices. They were no longer simply clever products of Yankee ingenuity but proof of and measured by "scientific principles."[10] The pres-ence of labor-saving devices became more and more essential to middle- and upper-class homes as the use of servants dwindled and women took on more of the housework themselves. (Women of lesser means had always performed this work.)

In her landmark 1915 book *Household Engineering*, Christine Frederick wrote at length about the necessity of investing in the latest labor-saving tools, especially for women running "servantless households." Women needed to be smart consumers, buying only "good tools and high-class equipment" and gaining an understanding of how those things worked. They should also avoid being tempted by all the "worthless equipment" then flooding the market and discern the difference between the two. Women risked "wrongly buying on cost only" and also needed to consider the num-ber of times a gadget would be used. A $1 cherry pitter, for example (along with pimple massagers, miniature printing presses, and the like), should give one pause. Although women might be "influenced" to buy such a device, they should proceed with caution. Quality, she wrote, was "the most difficult point on which it is necessary to be informed," especially given that women received most of their information from "the words of salesmen and descriptive circulars."[11]

By the first decades of the twentieth century there seemed to be as many cautionary tales about gadgets as gadgets themselves. Frederick and others enumerated the ways they were too good to be true. They might be too complicated to use. Some were of poor ergonomic design, "not shaped for the comfort of the hand"—too short, too long, flat rather than rounded—because of poor construction meant to minimize costs of materials and production. They could be poorly finished, like galvanized dish drainers with edges "so imperfectly soldered and so rough" that they cut people's hands, or the fireless cookers whose hinges, "so jagged," tore clothing."[12]

Plenty of gadgets might be truly labor-saving in their core function but generated more labor indirectly. Comprising several complicated and interlocking parts, many devices were too difficult to keep clean and too challenging to reassemble. The time and effort necessary to keep a gadget in good working order often negated its labor-saving benefits and counted, Frederick noted, as *part of the total time that the device is being used."* Because of the surfeit of such goods—and their increasing affordability—it was easy to misjudge new products on these and other grounds. Frederick interviewed one woman whose hulking kitchen cabinet, full of gadgets, monopolized valuable space. It was not uncommon for women to overstock their kitchens with "badly chosen" devices for no other reason than "she *had to have them.*"[13]

Women had so many new kitchen tools and appliances from which to choose that *Good Housekeeping* could no longer accommodate product reviews in the magazine itself and had to issue separately published pamphlets. Like Frederick, writers and editors of women's magazines appreciated readers' mounting need for reliable noncommercial information about new consumer goods—and gadgets in particular—because they had "sprung up in prodigious numbers." Of the 1,015 devices the Good Housekeeping Institute reviewed in one year alone, 412 (about 40 percent) did not receive its approval. Consumers "suffered disappointment" and grew "disgusted" because of all the "cheap," "useless," and "poor" articles available to them. Many gadgets could not meet manufacturers' "exaggerated claims";

buyers "literally" believed them and were therefore truly "dis-illusioned"—
that is, deprived of the magic.[14]

Despite people's occasional disillusionment, gadgets remained popular.
Both despite and because of "America's prodigal consumption," consumers
were not quite able "to distinguish between devices that save labor and
those that merely kill time."[15] The persuasive pull of marketing continued
to convince people that gadgets could do miraculous things. One contemporary wondered whether "any mechanical item can be advertised successfully without using such terms as 'Magic Brain,' 'Electric Hand,' 'Wizard's
Eye,' etc." Probably not, since the sorcery of efficiency and ease that such
labels suggested was what consumers were actually paying for. "Any gadget
that requires attention while operating is passé these days; people have
been *taught to expect* labor-saving performance from machines," declared
one observer.[16] Much to the dismay of domestic scientists and product
testers, average consumers did not greet innovations with the dispassion
of rational skepticism but with credulousness, "innocent wonderment,"
"curiosity," and "astonishment."[17] It did not matter that experts repeatedly
pointed out to the buying public that "gadgets frequently fail"; that was
almost entirely beside the point.[18]

Mechanical Madnesses

Many gadgets seemed patently absurd because they were. Devices in
search of problems, gadgets turned their extravagant futility into a compelling marketing gambit by making the vicissitudes of daily life seem corre-
spondingly complicated, confounding, and labor-intensive—which it often
was in industrialized America. As a result, consumers—and often critics,
too—had difficulty distinguishing visionaries and their useful inventions
from cranks and their silly contraptions. Often the difference had nothing
to do with the utility of the thing itself but lay in its ability to gain traction
in the market.

Rube Goldberg's cartoons perhaps best captured the spirit that at once
critiqued and celebrated the audacity of gadgeteers and their way of complicating even simple tasks. First appearing in 1914, Goldberg's cartoon
contraptions, in the words of one historian, "dramatized both the social
pain and the enjoyment felt by his generation as it learned to live with
automation." Like so many Americans, Goldberg's subjects were caught

Figure 4.2. Although there was much to mock about the never-ending supply of new-fangled gadgets, people continued to buy them. Rube Goldberg, "Automatic Sheet Music Turner," n.d. Artwork copyright and trademark, Rube Goldberg Inc. All rights reserved. "Rube Goldberg" is a registered trademark of Rube Goldberg Inc. All materials used with permission. rubegoldberg.com.

between the past and the future, between the "life-giving harmony" of man and nature on the one hand and technology's new frontier on the other.[19] Like real gadgets, Goldberg's cartoon machines suggested baroque solutions to problems that were either wholly illusory or brought on by modernization itself (fig. 4.2). By turns fanciful, crazy, and half-baked, they were nearly indistinguishable from the countless "wizard thingamajigs" that were the objects of consumers' never-ending "little prayers and desires."[20]

Other humorists also recognized and lampooned the absurdity of new gadgets and consumers' willingness to buy them. One wag urged inventive types to "put out a whatchamacallit to fit over doorways for use in cases where people stand around saying good night for an hour before leaving." The device could dump a bucket of water, throw a custard pie, or plop a carton of eggs on the offenders. Annoying door-to-door salesmen "who doorbell you out of the bathtub" (probably selling gadgets) could be automatically dispatched with a padded bowling ball or a large bag of cement dropped from overhead. Everyday annoyances inspired any number of innovations:

(A) Gadget for people who muss your hair and step on your feet at the movies.

(B) Doohickey to stop people from beefing about liquor prices.

(C) Invention to prevent waiters from putting checks face down on the table.

(D) Device for eliminating conversation about the weather.

(E) Attachment for radio to filter out advertising talk, Eddie Cantor, and advertising talk.[21]

In a 1934 column entitled "This Month's Madnesses," one magazine described a number of new devices, including the Fiz-It, a contraption to open bottles "which calls forth a few curses till pressed hard enough for an opening in the metal cap to be punctured." A lever "pressed beseechingly" would carbonate the contents "if you can make it work." Another "madness" was the Ritz Friller, which "will curl and wave your vegetables," as if the world needed "an undulated bit of parsnip, or a marcelled beet."[22] One well-intentioned man who wanted to prevent his wife from becoming a "drudge fiend" told of adding to his household a Getzall orange squeezer, a Mixum egg beater, a Papa's Pal razor blade sharpener, two different potato peelers, and a No-Squish milk bottle opener; all of them were useless—and difficult to distinguish from humorous fictions. (The Fiz-It and Ritz Friller were real, by the way.) That they were equally (im)plausible showed the extent to which Americans were now giving themselves over to the new-fangled and leaving the old-fashioned behind.[23] In the process of choosing the oil lamp or light bulb, the horse or automobile, the paring knife or the rotary apple peeler, consumers had to decide whether to look toward the future or remain stuck in the past (fig. 4.3).

The Next Greatest Thing

Because modern life brought an ever-new supply of vexations—many ginned up by gadgeteers themselves—the market in new-fangleds was perpetually renewable. Rather than learning from the countless devices that had failed them, consumers remained forever optimistic that the next greatest thing would make life's annoyances and burdens at last disappear. Writing in 1942, marketing expert Neil Borden observed that the expansion of consumer products had come not from true innovation but from the endless creation of "'meaningless' or 'inconsequential' product differences."[24] It was not simply that "minor differences are built into products" but that "advertising writers frequently seize upon these small differences and magnify them beyond their due."[25] As a result, every new-fangled gadget could test a consumer's ability to make the right buying decision.

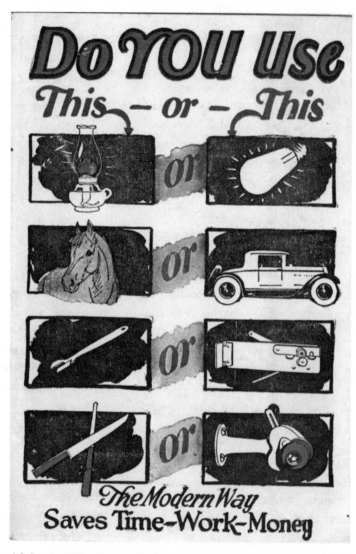

Figure 4.3. People who did not adopt the latest gadgets, marketers claimed, might as well be living in the past. The Speedo line of can openers and knife sharpeners was as modern as electric light bulbs and automobiles. "Do You Use This—or—This?" Speedo Collection of kitchen gadgets, 1934. Hartman Center, Rubenstein Library and University Archive, Duke University.

For consumers, embracing endlessly novel gadgets, whether practical or outlandish, was a way to shed their "old-fogeyism" and look perpetually to the future. Thus novelty triumphed over practicality time and again; it was the reason even the most improbable gadgets enjoyed viable markets in good times and bad. Even "thrifty housewives," who stood to benefit from labor-saving devices in their homes, could not help but be seduced by machines that promised to "perform with almost superhuman intelligence," whether an improved pea sheller, an automatic refrigerator, or "an electric toaster and percolator combined which also sets the breakfast table."[26] Critics of materialism wondered "whether people have sought happiness in a multiplicity of gadgets and have forgotten the art of simple living."[27]

Complicating matters was that even truly useful and time-tested gadgets existed in a state of perpetual obsolescence; the new-fangled became the old-fashioned at an ever-quickening pace. Door-to-door salesmen, whose livelihood depended on quick sales, were all too eager to demonstrate the latest "tricks." One woman lamented that her state-of-the-art vacuum cleaner, only six years old, had already been superseded by a "new ultra-scientific model" whose improvements included picking up cigarette ash, pulling threads out of cracks, polishing the floor, and killing moths. It would even dry her hair and, she sardonically noted, "transfer all the feathers from one pillow to another."[28]

Selling the latest "tricks" proved much more difficult during the Great Depression and Second World War, since families were struggling just to make ends meet and had little patience and even less disposable income to expend on them. However, nothing completely dampened the spirit of innovation and its promises. When consumer markets rebounded after the war, gadgets took their place among the various decorative and utilitarian material trappings of the suburban home. And many gadgeteers touted their devices as assisting with the efficient operations of the postwar home and the gender-prescribed roles within them. Automatic Toothpaste Dispensers were "swell for Dad's shaving cream," Neel-Ezpads prevented "Housemaid Knees" caused by waxing floors and scrubbing tubs, and Shower Chapeaus enabled women to put on their makeup before showering as they waited for "Hubby" to go first (fig. 4.4).[29]

All the while, the market for upscale gadgets, which were even more expensive and unnecessary, continued without interruption. The well-off, it seemed, were perpetually interested in the new-fangled and superfluous,

Come Clean in A Shower Chapeau

We're singing in the shower . . . about this wonderful new veiled cap that keeps your hair and face completely dry in the shower. Incredible, but true — you can actually bathe without dripping ends marring your pretty hairdo. And, if Hubby wants to go "first", don makeup while waiting; Chapeau keeps it perfect. Pastel plastic with "see thru" front.

B 7075 . . . . $1.00

Figure 4.4. Gadgets like the Shower Chapeau enabled postwar households to operate efficiently and maintain gender-prescribed roles. *Bancroft's Out of this World Selections*, ca. 1950s.

and the more obscenely outrageous and impractical, the better. For instance, rather than scale down operations in the 1930s, Hammacher Schlemmer, the vaunted hardware store, greatly expanded its line of inventions and luxury gadgets. The company was among the first to offer needless "wonders" like pop-up toasters, electric toothbrushes, and telephone answering machines.[30] While some now seem mundane, at the time they were testaments to excessive utility and pointless progress.

Making and selling gadgets for rich people turned out to be wildly profitable, helping companies like Hammacher Schlemmer not only survive the Depression and war years but flourish. In 1962 the company launched Invento Products Corporation, a subsidiary that encouraged inventing and oversaw product development, becoming "a clearinghouse for the new and unusual," including items from around the world that could be sold under the Hammacher Schlemmer name. Among the highly specialized gadgets solving rich people's problems over the years were tongs that cut sugar cubes in half, scissors that lopped the very tops off boiled eggs, grabbers for single stalks of asparagus, a French bean slicer, a pocket pepper mill, a parsley mincer, an electric chocolate grater, a device to measure the freshness of eggs (The Eggs Ray), a deluxe automatic natural yogurt maker ("new, delightful and easy"), and something called The Baconizer.[31] The company also offered personal, domesticated versions of items typically found only in professional or public settings, such as massage tables, whiskey casks, electric pants pressers, saunas, and even breathalyzers (fig. 4.5).

As Seen on TV

For a number of reasons, gadgets began once again infiltrating middle-class American homes after the Second World War. Technologies developed by and for the military were adapted and put in the service of the household; this retooling helped major manufacturers transition back into peacetime production. More rural American households were being electrified in the early 1950s, as were more appliances. In addition, there was considerable pent-up consumer demand dating from the Depression years. Finally, many Americans had disposable income to spend once again. Of all the new consumer goods, appliances were the things women most wanted to purchase, from washing machines and electric irons to radios and vacuum cleaners.[32] What was more, acquiring new furnishings to fill up their freshly built suburban homes became an important part of postwar women's roles as they left factory work.

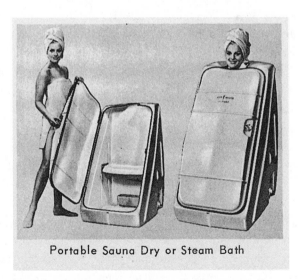

Portable Sauna Dry or Steam Bath

Figure 4.5. Among other items, Hammacher Schlemmer offered the Portable Sauna Dry or Steam Bath for $265 in 1967, equivalent to about $2,000 in 2019 dollars.

They bought televisions, too, a medium whose content consisted not only of newscasts, comedy shows, and dramatic serials but also of advertising. The glowing box, in fact, was perfectly suited to promoting gadgets, because they were "products that demanded context, explanation, or demonstration."[33] The intimate medium of television was able to do for gadgets what earlier forms of promotion had not, putting in front of countless viewers vivid, hypnotic, and action-packed demonstrations. The number of TV sets grew exponentially during the postwar era—expanding almost threefold in 1950 alone, to well over ten million sets. The number of broadcasting stations similarly increased during this time, as did the total number of advertising minutes sold. Broadcasters agreed that "any product that can be demonstrated in use and all goods that lend themselves to counter and show room display" were well served by televisual appeals.[34]

Such was the case with the Vita-Mix mixer, the subject of the very first television infomercial. Broadcast in 1949, it featured the strident yet mesmerizing sales pitch of onetime boardwalk barker and "food specialist" William G. "Papa" Barnard. In one half-hour block alone, for which he paid $270, Barnard was able to sell nearly three hundred mixers at a fairly substantial $29.95 each.[35] His device could grind corn, wheat, and soy into meal and flour; mix batter for waffles and pancakes; churn butter and whip cream; prepare pie fillings, omelets, or alcoholic drinks; and, of course, make "health cocktails." Staying true to the ideology of gadgets, the Vita-Mix, a panacea in mechanical form, promised to do many things superlatively, including dispensing "perfect health to everyone" (fig. 4.6).

Gadgets like Vita-Mix were tailor-made for television, where performances could be choreographed, edited, and reshot to accentuate products' real abilities and false promises. In addition, advertisers could purchase bargain slots during "graveyard" hours (between 11:00 p.m. and 9:00 a.m.). Bored insomniacs made for captive audiences, and infomercials were particularly stimulating during the nocturnal quiet. Late-night television provided the perfect opportunity for small entrepreneurs—"fledgling companies . . . door-to-door salesmen and garage inventors"—to hawk their products. Potential viewership was virtually unlimited. This presented opportunities for innovators and their strange new devices, people like Vita-Mix's Papa Barnard, who could produce longform commercials lasting upwards of thirty minutes and air them during cheap airtimes devoid of other programming, the wee hours and weekends. Although after

Figure 4.6. William G. "Papa" Barnard is credited with developing the modern infomercial when he broadcast a demonstration for his Vita-Mix Mixer in 1949 on television. The appliance came with a stainless steel jar, but he used glass so that viewers could better see the device in action.

the quiz show scandals of the late 1950s the FCC limited how much commercial airtime could be sold per hour, longer infomercials still aired on cable networks, where by the late 1970s and 1980s they flourished.[36]

In addition to promoting specific products, infomercials as a genre helped reinforce the prevailing celebration of American enterprise, championing the new-fangled over the old-fashioned. Even though products sold through early infomercials met with success only 50 percent of the time, hopeful inventors continued to try their luck.[37] The Rembrandt Automatic Potato Peeler was the perfect example of gadgetry designed for the televisual world. A veritable work of art claiming to be "the most revolutionary appliance in years," it made traditional potato peelers and paring knives—and the housewives who used such primitive tools—seem inferior by comparison. The Rembrandt's deep plastic bowl rested on suction-cup feet that anchored it to a counter. A hose connected the outfit to a faucet. Users placed their vegetables inside, locked the lid down, and turned on the water. In theory, the hydraulic pressure would pare, wash, and prepare for cooking any number of vegetables, all in less than a minute. It required no cleanup, since peels were "so finely pulverized they run freely down the drain and cannot clog." The finely calibrated process removed "only the thinnest layer of skin," leaving valuable nutrients intact. Rendering all

Figure 4.7. Often, gadgets invited more problems than they solved, and made relatively simple tasks more complicated and inconvenient. Rembrandt Automatic Potato Peeler advertising circular, ca. 1958.

other kitchen tools obsolete, the Rembrandt enabled beleaguered women to at last dispense with all their "old-fashioned vegetable peeling gadgets." It was "Science's newest contribution to greater convenience, leisure and economy" (fig. 4.7).[38]

Yet, true to the fundamental logic of gadgets, the Rembrandt actually created work, performed its designated tasks much less effectively, did not operate on "scientific principles," was not a "contribution" to the kitchen, and failed to make drudgery disappear at all, and certainly not "as if by magic." (One of the Automatic Potato Peeler's inquisitive customers was my grandfather, whose occupation during the 1950s took him on the road. He often found himself alone and awake in unfamiliar hotel rooms watching late-night gadget infomercials. The Rembrandt Automatic Potato Peeler was one of many devices he ordered, often without my grandmother's knowledge and never with her approval, she being a practical woman and rational consumer in the Christine Frederick mold. Ever the optimist, his much-anticipated test drive of the Rembrandt did not go as he'd hoped. His raw potatoes transformed magically not into delicate oven-ready morsels but instead into a starchy, macerated spray that stuck stubbornly to the kitchen walls. And to the ceiling. And to the floor. It took hours for him, abetted by his son-in-law, to scrape away the evidence of his trial before my grandmother came home. Certainly, the device *did* blow the lid off potato peeling, but not in the way he had expected. Also hewing to the nature of gadgets, the offending device was banished to a remote part of the garage and discovered only decades later.)

To be sure, my grandfather wasn't the only one fascinated by gadgets. The Gadget-of-the-Month Club (est. 1948) had quickly become a multimillion-dollar industry boasting membership in the hundreds of thousands. For $5 a month, each member received a new-to-market "whingding," "thingamajig," or "thingamabob." According to one account, "The package may contain anything, from a combination letter opener-letter weigher to the latest in car washing accessories."[39] Not merely a way for consumers to amass more stuff, the GMC also provided an important way for inventors to test-market and publicize their innovations. Anything the GMC's "impartial jury" thought filled a need would be manufactured in a hundred thousand units at minimum, and the inventor was promised royalties on each sale.

It was at this time, too, that the Popeil brothers, Raymond and Samuel—gadgeteers par excellence—got their start in the business. Like Vita-Mix's founder, they capitalized on surging interest in household innovations. Raised in a family of pitchmen, the Popeils knew as well as their predecessors that pointless innovations could nevertheless be highly successful, especially if effectively demonstrated. Rapt audiences, drawn in by "special offers" and "limited supplies," could then witness the peak performance of man and machine working in harmony. Such "effortless work" was just one purchase away. People bought, and bought into, the demonstrations rather than the devices. According to Samuel Popeil, "Making anything look easy is half the battle. If you're clumsy, the customer will walk away."[40]

Not only did they show the facility of a gadget, but demonstrations also prompted potential customers to engage with the product and its lively commercial representative on an emotional rather than rational level. One midcentury manual on salesmanship recommended that agents "dramatize" a product in order to "paint a glowing picture of it with a show of excitement and enthusiasm." This would make a prospective buyer less likely to think about price. A good demonstration would "fire his imagination with quiet intensity . . . shut[ting] out other thoughts from his mind," like whether his wife needed a $5 hydraulic potato peeler.[41]

The Popeil brothers were particularly successful at bringing to market simple gadgets of metal and plastic that could be produced cheaply and lent themselves to flamboyant pitches. A low price point—most of their devices sold for under a dollar—allowed consumers to take an easy chance and mitigated any disappointment if they failed to meet expectations. The Popeil Chop-O-Matic, demonstrated in early years by Samuel's son, Ron, was one of the most successful.[42] It appealed to the countless consumers who wanted new, labor-saving appliances in their homes but could not afford deluxe electrified versions and instead amassed a "substantial array of subsidiary gadgets" in addition to basic appliances.[43]

New gadgets stimulated the imagination as much as they spoke to rational "scientific" advancement. The Buttoneer, Miracle Broom, Tidie Drier, Kitchen Magician, Pocket Fisherman, and Speed Tufting Kit—to say nothing of the countless "O-Matics"—turned household drudgery into heroic feats and their users into wizards with extraordinary powers. In this way, gadgeteers reached back to earlier forms of commercial sorcery to help propel postwar nuclear families toward the ever-elusive goal of achieving the effortless, scientific, gadgetized life. Popeil and other companies helped promote this cultural zeitgeist through televisual appeals touting the latest space-age proprietary technologies. All told, consumers purchased over eleven million units of The Veg-O-Matic, for instance, which was hawked mostly through infomercials.[44]

Gadgeteers Master the Airwaves

Gadget master Ron Popeil most successfully harnessed the magic, showmanship, and pseudo-scientific marketing that had been so winning for itinerant barkers and hawkers for well over a century. Popeil's Ronco Teleproducts began telemarketing campaigns in 1964; by 1973 its net sales topped $20 million a year. Ronco's success was built on the company's strategy to market its never-ending procession of gadgets via commercial presentations that allowed people to see and judge for themselves and to imagine themselves as Popeil, able to perform their own miraculous feats. It was one thing to assert, for instance, that London Aire hosiery did not snag or run, but it was quite another to see pairs of the hose being subjected to the depredations of nail files, scouring pads, and cigarette lighters. So masterful were Popeil's demonstrations that *other* companies partnered with him to turn their slow sellers like the Buttoneer, Seal-A-Meal, Hula Hoe, and Miracle Brush into profitable product lines (fig. 4.8).[45]

The most attractive gadgets *performed* in the theatrical sense, and their value derived from the acting and the seeing rather than the doing. Gadgeteers sold their customers "the dream, the magical transformation." Their driving ethos was "State the before, but give them the after."[46] Infomercials enthralled audiences by exaggerating effects, creating convincing displays of facility, and emphasizing their performative aspects most of all. Postwar gadgeteers could credit much of their success to television, and vice versa. By 1996 gross sales of products sold through infomercials reached $1.2 billion; in 2015 the figure was $250 billion—accounting for one percentage point of the US GDP.[47] "Program-length commercials" crept into daytime slots and were used to generate direct sales, conduct beta market testing,

Figure 4.8. Gadgeteer extraordinaire Ron Popeil perfected the infomercial medium, using it to sell everything from spray-on hair to food choppers. This is a still from his infomercial for Ron Popeil's™ 5in1 Fryer™, a small fryer that does big things.

and create awareness about products that people could then purchase in retail settings (known as "retail driving").[48]

Inherently theatrical, infomercials created compelling narratives of transformation that were, of course, simply effective fictions. Infomercial presentations were often faked "for visual or dramatic effect." Producers "gaffed" their demonstrations to make products seem to work better than they actually did, and sometimes to make them work at all. One infomercial professional admitted that direct response marketing typically "involves degrees of manipulation."[49] Magic Wand hand mixers used precrushed pineapple that looked like whole pieces to demonstrate its literally unreal pulverizing abilities, and showed its amazing whipping prowess by using whole cream but claiming it was skim milk.[50] Timothy O'Leary, who produced infomercials for the renowned Ginsu knife, admitted, "We have to walk a fine line between entertainment and deception."[51] But did they?

Promotional "manipulation" was effective because it tapped into the needs, desires, anxieties, hopes, and fears that already resided, however inchoate, within the hearts and minds of consumers. People were sometimes hopelessly bedazzled by impressive product demonstrations, convinced that they, too, could battle growing waistlines, make perfect julienned potatoes, or deftly clean fish. They might be even more credulous since the myriad products, from power washers to yard weeders, pants fasteners to nonstick cookware, offered facile solutions to many problems. Just like other marketers, infomercial professionals became adept at problematizing consumers' lives and then obligingly offering them solutions, delivered in just a few easy payments (plus s&h). One late-century professional insightfully observed:

> As marketers, we identify an individual's *desire for more* as needs; needs that motivate and initiate the buying impulse. Need for more security (power, control, confidence, self-preservation, fear-of-loss); more wealth (greed, acquisition, something-for-nothing, bargain hunger); more love (vanity, glamour, self-esteem, exclusivity, conformity, guilt) and, of course, the need for more pleasure (sex, pain relief, power tools, and kitchen gadgets!). Such needs seek fulfillment and we are here to present solutions and, in the process, establish a relationship.[52]

If anything defined American consumers, it was their desire and need for *more*.

This helps explains both the enduring appeal of the entire genre of gadgety crap and the success of even the most improbable examples. Take, for
instance, the ThighMaster. Simply constructed—merely "two wire loops
covered with foam with a spring in the middle"—it was easily and cheaply
manufactured and sold at a significant markup. (Markups of 400 percent
were "not unusual," and products like the ShamWow cleaning cloth, "scraps
of cast-off industrial rayon and polypropylene," which sold wholesale for as
little as one cent each, retailed for over $5 each.)[53] Convincing puffery pre-
sented many possibilities for the improbable contraption: "The loops make
good handles. You can tuck it under your arm and into the curve of your
stomach. You can put it between your knees and work on your thighs."[54]
Aspirational personal narratives testified to the physical improvements
effectuated by the ThighMaster. And, like multi-tools and other combination gadgets, it was not one thing but many: "We're pitching it as a gym-
in-a-bag," its marketing representative explained.[55] The ThighMaster also,
crucially, benefited from the enthusiastic endorsement of Suzanne Somers,
a glamorous yet relatable celebrity. Within five months of launching the
product, the company was selling seventy-five thousand units a week; it
sold over six million in its first two years, becoming a multimillion-dollar
business.[56] A mere piece of metal covered in foam, the ThighMaster was
so popular that knock-offs quickly appeared. George Foreman, similarly,
was able to resurrect a tabletop grill its actual inventors had no hope of
selling and he himself was at first unenthusiastic about; he went on to
make $200 million from sales of his eponymous grill.

It wasn't just the masses who remained enamored of crappy gadgets.
The elite, who had developed enthusiasms for Baconizers and specialty
asparagus tongs in decades past, continued to purchase ever more outrageous and baroque contraptions. They were not, however, the "gimmicky"
products most often represented in infomercials but were instead more
exclusive items sold on the floors of upscale showrooms and within the
glossy pages of specialty catalogs.[57] These seemingly more refined, less carnivalesque forms of persuasion seemed better suited to the elite's perceptions of their more refined sensibilities and flattered their pretensions of
extremely conspicuous waste.

To members of the upper classes, outfits like Brookstone, Sharper Image, and Hammacher Schlemmer offered quasi-medical gadgets (infrared
pain wraps, facial nanosteamers, hand reflexology massagers); ostenta-
tiously specific devices (Wi-Fi pet treat dispensers, digital tape measures,

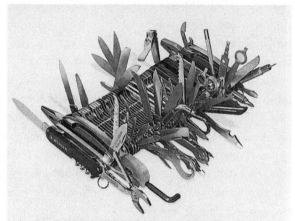

THE ONLY COMPLETE SWISS ARMY KNIFE.
This is the largest Swiss Army knife in the world, holder of the
Guinness World Record for "The Most Multifunctional Penknife,"
with 87 precision-engineered tools spanning 112 functions. Made
by Wenger, crafter of genuine Swiss Army knives since 1893, it uses
stainless steel for all parts and is hand-assembled by just two
cutlery specialists in Delémont, Switzerland, ensuring that every
knife meets exacting standards. It has seven blades, three types of
pliers, three golf tools (club face cleaner, shoe spike wrench, and ·
divot repair tool), 25 flat- and Phillips-head screwdrivers and bits,
saws, wrenches, and more. It also has a bicycle chain rivet setter,
signal whistle, 12/20-gauge shotgun choke tube tool, combination
fish scaler, hook disgorger, and line guide tool, cigar-cutting scissors,
laser pointer, tire-tread gauge, toothpick, tweezers, and key ring. See
hammacher.com for a complete list of tools. 3¼" L x 8¾" W. (2¼ lbs.)
HR-74670 $1,400

Figure 4.9. The Only Complete Swiss Army Knife was offered in the 2011 Hammacher Schlemmer *Gift
Preview* catalog. Although priced at $1,400 and described as "The Most Multifunctional Penknife," with some
112 functions, the device likely could perform none of them due to its extreme gadgetization.

cold butter graters); and novelties (Star Wars toasters, Superman capes for
office chairs, remote-controlled beach balls). Although more outrageous
and expensive, it was still crap. Many upscale devices resonated with people
whose income afforded them the dream of effortless work performed by
dirt-detecting robotic vacuums (a bargain at $700) and Bluetooth-enabled
3-D full-body heated massage chairs (at a fairly modest $4,299). At the
same time, there was something intriguingly creative, optimistic, and awesome about all the new whats-its and thingamajigs that, reliably, appeared
on the market, season after season, year after year (fig. 4.9).

Whether it was a modest product like the aerosolized bald spot concealer
Hair in a Can or a $58,000 golf cart hovercraft, the new-fangled could be

simultaneously ridiculous and very, very cool. By the end of the twentieth century, extreme gadgetization was the logical result of American consumers' desire for more: more functions, more features, more tasks performed more efficiently, more effortlessly, and more entertainingly. So Americans *did* get more: more expense, more waste, more labor, more futility, more disappointment, and, perhaps, more entertainment, more hope, more optimism. And because it was the *more* that mattered most, gadgets ultimately did live up to their many outlandish and empty promises.

PART 3
Land of the Free

5

GETTING NOTHING FOR SOMETHING

Something curious happened with the rise of commodity capitalism in the early nineteenth century. As business enterprises became even more motivated to maximize profits, they also started giving stuff away. In thinking about consumer culture, we often, quite sensibly, focus on transactions between buyers and sellers, when goods are bartered, exchanged for cash, or purchased on credit. But a lot of merchandise has made its way into Americans' homes because they got it for free. These "gifts," "inducements," "prizes," "rewards," and "incentives"—call them what you will—have been both an incredibly successful sales gambit and an efficient way for recipients to encrappify their lives.

The Alchemy of Free

Those who covet rewards miles and sign up for free t-shirts today descend from generations of consumers who also embraced various kinds of free things. As soon as crappy goods entered the American marketplace, entrepreneurs were not just selling it but offering it for free so they could sell other stuff. As early as the 1820s, the publisher of the *Christian Advocate* magazine was giving a free subscription to its traveling ministers for every six they secured. Even selling the word of God, apparently, needed to be incentivized.[1] A few decades later, Benjamin T. Babbitt—who pitched his product from a traveling wagon (and is credited with coining the phrase "get on the bandwagon")—recognized that he, too, needed to goose his sales pitch for baking soda, a necessary but unsexy product. For every box she purchased, a customer would received a cheap lithographic print.[2] The prints lured people to Babbitt's show and helped him sell more boxes of his conveniently packaged and distinctively branded product.

Realizing the effectiveness of Babbitt's strategy, many others soon used free giveaways, known in the trade as retail premiums, to make sales. One of them, soap seller Hibbard P. Ross, dubbed himself "Major Ross, the World-Renowned Soap Man." Ross's soap, also a generic product, was not even considered a necessity at midcentury; since soap did not sell itself, Ross staged a traveling show, often lasting several hours, while dressed in "pointed shoes, shorts, flapped-waistcoat, ruffled shirt, and peaked hat." Described as "athletic and spunky," he was known as "one of the most re-markable peripatetics in New England" (fig. 5.1).[3]

We can imagine Ross enticing prospective buyers with an entertaining performance and then sealing the deal with a "schedule of presents" dis-tributed with every purchase. At the time, Ross was not only competing with other equally charismatic traveling pitchmen but also facing a recal-citrant public slow to appreciate the benefits of using fine soap for personal hygiene.[4] So he offered nearly thirty different kinds of incentives—free things he claimed were worth from twenty-five cents to $500—for every dollar's worth of soap (ten bars) people bought.[5] It might be a linen hand-kerchief, a random issue of the *Illustrated Magazine of Art*, a gold pocket watch, or, too good to be true, a choice plot of land near a railroad station. "There is no gammon or soft Soap in this affair . . . there is no humbug," he assured. Although the Major's customers *did* receive a "present" for pur-chasing in bulk, chances of getting a house or even a fine piece of jewelry were long indeed. (Perhaps one in twenty thousand, if the prizes actually existed.)

Early promoters like Babbitt and Ross rightly surmised that the mere *prospect* of getting something for free, even if only an inexpensive handker-chief, spurred people to buy products they would not have otherwise, and in greater quantities than they needed or wanted. Consumer psychologists in the next century unpacked how the alchemy of free worked. Getting free stuff creates positive feelings of hope, desire, anticipation, and "good-will," incentivizing customers to do what merchants want—to buy their things—and to make them feel rewarded for it. The language itself is in-tended to kindle these positive feelings: "gift," "present," "prize." But sellers are in the business of making profits, not friends, and so the fundamental contradiction at the heart of free is that it comes at a cost. Sometimes it is

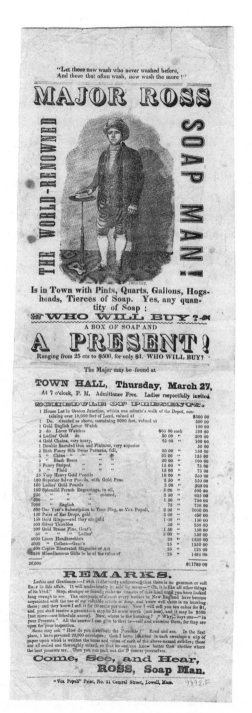

Figure 5.1. Traveling soap seller Hibbard P. Ross offered free prizes with every purchase; his "schedule of presents" was prominently featured on his advertising broadsides. *The World-Renowned Major Ross Soap Man!*, [1856]. Library Company of Philadelphia.

a monetary cost, as when people purchase more than they need or something they don't want, like ten bars of soap, or when they pay more for one thing in order to get another gratis. People pay emotionally as well, since all those "gifts" and "presents" that purport to engender "goodwill" are fundamentally insincere, designed not to reinforce bonds between family, friends, and neighbors but to create commercial obligations between sellers and buyers, sales agents and business associates. Perhaps the most pernicious cost of free things has been the way they have enabled the world of commerce, in the form of crappy goods, to insinuate itself into people's homes. This, then, helps explain why giving things away was not antithetical to the capitalist enterprise but an increasingly important part of it.

By the antebellum era retail premiums had become a popular and effective strategy for producers and distributors to reach customers. As important, giving away "free gifts" provided a viable market outlet for surplus and otherwise unsaleable goods: it was an ingenious way to create value where it did not exist. One early example was the "prize package," a sealed assortment of cheap stationery-related products accompanied by a "prize," sold by peddlers in the 1860s. The "great, original" S. C. Rickards prize packages, for instance, included "Writing Materials, Engravings, Fashion Plates, Fancy Articles, Yankee Notions, Games, Recipes, Many Ways to Get Rich, Rich Presents of Fine Jewelry, &c. The whole worth several dollars if bought separate. Price only 25c."[6] There were also Opposition Prize Packages, Valentine Packages ("new and very attractive . . . for soldiers"), Eureka Prize Casket Packages ("the largest ever sold"), and many more.[7]

Despite the boasts, these packages contained crap that had no value in the retail market and were merely masquerading as secret treasures. W. H. Cately & Co.'s Dime Panprosphosium Prize Package consisted of a few sheets of writing paper with mismatched envelopes, an outdated calendar, and miscellaneous magazine pages—publishers' overruns whose only value was in their rag content.[8] The "valuable prize" might be a pin, ring, sleeve button, or any number of other notional goods produced by the hundreds of thousands and costing only pennies wholesale.[9] For instance, at his "Head-Quarters for Cheap Jewelry," J. S. Andrews sold hundred-piece lots—mostly destined for cheap prize packages—for a mere $4.[10]

Remaindered books languishing on booksellers' shelves also found new market value when combined with other crappy goods. In addition to auctioneers, peddlers, and secondhand booksellers, a new kind of midcentury reseller, "the gift-book people," capitalized on cheap, discounted books. This business model rested on selling cheap books by offering a cheap prize determined by a random number inscribed inside the back cover. Rather than discounting the remaindered books, though, sellers actually marked them up, because people were willing to pay more for a cheap book that came with a "gift" or a "present," since it was the present that they were really after.[11] This motive force kept remaindered books circulating in the market, rather than being relegated to the pulper, and endowed these otherwise unsaleable things with value amounting to tens of thousands of dollars

People came to gift-book showrooms not for the books but the prizes.[12] George G. Evans, one of the most successful operators, acknowledged that "many," but certainly not *all*, of his customers made purchases "because they want the books." He admitted, "We think there is not much doubt that every individual who orders a book of us, entertains, at least, a secret hope of securing a valuable present." Evans's printed mail order catalogs routinely put the substantial "Watch and Jewelry Catalogue" of "inducements" in the front. Taking up more pages than the books themselves, the catalogs enumerated almost fifty different "classes" of "gifts" worth from twenty-five cents ("miscellaneous articles" such as thimbles, pen knives, and "articles for the toilet") to $100 ("Patent English Lever Gold Watches").[13] A contemporary court case charging a gift-book operator with running an illegal lottery scheme stated emphatically that gift-book establishments sold books "above their real value," and that the defendant purchased a book for a dollar that was worth much less, paying as much for the chance to win a prize as for the book itself. (Ironically, one of his purchases was *The Life of Barnum*.)[14] People were quite happy to pay for free (fig. 5.2).

Gift-book proprietors themselves readily acknowledged the shoddiness of their prizes. Albert Colby, for instance, described his "gold" jewelry prizes as "slightly brassy."[15] Responding to claims that his prizes were "too good to be true," Evans explained that he sometimes paid in cash and bought in bulk, which allowed him to *"purchase at less than one-half the cost to manufacture."* He also boasted of purchasing unsold stock from failed businesses and production overruns "made *in excess of demand*." None of it, in other words, was actually very good.[16] As long as crap was free, though, it worked its alchemy; items that individually had no viable market became

Figure 5.2. People milled about showcases contemplating the prizes they might win if they purchased a cheap book. Edward Sachse, *Interior View of Evans' Original Gift Book Establishment*, 1859. Historical Society of Pennsylvania.

desirable, especially when offered with other crappy things. Consumers who thought they were getting something for nothing along with an incidental purchase were able to doubly partake of the exhilaration of the nineteenth century's burgeoning consumer culture.

Since many of these items were new to them, consumers might be unable to judge their true quality and hence be quite pleased with their fortuitousness in receiving the "gifts" of market bounty. In the late 1850s residents of Stone Mountain, Georgia, for instance, purchased copies of relatively common books like *Uncle Frank's Pleasant Pages for the Fireside*, Samuel Mitchell's *School Atlas* and *Geography*, and T. S. Arthur's *The Angel and the Demon* from their neighbor G. R. Wells, who, acting as an agent of G. G. Evans, was himself promised free things if he sold volumes in bulk. The prizes they received included a pair of men's engraved gold studs (ostensibly worth $2.50), a lady's plain gold pin, "new pattern" (also worth $2.50), a silver-plated butter knife ($1), men's sleeve buttons ($2.50), and a man's gold-plated pen ($2).[17] People in the hinterlands might have been delighted with these petty urbane luxuries and would have little way of knowing whether they truly were worth what Evans claimed; perhaps they did not care.

In addition to appealing to consumers' rational selves by promising something for nothing, retail premiums tapped into emotional yearnings, despite their triviality and cheapness. Presented with the opportunity to purchase a typical prize package, one woman described being "overtaken by the desire to buy." Intensely scrutinizing the outside of the sealed box, she "fairly hankered for twenty-five cents with which to test the delusive promise of a possible one-dollar greenback within, not to mention 'attractive articles' of jewelry and unlimited stationery."[18] Fully aware of the prize box's "delusive promise" and the dubiousness of the "attractive articles" (the sardonic quotation marks were her own), she was nevertheless seduced by the mystery. Purveyors of free things understood that these crappy goods could stoke acquisitive passions: they often referred to them as "inducements" and "incentives," words with roots meaning, respectively, "to lead" and "to smolder" or "to burn" (i.e., to incinerate). Incentives made (and make) people *feel* a certain urgent excitement that induced (and induces) them to *do* something in response.

Free Becomes Systematized

By the 1870s countless enterprises across the country, large and small, began adopting inducement strategies. The Atlantic & Pacific Tea Company (A&P), for one, implemented a club system during the mid-1860s. Its expansive mail order program, which offered free boxes of tea to people who organized groups that would purchase in bulk, was crucial to the company's success.[19] Other businesses followed suit, like the Larkin Company, which instituted its "Clubs of Ten" system in the 1890s to encourage rural women to act as sales agents by leveraging familial and social networks to sell soap. Instead of receiving wages or commission fees, club organizers earned product discounts and premium gifts if they met sales quotas. The system was so successful that it enabled Larkin to eliminate all middlemen and sell directly to customers through mail order. They purchased premiums cheaply in bulk and even manufactured some of their own in order to further maximize profits. By the early twentieth century Larkin was offering its club organizers a choice of over 1,600 premiums, a more dazzling array, to be sure, than the modest variety of soaps the company actually produced.[20]

Giveaways became increasingly important in companies' advertising literature, often taking up more space than the primary products themselves. As marketing expert Henry Bunting noted in his seminal book of the time, *The Premium System of Forcing Sales*, "People buy the goods because of the premium. The premium, *not* the goods, is the inducement. . . . The way to sell the goods is to make public your premium offer."[21] The Boston-based Great London Tea Company's 1891 illustrated price list, for example, devoted a scant ten pages to descriptions of its teas and coffees but over a hundred pages to its giveaway silverware, pocket watches, bronze figures, lamps, mantel clocks, serving trays, and buttons. Should they not wish to buy the tea to use or resell, people could pay a "cash price" for these things, ranging from $1.25 for a hanging match safe to $20 for a 130-piece china dinner set. The dinner set came free with a $60 tea order, meaning sixty to a hundred pounds of tea depending on quality; the match safe required a more modest $5 order. Either way, that was a lot of tea—more than a lifetime's worth in some cases. The items were offered "simply to induce people to order in larger quantities and to get others to join with them in ordering," the company explained.[22] In this way the outfit was able to sell, by orders of magnitude, more tea than there were mouths to drink it. Other businesses, such as magazine publishers, used premiums to get people to sell subscriptions on their behalf, paying wages in the form of an often trifling object. Haverfied & Givin, one among many, offered a $15 "silverine" pocket watch to any sales agent who sold sixty subscriptions to *Home and Youth*, "in order to induce you to work more earnestly for us" (fig. 5.3). And mass merchandisers incentivized their traveling agents by offering free things as rewards for meeting sales marks.

M. W. Savage gave over so much of his advertising space to describing his premiums and detailing the complex system for earning them that it was hard to know exactly what he was actually trying to sell. Apparently it was something called International Stock Food Tonic, a product not even mentioned until page 20 of his 1914 promotional booklet *Savage's Free Premiums*. The pages and pages of stream-of-consciousness text explaining the free premium plan made no apologies for the fact that his gifts came not from the heart but from self-interest:

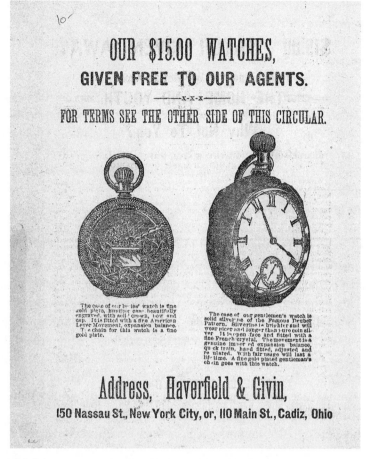

Figure 5.3. Sales agents were often incentivized by the idea of free. Circular advertisement for Haverfield & Givin, publishers of *Home and Youth*, ca.1880s.

I give Free Premiums BECAUSE they Help Me to Largely Increase My BUSI-NESS and this Helps Me to make more money at the end of the year. A Larger Volume of Business Greatly **Decreases** the Overhead Expense of Any Business. . . . I have found that I can secure Increasing Sales for Less Money, **with Free Premiums,** than I can in Big Newspaper Advertising **One** or the **Other** being absolutely necessary, and my Free Premiums, GIVE Consumers, this necessary **Business Increasing Expense,** instead of giving it to other people. . . . I believe this is Practical, Every Day, Fair and Square CO-OPERATION because

this Big Cash Saving, goes Directly into the Pockets of the PEOPLE who Help
Me Increase My Business, simply by using My Preparations or Products,
when needing such goods. My Free Premiums, are actual—**Cash Premium
Dividends**—returned directly to YOU. . . .[23]

Savage offered an eye-popping array of goods, from cut crystal bowls to
diamond rings, leather change purses to pocketknives—all described in
obsessive detail in dense blocks of 8-point type (fig. 5.4a–b). His systems
were several and baroque: not only Regular-Extra Quality Free Premiums
but Extra High Quality Premiums, Splendid Double Free Premiums, and
Cumulative Premiums.

The advent of trading stamps and coupon systems modernized, systematized, and spread the use of retail premiums further. Trading stamps
were first issued in 1892 by the Milwaukee-based Schuster's Department
Store with its Blue Trading Stamp System. Shoppers received a certain
number of stamps with each product purchase, to be dutifully pasted into
specially designed booklets; since Americans were already engaged in a
scrapbooking craze during this time, they seemed quite amenable to all
this licking and pasting.[24] Each booklet, containing five hundred stamps
and representing $50 in retail purchases, could be redeemed for a dol-
lar in merchandise or seventy cents in cash, the equivalent of a 2 percent
discount for merchandise and 1.4 percent for the cash (similar to today's
cash-back rewards programs).[25]

Over time, retail premium systems became even more elaborate. While
many companies incorporated trading stamps, others offered "profit-
sharing" coupons carrying points to accumulate toward free things. Coupons sometimes became integrated into the packaging itself, obliging purchasers to save cigar bands and fruit wrappers, clip the lids of tin cans, or
cut the backs off cigarette packs. Sometimes companies inserted printed
cards or slips of paper inside products to be fished out and saved for later.
And sometimes premium coupons were untethered from specific products
and distributed by retailers in exchange for minimum purchases of a generic group of goods.

Placing the onus of tracking and tallying points and purchases on consumers saved companies money. As important, the very practice of amassing, pasting, and archiving encouraged consumers to feel more actively
engaged in prize-getting, increasing their emotional and economic impe-

tus to purchase products. What was more, it helped raise brand awareness among consumers, who by the end of the nineteenth century were beginning to learn a new commercial language centered around commodities and their associated characteristics. Trading stamps and coupons encouraged consumers to more closely scrutinize packaging and labels and to be more discriminating buyers; ideally they would become loyal to a single brand or specific suite of products.[26] Improved printing and packaging technologies helped make this possible, enabling marketers to turn otherwise undistinguished goods into individuated products with distinctive packages and striking labels, giving them unique identities and personalities to which consumers could affix loyalty, as if they were people.

Shifts in retailing and manufacturing also helped free stuff to flourish at the end of the nineteenth century. The expansion of the postal system into far-flung areas and cheaper shipping rates encouraged the growth of mail order; buyers could finally break their dependence on traveling sales agents and even local retailers. In addition, mass industrialization, especially among American manufacturers, delivered more and cheaper goods to the marketplace. It was easier than ever for companies to offer free premiums without losing any money, since these items were now ubiquitous. What was more, they were no longer simply commercial overruns, irregular goods, or last year's models but items made specifically to be given away. Entire manufacturing sectors developed solely to make free crap.

The Cost of Free

The "fate of whole branches of industry," according to one expert, relied on the robust use of premiums.[27] This might explain why, by the first decade of the twentieth century, a veritable "premium craze" swept the country. "Scarcely a middle-class or wage-worker's family may be found, at least east of the Mississippi," remarked political economist I. M. Rubinow, "where some kind of coupons are not saved and some kind of a free prize not expected."[28] Writer Lucy Salmon concurred, describing in 1909 the many "special enticements" available to consumers "in the form of prizes of every conceivable device." Free was creeping deeper into the market, from dry goods merchants offering "prizes of lace handkerchiefs" for buying

SAVAGE'S EXTRA HIGH QUALITY, BIG CASH SAVING, FREE PREMIUMS WILL BE GREATLY PRIZED IN YOUR HOME. IN REALITY THEY ARE 15% TO 20% DIVIDENDS FOR YOU, IN RETAIL VALUES.

THEY ARE SECURED,—ONLY THROUGH CONSUMER RETAIL DEALER PURCHASES OF M. W. SAVAGE'S PREPARATIONS,—BOUGHT DIRECT BY RETAIL DEALERS FROM INTERNATIONAL STOCK FOOD CO., OR FROM ANY OTHER OF M. W. SAVAGE'S SPECIFIED COMPANIES.

DESCRIPTION OF MY—REGULAR—EXTRA QUALITY FREE PREMIUMS.

No. M101. German Silver Mesh Bag. The Frame of this Handsome German Silver Mesh Bag is Deeply Engraved with a Very Artistic Design and finished in the Beautiful French Gray Silver Finish. The Chain is Especially Strong and composed of Hand Soldered Links. The Clasp is Strong and is Soldered to Frame and the Mesh is of Fine German Silver, and is Especially Strong. This Beautiful Mesh Bag is Splendidly Lined with a Fine Quality of White Leather. Free with Total Purchases of $45.00.

No. M102. "Tango Style" German Silver Mesh Bag. It is made of Genuine Silver Mesh and the "Tango" formation makes a most Popular Novelty. Frame and Catch are made with the Bright Silver Finish and the Catch is Especially Strong. The Strap is made of the Same German Silver Mesh as Used in the Bag Itself. Length 10¾ inches, and width 6 inches. Free with Total Purchases of $55.00.

No. M103. Best Rolled Gold Gentlemen's Watch Chain. Every Link is soldered by hand and this Fine Watch Chain will Give Many Years of Satisfactory Service. Made of Best Rolled Gold and Guaranteed by the Manufacturer. A Splendid Quality Watch Chain of Attractive Design. 11¼ inches long. Provided with Bar and Swivel and Ring for Watch Charm. Free with Total Purchases of $55.00.

No. M104. Gentlemen's Watch Chain. Here is a Watch Chain of Splendid Quality. Its Extremely Heavy Boston Links are Hand Soldered and of Positively Guaranteed Material. This is an Exceptionally Heavy Chain about 11½ inches long and will wear Splendidly. Free with Total Purchases of $48.25.

No. M105. Gold and Silver Finish Vanity Case. This Vanity Case is finished in a Rich Gold Striped Design, which contrasts Beautifully with the Bright Sterling Silver Finish of the Balance of Case. Panel in the front is Surrounded with an Artistic Scroll Design and will Admit Engraving an Initial or a Monogram. Back is finished in Bright Sterling Silver Finish and is Extremely Rich in Appearance. Catch is Ornamented with a Beautiful Imitation Jewel. Coin Holders will hold Coins of Two Denominations. Mirror forms Cover of Powder Receptacle and inside cover contains Memorandum Panel. This Receptacle is Provided with Dainty Powder Puff. Card Holder has Splendid Spring of Especially Attractive Design. Chain is Very Strong. Free with Total Purchases of $36.25.

No. M106. Vanity Case of Highest Quality. This Vanity Case is finished in French Gray Silver with an Extremely Attractive Border Deeply Engraved. The Back has the Beautiful Lustrous Appearance of Genuine Sterling Silver. Catch is Beautifully Ornamented with an Imitation Jewel. It has Two Coin Holders, Powder Puff, Celluloid Memorandum Panel, Visiting Card Clasp and Mirror. The Chain is made of Soldered Links of Very Attractive Pattern. Free with Total Purchases of $32.75.

No. M107. Fine "Mayfair" Adjustable Bracelet. This Bracelet will Fit practically Any Sized Arm. Made of Fine Gold Filled Stock and Positively Guaranteed. In Design it is a Beauty, each Alternate Link being Deeply Engraved. You will Surely Appreciate this Beautiful Bracelet. It is Packed in a Handsome Velvet Box. Free with Total Purchases of $26.75.

No. M108. Gold Filled Bracelet. Here is a Magnificently Engraved Genuine Gold Filled Bracelet. It is five-eighths of an inch wide and has Safety Bar which prevents loss and also prevents breaking bracelet when opening. Finished in Bright Gold finish and incased in Handsome plush lined box. This Elegant Bracelet will Surely Appeal to you. Free with Total Purchases of $48.25.

No. M109. Adjustable Bracelet. This Patent Adjustable Style will Fit any ordinary sized Arm. Made of the Best Rolled Gold. Has a Beautiful Bright Gold Finish and the Links are Most Attractive. Front Ornamented with Very Stylish Twisted Wire Design and Signet upon which initial or monogram can be Engraved. Signet is bordered by a Rich Beaded Design. Substantially Made of Splendid Material. Free with Total Purchases of $32.75.

No. M110. Watch Fob. The design of this Fob is Extremely Artistic. Except the four upright center pieces, it is finished throughout in Bright Gold Finish. These center pieces are Beautifully Engraved. Seal pendant is Bright Gold Finished and has Signet at bottom. Has safety guard and chain and is 4¾ inches long. Free with Total Purchases of $36.25.

No. M111. Watch Fob. The upper and lower Panels of this Fob are Engraved in a Beautiful Scroll and Leaf Design. The outer chains are of Curbed Pattern and are Bright Gold Finished, while center Chain is Engraved and has a Roman Gold Finish. Locket holds Two pictures. Design of Locket is Beautiful. It has a round panel which can be used for Engraving initials or Monogram. Has Strong Safety Guard and Chain 4½ inches long. It is Surely a Beautiful Fob. Free with Total Purchases of $65.50.

No. M112. Watch Fob. This Handsome Fob is of Seamless Gold finish Wire Ribbon and is 1¼ inches wide by 5 inches long. This Gold Finished Wire Ribbon is Extremely Attractive. It is Ornamented at top by round panel and in center by oblong panel, both Engraved in Artistic Scroll Designs. Seal Pendant is of Bright Gold Finish. Has Strong Safety Guard and is 4⅜ inches long. Free with Total Purchases of $36.25.

No. M113. Ladies' Watch Fob. This Lady's Fob has a very Dainty Appearance. Its Finish is the Bright Gold Finish and its Design is Appealing in its Beauty. Center composed of Five Strands of Polished curbed link chains, while Pendant is of Seal Pattern and bears Signet for Engraving Initial or Monogram. Has Safety Guard. Free with Total Purchases of $52.75.

No. M114. Watch Fob. This Gold and Platinoid Fob is composed of Chains of Two different patterns. Outer and center chains being formed of Oblong Links, each alternate Link being of Gold and Platinoid. The remaining two chains are of Rolled Gold in a Beautiful Scroll Pattern. The Pendant Charm is both Gold and Platinoid, center circle beautifully Engraved with figure of a Thoroughbred Horse. Outer Circle is of Platinoid, Richly Engraved. Has Safety Guard and is 4⅞ inches long. Free with Total Purchases of $65.50.

No. M115. Ladies' Fob. This Dainty, yet Most Serviceable Fob is made of very Best Materials. It is very Beautiful in its Oddity of Design. Finished in the Bright Gold Finish and portions are Handsomely Engraved. The Charm is very Handsome and Beautifully Finished and has places for Two pictures. Has Safety Guard and is 3⅛ inches long. Free with Total Purchases of $36.25.

No. M116. Scarf Pin. This Elegant Scarf Pin is made of Ten Carat Solid Gold and the Setting is a most Beautiful pink and white Cameo, bearing a head of a Beautiful Woman. Particular Attention is called to the fact that this Entire Scarf Pin is made of Ten Carat Solid Gold. Free with Total Purchases of $36.25.

No. M117. Scarf Pin. This Exquisite Scarf Pin is made of Ten Carat Solid Gold with Beautifully Engraved Design, finished in popular Bright Gold finish. This Scarf Pin is set with an exquisite Iridescent Stone of Beautiful Coloring. Do not overlook the fact that this Scarf Pin is made of the Finest Ten Carat Solid Gold. Free with Total Purchases of $22.50.

No. M118. Brooch. Here is a most Charming Brooch made of Ten Carat Solid Gold. It is made of Bright Gold except the Beautiful Leaves which are made of Exquisitely Tinted Solid Gold. The Setting is an Exquisite, Iridescent, Round Stone, Gleaming in all Colors of the Rainbow. A beautiful Pearl graces each end of this Valuable Solid Gold Brooch. Free with Total Purchases of $38.75.

No. M119. Cross Pendant. This Genuine Gold Front Cross Pendant is very Odd and Attractive. Its Genuine Gold Front is Ornamented with a Very Attractive Design in Hard Black Enamel. This Cross is most Attractive. It will surely be Greatly Appreciated by its wearer. Free with Total Purchases of $36.25.

No. M120. Charm. Both front and back of this Beautiful square shaped Charm are finished in Bright Gold and Entire Locket is made of the Very Best Gold Finished Material. The Exquisite Scroll Design is deeply Engraved and the Leaves, as shown in the Illustration, are made of Genuine 14 Carat Green Gold. The inside of this charm is fitted with two places for pictures. Such a real Gold Finished Charm with Genuine 14 Carat Gold Decorations is indeed Most Desirable. Free with Total Purchases of $29.50.

No. M121. Charm. This Exquisite round Charm is made Throughout of Bright Finished Rolled Gold and the front is Exquisitely Ornamented by beautiful Striped Engraving surmounted with a conventional Floral Design of Rare Perfection. In the top are set Fourteen Brilliants. These Brilliants are splendid imitations of Genuine Diamonds. The center is also Set with Two Large Sized Brilliants and Four imitation Emeralds. The inside is fitted with places for Two photographs. Free with Total Purchases of $48.25.

No. M122. Cross Pendant. This Exquisite Gold Front Pendant Cross is made of material of Extremely Fine Quality, Finished in the Rich, Bright Gold Finish and Ornamented with a Beautiful Scroll Design. For Quality and Beauty this Exquisite Gold Cross Pendant is Difficult to match. Free with Total Purchases of $32.75.

No. M123. Cross Pendant. This Elegant Cross Pendant is made of fine, Hard, Black Enamel ornamented with a Gold Edge in an exquisite beaded effect. It is a Solid Gold piece of Jewelry and is set with 20 Fine imitation Pearls. The Pearls contrast with the beautiful Gold and Black Finish making it a most Attractive piece of Jewelry. Free with Total Purchases of $45.00.

No. M124. Watch Chain. This Very Handsome Chain is made of the Best Rolled Gold, Positively Guaranteed. Links are Soldered and of a very Desirable oblong Design. Has Splendid bar and swivel and a ring for watch charm. To those who Admire a Chain of medium heaviness this one will Surely Appeal and it will Give Splendid Wear. Free with Total Purchases of $42.50.

No. M126. Coat Chain, Scarf Pin and Scarf Holder. This Set consists of a Solid Gold Front Coat Chain, Scarf Pin and Scarf Holder. They are Magnificently Engraved with a Beautiful Scroll Design, bearing an Exquisitely finished panel. Coat Chain is 9 inches long and has alternate links of bar and twisted design. Button of Coat Chain and tops of Scarf Pin and Scarf Holder have Solid Gold Fronts and Coat Chain is Bright Gold Finished with Soldered Links. Free with Total Purchases of $38.75.

No. M128. Secret Locket Pendant. This Secret Locket Pendant has a place for a Picture, the location of which is a Secret from Everybody but the Wearer. This Beautiful Pendant is set with Two Exquisite imitation Rubies. Has a Bright Gold Finish and has a very Elaborate design and Two handsome Stone Settings. This Secret Locket is a Very Handsome one and will be Worn with Pride. Free with Total Purchase of $38.75.

No. M129. La Valiere. This La Valiere is Solid Gold and Truly Beautiful. The Design of the Pendant is Very Artistic and the Coloring of the Bright Gold harmonizes with the Garnet and Pearl Settings. The central Setting is an oblong imitation Garnet of Rare Beauty. The Pearls consist of three gleaming round Pearls with two Genuine Slug Pearls at top and bottom. The Chain is Solid Gold with Soldered Links. Free with Total Purchases of $46.75.

No. M130. La Valiere. This Solid Gold La Valiere will be Greatly Prized by its owner. Chain and Pendant are made of Solid Gold and Pendant is Set with a Beautifully cut imitation Ruby, an imitation Pearl and a Genuine Slug Pearl. We cannot speak too Highly of its Quality and Beauty. Free with Total Purchases of $39.75.

No. M131. Diamond Set Cuff Links. These Gold Filled Cuff Links are Particularly Desirable as each one is set with a Genuine, Rose Diamond. These Diamonds are Not imitations but are Genuine Chip Gems. They are Beautiful in their star shaped setting, carved in the Exquisite Roman Gold Finish of the Cuff Links. Free with Total Purchases of $44.50.

No. M132. Diamond Scarf Pin. This Scarf Pin is the Very Finest shown in our premium book. It is made of Genuine Ten Carat Solid Gold and is Set with a Genuine Hand Cut Blue-White Diamond and a Beautiful Pearl. This Diamond is Cut in the same manner as Diamonds which Cost Hundreds of Dollars. The design of this Extremely High Grade Scarf Pin is of the Popular La Art Noveau Pattern. You will Surely make no mistake in selecting this fine Genuine Diamond, Solid 10K Gold Scarf Pin as it is Surely a Beauty. Free with Total Purchases of $65.50.

No. M133. Diamond Scarf Pin. This is one of the Finest Scarf Pins in our premium book and is made of 10K Solid Gold, set with a Genuine Rose Diamond. This Rose Diamond is Not an imitation Diamond but is a Genuine Chip Diamond. The Design of this Stick Pin is most Attrative, the Crescent is finished in the Beautiful Roman Gold Finish, while the Star is finished in Bright Gold Finish. Free with Total Purchases of $26.75.

No. M134. Secret Locket Pendant. This is a Very Beautiful Secret Locket Pendant. It is finished in Bright Gold Finish. The Center Setting is a Splendid Diamond shaped imitation Amethyst surrounded by Eight Handsome imitation Pearls. A Genuine Slug Pearl droops from the bottom. This Secret Pendant has space for one photograph, the location of which is known Only to the Wearer. Free with Total Purchases of $52.75.

M. W. SAVAGE OFFERS THE MOST EXTRAORDINARY LINE OF VERY HIGH CLASS, FREE PREMIUMS IN THE WORLD, FOR SUCH SMALL SIZED PURCHASES, BECAUSE HE BELIEVES IN FAIR AND SQUARE CO-OPERATION AND IN THIS WAY HE RETURNS A PER CENT OF HIS PROFITS DIRECT TO THE PEOPLE WHO HELP HIM INCREASE HIS BUSINESS,—SIMPLY BY USING ANY OR ALL OF HIS LARGE NUMBER OF DIFFERENT PREPARATIONS AND PRODUCTS, WHEN THEY NEED PREPARATIONS AND PRODUCTS OF THESE DIFFERENT SPECIFIED KINDS.

Page Fifteen

Figure 5.4a–b. Premium promotions often overshadowed the actual products being sold, as these pages describing the "elegant, extra high quality free premiums" from M. W. Savage's 1914 advertising pamphlet for International Stock Food Tonic show.

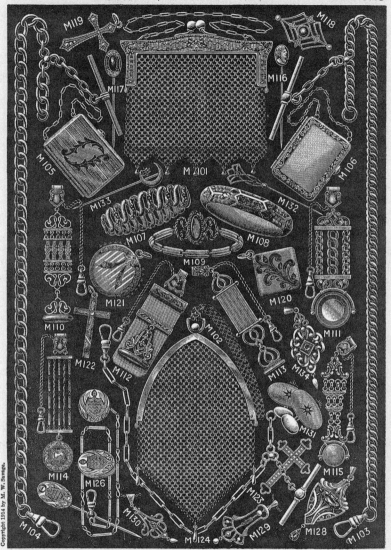

ARE BASED ON FAIR AND SQUARE, PRACTICAL CO-OPERATION.

canned goods to newspapers promising free excursions to readers who participated in surveys.[29]

Not everyone, however, was so enamored with free stuff. Many observers felt there was something fishy about the prospect of getting something for nothing. Rubinow, for one, called retail premiums a "moral epidemic." At minimum, premiums persuaded people to buy more of a product than they might need. Through another bit of alchemy, premiums bound together two unrelated things, thereby creating ambiguity about the value of each. Rubinow was at a loss to fully explain this economic "aberration," writing, "The phenomenon of a combined price for two commodities, entirely different, presents itself to baffle the most careful inquiry into the nature of value."[30] This "bafflement" led people to overly prize merchandise that was only of negligible value and to purchase the primary products at inflated prices. In addition, putting unlike with unlike curiously made them *both* more desirable.

Companies also tantalized consumers by implementing different kinds of exchange systems simultaneously, offering premiums for free with minimum product purchases, for cash plus coupons, and for cash alone. Cash and part-cash plans had special appeal for coupon savers who lacked the time and patience to "get up" enough to redeem for what they wanted. Having accumulated coupons toward possession, these buyers had already made psychic investments in the premiums (investing them with emotional value). What was more, and contradictorily, because premiums could come free with purchases, they seemed like good bargains (and less valued monetarily). And so, many companies gave consumers the opportunity to purchase premiums outright, or to pay the difference between their alleged market value (always inflated) and whatever the buyer had already accumulated in coupons or stamps. People continued to find free premiums enticing even if they had to pay for them.[31]

Over time, more and more retailers adopted trading-stamp and coupon premium systems in order to match competitors who were already garnering patronage with them. Competitive pressures meant that free was actually expensive for retailers, too, because consumers no longer remained loyal to particular stores but could—and did—shop around. During a 1914 congressional investigation into whether tobacco companies' use of coupon-based premium systems constituted a monopoly, for instance, one witness testified that free stuff, rather than cash discounts, "brings the customer back for more goods so as to get more coupons."[32] Tangibility prevailed over monetary value, transforming mundane premiums into compelling retail hooks. As Henry Bunting observed, the premium took the "invisible vapor" of the cash discount and transmuted it through "concentration and condensation" into "a concrete parcel of actual property which the consumer can feel with his fingers."[33] It really *was* a form of alchemy.

While premiums helped encourage customer loyalty to particular retailers, they also pressured retailers to be loyal to specific premium outfits. Because consumers had come to expect free stuff but retailers typically did not have the means to implement their own premium systems, many businesses felt pressured to hire the services of companies specializing in offers of free—and often lost money in the process.[34] Outfits described their proprietary premium systems in specialized publications meant to convince retailers that their businesses could not survive without offering giveaways, and touting the benefits of their systems in particular. Retailers themselves did not profit from premium systems but hoped premiums would "fix the place of purchase" for the "establishment of habit" among shoppers. The Buffalo-based Penfield Merchandise Company, for instance, explained that its system was intended to "induce" people "to do all your trading at this store."[35]

In other words, premium systems created market-defined and -administered customer loyalty. What consumers received in return was, in Rubinow's words, "cheap and useless articles, namely bric-à-brac, the very production of which in such enormous quantities is an enigma." The ethos of free showed quite clearly that patronage was a commodity bought for cheap, had its foundations in materialism, and was forged and reinforced through commercial transactions. Within capitalism, this is what loyalty looked like.[36]

In addition, free came at a material cost. Premium companies insisted, often quite elaborately, that their free merchandise was useful, of the highest quality, and beautiful. This was indeed true for many of the best premiums. All those pianos, diamond jewelry sets, silver tea services, fur coats, and Victrolas were at least theoretically attainable, but they took years' worth of purchases to get—assuming that the redeeming outfit hadn't gone out of business in the meantime. In 1911 A. J. Brown placed a notice in the *Railroad Telegrapher* asking his fellow railroadmen for twenty thousand Central Union Smoking Tobacco labels, which would earn him two artificial legs. Later that year, he was still eighteen thousand short—many of the labels people sent to him, representing other tobacco brands, were "of no value." In 1912 F. E. Pomeroy made a similar appeal, as he, too, "was badly in need of an artificial leg." Two years later, he still needed a thousand to reach his goal.[37]

Free and Easy

Unlike high-quality premiums, most premiums were truly cheap crap that flowed freely in the market. "The customer is blinded by the attraction of getting something for nothing," observed Lucy Salmon at the dawn of the twentieth century. It didn't matter whether the giveaway was a cheap gilt frame with a cheap color print, or a tinplate dime store saucepan. Some critics called premiums, cynically, the "something-for-nothing" idea. Others argued that both the primary product *and* inducement were of "inferior quality," since neither could be sold on its own merits. The "innocent, inexperienced" buyers "who know nothing of quality, value or price" were particularly susceptible to their allure.[38] Yet despite the opprobrium leveled by consumer rights advocates and economists, consumers still loved premium schemes. An estimated ten million families had at least one "stamp collector" in their home in 1917. Why? Because, according to one observer, "the *buying public does not think straight* or, thinking straight, does not act as it ought to act."[39]

That free thwarted the decision-making capabilities of the rational consumer was precisely the point. Because consumers reveled in the carnivalesque excesses of free merchandise, manufacturers, wholesalers, and retailers were able to profit from cheap and often unsaleable goods disguised as desirable gifts. Many merchandisers were influenced by the psychology of the carnival midway itself, and they not only observed the increasing popularity of traveling carnivals in the first half of the twentieth century but recognized the immense profits carnies were earning from their gaffed games and crappy prizes.[40] As simple entertainment, playing carnival games counted as money well spent. But many fairgoers who tried to shoot balls through hoops or knock down milk bottles were surely disappointed, since they rarely won the "teaser" prizes and instead left the booth with cheap trifles. The prizes—"slum" and "garbage" in the unadorned parlance of the carnival—were worth much less than the dime it cost to play the game.[41]

Nevertheless, they kept people coming back for more. Some games, "creepers," enticed players to spend more money just to get a free prize. In an exposé about carnival culture, the writer Harry Crews explained that carnies often gave out a crappy prize to "cool the mark" and keep him playing. He recalled a typical scene: "I watched the mark finally get thrown a piece of plush, in this case a small, slightly soiled cloth giraffe. The poor bastard had paid only $12 for something he could have bought for two and a quarter out in the city."[42] Other games, like roulette wheels, always paid out, but the prizes were worth much less than the cost of a spin; operators deemed them "perfect for Slum."[43] The compartments of the Country Store Wheel were "laden with clocks, thermos bottles, kewpie dolls, and other desirable articles" that were unwinnable. What was more, its very action seduced more people to try their luck: "The sight of the whirling wheel creates what the gamesters call a 'flash,' and attracts more customers." Even fairgoers who knew games were gaffed still could not resist the chance to win a prize. Much like mystery prize boxes, games provoked a "tantalizing fascination," which made players "come back for more" (fig. 5.5).[44]

The power of free motivated people to spend more money, whether they were purchasing a hundred pounds of tea or playing gaffed games on the midway. Retailers and carnival barkers alike helped funnel cheap goods to the masses and, in the process, create viable opportunities for crap's many producers and distributors (fig. 5.6). Some businessmen, like Samuel Pockar, made their living dealing in nothing but slum—in his case, crappy jewelry, "the cheapest grades of gaudy baubles," which carnies in the 1930s and 1940s bought by the gross for $2 or $3. (That might seem cheap, but just a few years earlier slum was going for sixty to seventy cents per gross.) Pockar's vast inventory of cheap stuff included brass rings falsely marked 10K on the band, $1 pocket watches "that looked like a $25 solid gold model," close-outs, "broken merchandise," and other things that had no other market.[45]

Creating value out of thin air, admen and carnival barkers knew marks when they saw them, happy to separate people from their money by selling them free. A promotional catalog from the Lee Manufacturing Company inadvertently revealed the pedestrian nature of its premiums. The catalog featured full-color illustrations of Lee's dime-store-grade lines of

Figure 5.5. People who played gaffed games on the midway often spent much more money to win prizes than they were worth. Girls at a carnival with their Kewpie doll prize, ca. 1920s.

face powders, hair tonics, and beauty creams alongside the array of available free merchandise—sets of chinaware, pieces of glassware, and the like. Highlighting the scale and volume of the company's business, interior shots of the enterprise reveal hives of industrious employees sorting mail, answering orders, and working the packing floor. In a view of the order-filling department, which took up an entire floor, workers can be seen diligently satisfying premium orders, pulling china plates, cups, platters, and other goods from large bins: the seemingly unique, high-quality pieces lushly rendered in the catalog were actually quite the opposite—anonymous products of industrial production (fig. 5.7, plate 4). Stacked one on top of another and typical of what could be found in every cheap variety store, they were not even worthy of special care and handling.[46]

Businessmen capitalized on the way emotions can overtake rational decisions, even when consumers themselves realize that what they *feel* they

SOFT-STUFFED PLUSH CARNIVAL DOLLS

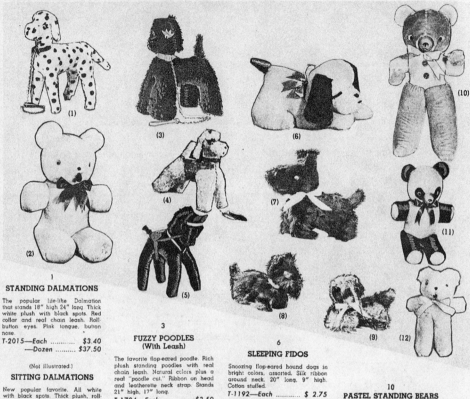

1

STANDING DALMATIONS

The popular life-like Dalmation that stands 18" high 24" long. Thick white plush with black spots. Red collar and real chain leash. Roll-button eyes. Pink tongue. button nose.

T-2015—Each $3.40
  —Dozen $37.50

(Not illustrated.)

SITTING DALMATIONS

New popular favorite. All white with black spots. Thick plush, roll-button eyes. Red Leatherette collar and real chain leash. 21" high.

T-2014—Each $2.90
  —Dozen $32.00

2

PLUSH GIANTS

Giant bears and Pandas. Full 38½" tall. Assorted Pandas and Honey bears. Includes the new all white bears.
Silk matching ribbons. Roll-button eyes.

T-2020—Each .... $ 6.90
  —Dozen $80.00

(Not illustrated.)

PLUSH BEARS

A special! 25" tall plush bears in assorted colors. Roll-button eyes. button noses. Bright silk ribbon around neck. Cotton stuffed.

T-1217—Each $2.25
  —Dozen $24.00

3

FUZZY POODLES
(With Leash)

The favorite flop-eared poodle. Rich plush standing poodles with real chain leash. Natural colors plus a real "poodle cut." Ribbon on head and leatherette neck strap. Stands 21" high, 17" long.

T-1794—Each $3.50
  —Dozen $39.00

4

15" FUZZY POODLE

Identical to the above but a cuddly 15" high. Long curled plush, silk lined ears. Popular new pastel colors. Chain leash.

T-289—Each $2.75
  —Dozen $30.00

5

RED HORSE

Soft stuffed, all red plush horse with plastic saddle and trappings. 19 inches tall.

T-4956—Each $3.25
Dozen $36.00

Same as above in assorted colors.

T-4956—Each $3.25
Dozen $36.00

6

SLEEPING FIDOS

Snoozing flop-eared hound dogs in bright colors, assorted. Silk ribbon around neck. 20" long, 9" high. Cotton stuffed.

T-1192—Each .... $ 2.75
  —Dozen $30.00

7

FUR SCOTTY

Life-size Scotty dog with a long-hair natural looking fur coat. Ribbon bow around neck. All black or assorted black and brown. 16" tall, 18" long.

T-2310—Each $2.75
Dozen $32.00

8

FUR SCOTTY PUPS

Same as above except 9½" tall, 12" long.

T-1876—Each $1.50
Dozen $16.00

9

FUR PEKINESE

Cuddly, life-size fur Pekes in assorted colors. Ribbon bow around neck. 10" tall, 9" long.

T-1875—Each $1.50
Dozen $16.00

10

PASTEL STANDING BEARS

Standing bears and Pandas in the new pastel colors. Colors assorted. Cotton stuffed. Roll-button eyes. Plastic nose and mouth. Large silk ribbon.

T-2019—Each .... $ 3.60
  —Dozen $40.00

11

PASTEL CUDDLE BEARS

New colors for real flash. Soft-stuffed cuddle bears in bright new pastels. Assorted colors. Plastic noses and mouth for realism. Silk ribbon.

T-2006—Each .... $ 3.60
  —Dozen $40.00

12

CUDDLE BEARS

Soft-stuffed cuddle bears in assorted colors. 15" high. Roll-button eyes. button nose. Bright silk ribbon.

T-1509—Each .... $ 1.25
  —Dozen $12.00

Figure 5.6. Companies like Kipp Brothers provided cheap carnival goods—called "slum" in the trade—to midway operators. Kipp Brothers, *Carnival Catalog, no. 166*, [ca. 1940?].

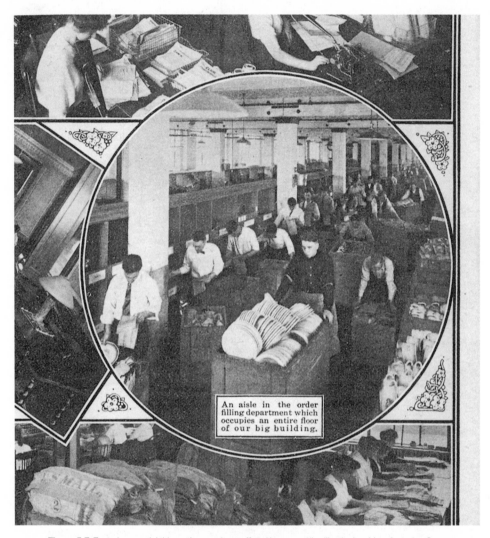

An aisle in the order filling department which occupies an entire floor of our big building.

Figure 5.7. Touted as special things, the premiums offered by companies like the Lee Manufacturing Company were really just cheap mass-produced items pulled from bulk bins. *Lee's Wonderful Catalogue of Easy Selling Goods and Premiums*, 1924.

are getting does not match what they *know* they are getting. While decrying the many perils of retail premiums, even economist I. M. Rubinow, who presumably knew better, admitted that he himself "was, for a time at least, a victim of this craze, and has quite a respectable piece of cut glass to show for it."[47] That "respectable piece" was likely pulled from yet another volume bin of cheap merchandise. Equally confounding and enticing, that was the nature of free.

Little Kids, Big Dreams

Easily incentivized, children could be even more bedazzled by the prospect of free than their parents. In the early nineteenth century they earned rewards of merit—small printed certificates embellished with pictures of animals—for getting good grades in school, correctly memorizing Bible verses, or simply doing what was asked of them. By the later decades of the century they no longer had to settle for flimsy pieces of paper but, like their parents, could obtain *real things* for free from the market. The authority of teachers, parents, and even God had been supplanted by commerce. Children didn't have to behave well or learn well. They just needed to spend well.

Long before Cracker Jack began inserting miniature prizes into boxes of caramel-coated popcorn in 1912, retailers realized they could entice kids with trifling things. As early as the 1870s candy shop owners began offering free toys with purchases of even the cheapest penny candies in order to cultivate "friendships" with their young customers. The Philadelphia confectionery company John M. Miller & Son, for instance, used all manner of inducements, from fireworks to Christmas goods, to sell its sweets. Among other items, the business produced over twenty kinds of Prize Boxes. Selling for $2 per dozen wholesale and retailing for twenty-five cents each, the Bon Ton contained cash and jewelry along with candy. The Centennial, retailing for just a nickel, contained jewelry and chromo prints along with candy. The United States Mint box included coins valued at anywhere from a nickel to a dollar. The International Prize and Fortune Box, "among the latest novelties out," came with a free fortune. Some prize boxes promised a piece of jewelry, a gold-plated pocket watch, or a $100 bill in every hundredth box.[48] "We predict for these an UNPARALLELED SALE," the company boasted (fig. 5.8).[49]

New industries emerged dedicated to make trifling giveaways for the kids who were now accumulating their own spending money by running errands, selling newspapers, scavenging scrap, and working in factories. Manufacturer W. C. Smith produced several lines of toys for "penny goods"—cheap things to give away with other cheap things. Sold by the

A new candy prize package. Good for resale purposes. Very attractive because it is printed in very pretty colors. The box measures 6½x2¼x1 inches. Each box contains five pieces of Peanut Butter Kisses and a prize or toy that will please a girl or boy. The prizes used consist of many different styles such as rings, lanterns, pipes, tops, auto, etc. "Sweeties" are packed 250 to a carton; no less sold.
No. 12A1. Per thousand (4 cartons)............ **14.50** | Per carton (250 pkgs.)........................ **3.75**

Figure 5.8. Children were tantalized by candy offered with prizes. Each box of "Sweeties," for instance, contained "a toy that will please a girl or boy." N. Shure Co., *Shure Winner Catalog No. 121*, [1933].

gross to retailers, these wonderfully crappy items included "Puzzle Whistles, French Puzzles, Pop Guns, Tin Dishes, Stamped Spoons," and countless other kinds of merchandise appealing to the tastes and modest buying power of girls and boys.[50] Children could hold special prizes in their hands, play with or trade them, tuck them into their pockets as hidden talismans, covet them, and perhaps count them among their few possessions. The "Metal Novelties" and "Penny Prize Goods" made by Dowst Brothers of Chicago (future manufacturer of Tootsietoys) were tiny cast-metal charms and tokens re-creating, in miniature, the world of adults, stoking grown-up desires among those with child-size pocketbooks. Dowst's tiny replica of a fruit dish, "correct in every detail," was aspirational: it made for "an ideal doll house article," whether a girl actually owned such a thing or could only dream of it. The small lantern, outfitted with a transparent imitation glass globe "that looks as if lighted," was "probably the best novelty ever placed on the market." The lady's shoe, "a cute reproduction, showing every detail," was rendered "in perfect proportion," sported "the latest French heel," and was "finished on both sides."[51]

Free, then, incentivized consumption in the here and now and served as a gateway to future consumption, training a young generation not only *that* they should covet but *what* they should covet as grown-ups. *Novelty News* noted in 1909 that "stores whose proprietors invested liberally in such things as tops, whistles, and other peace breakers, kites, wagons, and similar playthings" saw increased sales.[52] Retailers were able to gain newer and ever-younger customers. "There is an unlimited field for the exploita-tion of toys, playthings, games and mechanical devices," wrote Henry Bun-ting. Kids' freebies could be called forth to sell cereal, soup, salt, and even coffee. "Good results have been recorded," Bunting noted, "in every case where toys were intelligently used to supplement space publicity," in other words, in addition to traditional advertising.[53]

At the same time, toy manufactures, like suppliers of other premium goods, were able to increase their output to meet the growing desires for cheap stuff among these little consumers. What was more, because chil-dren were relatively easy to please and responded more to novelty than to quality, makers of crappy toys had no reason to make them any bet-ter, since kids only cared about "the possibility of possession without the cost."[54] In one case, a toymaker early in the century found himself with eleven thousand unsold toys after a failed Christmas season and resigned himself to the idea that "the stuff won't sell, and that's all there is to it." To avoid losing money, the manufacturer "worked his thought emporium overtime," finally realizing he could market the worthless toys as premiums for soap, magazines, and tea. Sure enough, in just two months, "the line was cleaned up, the money was in the bank, and the toy maker was rid of a bad bargain."[55]

Trading cards in cigarette packs, trinkets and tokens in packages of caramel popcorn, and, eventually, secret decoder rings in boxes of cereal were some of the many retail premiums designed for children over the years. Cheap as they were, these tokens helped secure long-term customers at very little expense: children could be bought most cheaply of all. Not only were they able to influence their parents' purchasing habits, but their own commercial loyalties, imprinted while young, often continued into adulthood and could last entire lifetimes. Offering free premiums was an easy way for marketers to increase sales not just immediately but well into the future.[56] One 1922 article noted, "Children, particularly, are keen in patronizing places where they get a little extra for their money. They urge their parents to take them to the barber shops that have the most imposing hobby horses or that give them a celluloid ball or some other souvenir. . . . Any remembrance, no matter how slight, will win the favor of the child."[57] The Newton Manufacturing Company, another maker of giveaways, re-marked in 1923, "Children are impressionable—they remember favors and are loyal to their friends. If you could get the friendship of half the boys and girls in your trade territory you would be surprised at the influence which they have with the 'grown-ups.'"[58]

By the mid-1930s the new medium of radio enabled marketers to penetrate the children's market further still, especially if premiums were incorporated into the promotions.[59] Children liked collecting and, as it happened, also enjoyed sending away for free things through the mail. Undoubtedly, some of this impulse originated in Depression-era material scarcities and the desire to get something for nothing. But other psychological factors were at work as well, namely, the exquisite pain and pleasure of anticipation, two sides of the same coin. Kids spent the days and weeks waiting for the mail in a state of frustrated anticipation: like Christmas Eve, the wait seemed eternal, but allowed more time to fantasize about the package to come.[60] That exquisite anticipatory desire disappeared the moment the dream materialized into reality. But it would soon be reignited by yet another prize offer. Children's responses to a series of interviews with marketers in the 1930s are telling. Patsy met the mailman "every day" after sending in her box tops and decided "that the people were naughty to keep me waiting so long." Donnie "didn't like to wait" and told his mother he "thought they were lazy at that office," adding, "I thought [my money] was gone out the window—just lost forever. But finally the prize came." Having waited three weeks for his prizes, Eugene "got all up in the air about the delay—I didn't think it was so hot."[61]

Marketers also encouraged children to form clubs in order to trade their free things with each other, creating an even more fervent interest in collecting series of premiums and, of course, in purchasing the primary products that made them possible. In the process, kids were learning the social dimensions of being consumers, especially the role that one's possessions play in establishing membership and status within a group. In some towns, Boy Scouts cooperated and competed to "earn" merit badges by collecting entire sets of Wheaties miniature license plates. Thus, in truly American fashion, commerce seamlessly infiltrated children's milieus, as it had their parents', and fused civic-minded efforts with materialist pursuits.[62]

This strategy proved especially effective when promotions, from decoder rings to sheriff's badges and pictures of cowboys, were hawked by well-known radio characters and the actors who played them.[63] Listeners joined Ralston Purina's Tom Mix Straight Shooters Club, organized in conjunction with the *Tom Mix Straight Shooters* radio show of the 1930s, by sending in a Ralston cereal box top; the more box tops you sent in, the more free stuff you could show off to friends, from comic books to cowboy clothing. Evidence of the strategy's popularity, Ralston received thousands of box tops each week: that was a lot of cereal.[64]

Box top redemption promotions became even more popular in the 1940s and 1950s, especially as marketers entered the home through product tie-ins on television. By 1941 American companies were giving away over $450 million worth of premium merchandise (about $3.40 for every US resident); the "youth market" comprised the top recipients.[65] In 1946 over three million children sent in fifteen cents and a Kix box top to get the coveted Atomic Bomb Ring promoted by General Mills.[66] These efforts to "bribe" children, in the words of one contemporary marketing professional, resulted in "many commercially beneficial things."[67]

By the mid-twentieth century, youth-oriented premiums were more effective promotional tools than mainstream print, radio, and television advertising. Even kids who lacked their own spending money could have significant influence over their parents' purchasing decisions, pressuring them to buy particular brands and products based on the free promotion. One such study in 1957 concluded that despite being resentful of premium "gimmicks," mothers usually gave in.[68] According to another study in the early 1960s, 71 percent of children were aware of cereal brands and asked their mothers to buy specific ones; 90 percent of their mothers acceded.[69]

Mothers were right to be cynical, though, since free stuff for kids was especially crappy. According to one marketing expert:

> The fact that children's premiums . . . are likely to be greatly treasured for a short while and then broken or discarded has led a few users down the rather dangerous path of poor quality. "Action" rings that don't work, wheeled toys that don't function properly, whistles that blow weakly or not at all and many other poorly made or poorly conceived items have occasionally been offered to youngsters.[70]

At its height, the Jack Webb cereal premium—a plastic police whistle with a tie-in to the popular radio and television show *Dragnet*—was being manufactured at a rate of four million units *a week*. It cost less than two cents to manufacture each one, an astonishingly low figure considering the costs for design (including product liability research), fabrication, assembly of its four parts, sealing in cellophane, shipping to the plant, packing in cereal boxes, advertising, and administering distribution across the country.[71] We don't know if they were shoddy enough to create "premium backfire," which, according to one marketing professional, "is a well-known phenomenon in the trade."[72] Although children would take practically anything offered for free, they could also be tough critics and might equate poor-quality free things with the brands themselves.

As a result, companies did not offer better things, which would undercut their profits, but simply tried to manage expectations. Children, according to one expert, "tend to romanticize the idea of the premium and build it up during the anticipatory period before the item arrives." Although the trade literature cautioned marketers to tell their young consumers "the absolute unvarnished and not too flattering truth about the construction of your mystery gadget," kids would nevertheless "enlarge upon" the description."[73]

Professional cautions aside, most premiums were hyped as sensationally as possible, as the ad for Kix's Atomic Bomb Ring illustrates (plate 5). In order to sell the primary product, advertisers had to convince children that the premium "is worth going after." They did this by showing it in action, thereby creating a sense of excitement, urgency, and competition.[74] Thus, advertisers encouraged children to desire consumer goods both as individual things and in the aggregate. They were also teaching kids about the need to pay attention to advertising messages, instilling in them the importance having marketing literacy—above and beyond the specific messages themselves. Advertising became a new authority to be heeded as cultural advisor and commercial arbiter.

Children's tendency to not only believe advertising rhetoric but "enlarge" on it, as advertisers feared (and also hoped), showed the true power of modern marketing and the particular potency of free, even when it failed to meet expectations. Responses to marketing interviews bear this out. Patsy was quite pleased with the ring inscribed with her initials but was "disappointed" with her party kit, which she expected to be "a big box filled with cutouts and things." It was just a book. Jean wasn't able to see the numbers on her Decoder pin. Some of the other premiums she received, which cost her ten cents, "were too cheap" and "weren't worth the money." George wanted the birthstone ring "for one reason," because

he believed what the advertisers said about it. When he received it, however, he found "it wasn't very good—all the gold stuff came off."[75] Frances
thought most of the prizes were "just cheap and were not of any value—no
good." Eugene really wanted the identification tag but was "disappointed"
in it, since "they said it showed the number on the inside—but when I got
it I found that the number was simply on the back." He explained, "Their
description didn't fit with my idea of what it was." Disappointment did not
dampen desire but actually stoked it, by urging kids to seek out newer and
hopefully better kinds of free stuff and making them more discerning consumers in the process. A feature rather than a bug, unfulfilled expectations
merely led to more consumption. Crap begat more crap.

The most effective giveaways, in fact, promoted series of things that
children could collect over time to complete an entire set. Eventually, kids
harangued their parents to patronize specific gas stations offering free toys
with each fill-up, presaging the sets of "collectibles" offered with McDonald's Happy Meals. One of the most inspired was ARCO's Noah's Ark set
from the early 1970s, which included dozens of animals, plus Noah and
his plastic ark. Every few months, stations received new animal mates;
kids had to snatch them up before they ran out, meaning many trips to
the pump for the adults (fig. 5.9). By this time, children's premiums had
become so pervasive that the government worried they were taking hold of
young consumers' susceptible minds. In 1973 the FTC proposed banning
advertisements for premium offers geared toward children. According to
the American Marketing Association, supporters of the ban thought premiums could "confuse or bedazzle," "inhibit a judicious buying choice," and
encourage materialism.[76] That was precisely the point.

Postwar Premiums

For adults, free meant something a bit different, especially in times of material scarcity. During the Great Depression, people struggled to make do
and get by, taking little consolation in the prospect of getting things for
free, especially if they required extra effort like pasting stamps into books.
Giveaways, ironically, had to seem worth it.[77] Many companies recognized
this shift in consumers' attitudes and began offering premiums that were
more useful and durable and less crappy. In 1931, for example, an issue of
Novelty News promoted premiums ranging from blended wool blankets

Figure 5.9. Gas stations could credit many families' fill-ups to cheap free premiums they gave away in series over time, like ARCO's Noah's Ark premium campaign from the early 1970s. Tim Tiebout Photography, www.timtiebout.com.

("something every woman *wants* and *needs*") and flashlights ("well designed and strongly constructed") to "Quality Fountain Pens," "sturdily built" rubberized briefcases, and axes with a "perfect grip, perfect balance, [and] perfect *feel*" made of laboratory-tested "better" steel.[78] *Business Week* noted at the time that "premiums for the mature bear down hard on utility features."[79]

As the hard years of the Depression continued, many manufacturers could stay afloat only by integrating their products into other brands' premium lines. In other words, their merchandise could be profitable only if it was given away. Since consumers remained more responsive to free stuff than to cash discounts, newer companies had to adopt premium strategies and others had to implement them more aggressively. Many companies used premiums as a way to get around the National Recovery Administration's price control legislation, which sought to reduce price gouging in tough times and to equalize the purchasing power of smaller retailers in the face of chain store competition.[80] "Small-fry smarties" continued to manufacture crappy premium goods and survived because larger-profile businesses found incentives so necessary. Ironically, legitimate retailers' solvency relied on being able to give away crappy items: cheap luggage, kitchen utensils, china, and glassware could make or break a business.[81] At the same time, and perhaps even more ironically, well-established companies like Revere Copper, Corning, Zenith, Oneida, and Eastman Kodak were also trying to get their wares adopted as premium promotions.[82] In many ways, free kept a good part of the manufacturing and selling sectors afloat during the Depression. At the time, the annual wholesale value of premium products—items that "would not have been bought at all during an era when everyone was concentrating on naked necessities"—was estimated to be $200 million.[83] By 1938 American businesses were spending some $500 million on premium merchandise "intended to appeal to the good old human desire to get something for nothing."[84]

During World War II, people began enjoying more disposable income from the jobs they performed for the war effort. But their choice of consumer goods was greatly curtailed, and tracking purchases with booklets and stamps was reserved for rationing and not buying; people had better things to do. Many trading-stamp companies went out of business, and others cut back their operations. But by the late 1940s Americans started consuming with renewed vigor. Although one analyst writing in the late 1950s asserted that "trading stamps have reached the crest of the greatest boom in their 65-year history," he was wrong. Soon they regained popularity, particularly among grocery stores and gas stations. By the 1970s more than 40 percent of supermarkets were offering trading stamps, which owners believed had become essential to their stores' survival. Upwards of 80 percent of surveyed households—and as many as 95 percent in some neighborhoods—collected and redeemed trading stamps.[85]

By the mid-twentieth century, premiums aimed at adult consumers included much more merchandise of middling and better quality that households found truly useful, shying away from knock-offs, overruns, and remainders. While some consumers resented what they saw as "forced loyalty" to stamp programs, many others reported material and psychological benefits. Some relished the "licking and sticking" of stamps into books, and surely, too, the gratification of collecting and the satisfaction of completing each page and each book. Many also embraced their status as thrifty shoppers, being able to boast about getting something for nothing. And many others, of course, found pleasure in being able to choose free goods from lavishly illustrated catalogs, converting their stiff and

Gifts FOR EVERY HOME
(A) 17-237 FINE ARTS TOLEWARE OCTAGON TRAY. 2⅘ Books
(B) 17-236 FINE ARTS TOLEWARE SCOOP. 1⅗ Books
(C) 17-238 FINE ARTS TOLEWARE BOWL. 1⅕ Books
(D) 17-235 NELSON-McCOY COFFEE SERVER. 2½ Books
(E) 17-233 NELSON-McCOY SET OF NUT DISHES. 1 Book
(F) 17-232 NELSON-McCOY CASSEROLE—With wrought iron frame. 2⅗ Books
(G) 17-234 NELSON-McCOY CANDY DISH. 1 Book
EDWIN M. KNOWLES CHINA
(H) 19-122 TIFFANY—16 pc. Set. 3½ Books
(J) 19-123 TIFFANY—53 pc. Set. 11 Books
(K) 19-118 PARK LANE—16 pc. Set. 3⅘ Books
(L) 19-119 PARK LANE—53 pc. Set. 12 Books
(M) 19-120 SIMPLICITY—16 pc. Set. 3 Books
(N) 19-121 SIMPLICITY—53 pc. Set. 10 Books
(O) 17-185 17 PC. BEVERAGE SET. 2⅘ Books
(P) 17-184 9 PC. BEVERAGE SET. 1⅗ Books
CONTINENTAL SOLID BRASS
(R) 17-227 GALLERY HANDLED BREAD TRAY—Oval. 3½ Books
(S) 17-228 3-TIER SWING-AWAY SERVER. 7 Books
(T) 17-229 COMPOTE DISH. 2⅕ Books
(U) 17-231 COVERED CASSEROLE ON BALL FEET—With 1½ qt. ovenware insert. 7 Books
(V) 17-226 SILENT BUTLER—Fully lacquered, 6½". 3½ Books
(W) 5-138 BRASS BOOK ENDS. 1 Book
(X) 5-140 WASTE BASKET. 3⅖ Books
(Y) 5-144 LAMP TABLE—Black with brass trim, glass top. Planter container is removable. 6 Books
(Z) 5-139 MAGAZINE RACK—Luxurious style in brass and black. 3⅖ Books
(AA) 17-246 BUFFET SERVER. 3 Books
(BB) 17-245 BUTTER DISH. 1⅕ Books
(CC) 17-244 COFFEE SERVER. 2⅘ Books
(DD) 17-247 TIDBIT TRAY. 1⅗ Books
(EE) 17-243 GRAVY BOAT. 1⅗ Books
(FF) 17-242 SUGAR AND CREAMER. 2⅗ Books
(GG) 17-230 CANDY DISH. 1½ Books
(HH) 17-240 BONE CHINA AND BRASS PLAQUE. 1 Book
(JJ) 17-239 PAIR OF HURRICANE LAMPS WITH CUT GLASS CHIMNEYS. 6 Books
(KK) 17-241 BRASS PLAQUE. 1⅕ Books
5

Figure 5.10. Paper and paste could be converted into beautiful and practical household items, like these. Philadelphia Yellow Trade Stamp Company, *Yellow Trading Stamps: The Seal of Approval for 53 Years*, 1957.

sticky books of stamps—mere paper and paste—into beautiful and practical things (fig. 5.10). "Over one-third of stamp savers," according to one midcentury report, "plan in advance for their premium, and a majority report a sense of urgency to start saving again once the first premium is obtained."[86]

The lure of free offered opportunities and contradictions, keeping people enthusiastically locked into a consumerist mindset. Even when they found themselves comfortably at home, the market was never very far away. Spending masqueraded as saving. And people's shopping goals tended to be motivated, ironically, by what they could obtain for free along the way.

6

THE PRICE OF LOYALTY

Over the centuries marketers found it increasingly easy to sell the idea of free to a growing number of American consumers, especially because getting free stuff felt so good. Cloaked in the entertaining rhetoric of the carnivalesque, promotional giveaways were touted as prizes and rewards. What was not to love about getting something for, seemingly, nothing? Of course, as we've seen, free embodied many contradictions: giveaways were not exactly free; the rewards consumers amassed for making purchases were no real prizes themselves; the bargains often came with strings attached. Yet there were other dimensions to free, and different iterations of it beyond retail premiums. More insidious were free advertising specialties, which, more than scratching the itch to get something for nothing, tapped into deeper desires to be accepted and, perhaps, to be loved.

Paying for Goodwill

At the eighth annual convention of the Associated Advertising Clubs of America in 1912, adman Lewellyn E. Pratt spoke passionately to a rapt audience of tradesman about the benefits of a new form of advertising, which, unlike billboards and newspaper ads, "gives the personal, human touch to an advertising campaign." This new appeal, he remarked, "rises and falls with every pulsation of the buyer's heart, like the personal greeting, the handgrasp that singles the friend out of the crowd."[1] Pratt was talking about "advertising specialties"—a.k.a. free crap: metal trays and celluloid buttons and leather-bound diaries produced by the millions. This was free with a face, with a name, with feelings.

Innovations in printing and materials technologies at the end of the nineteenth century expanded advertisers' abilities to reach American consumers through free stuff. Sure, the buying public continued to encrappify their lives with the giveaways they received by collecting stamps, clipping coupons, joining loyalty clubs, and buying other products. But the advertising specialties Pratt described were different, functioning as gifts to curry favor rather than as rewards for a purchase. They were reminders rather than inducements. As advertising professional George Meredith explained, "A premium is related to *sales*—and by definition, it has important strings attached. The advertising specialty, on the other hand, cannot have strings attached—it is a piece of merchandise given freely, without condition, and usually bearing an advertising imprint. It is, in short, an *advertising medium*, whereas the premium is a *merchandising device*."[2]

In short order, advertising specialties became an important and powerful form of commercial currency, helping businesses establish seemingly intimate relationships with their customers as they curried favor, cemented brand loyalty, and made consumers feel appreciated and affirmed. Advertising specialties were intended to help establish interpersonal relationships between a tradesman and his customer, whether a consumer, such as a housewife, or a fellow business associate. The success of these kinds of goods rested on their ability to seamlessly elide money and emotion and concretize it, made explicit in the very names professional advertisers gave to them: Gift advertising. Business souvenirs. Advertising intimacies. Personal advertising. Goodwill advertising. And, among my favorites, and even more baroque: psychological moment publicity and closing argument publicity. (Today, we get more to the point and call it swag: "stuff we all get."[3]) Quasi-gifts, they signified particular kinds of relationships, inflecting the interpersonal with the commercial and vice versa. Although advertising specialties were minor, if not negligible, forms of crap, they did not simply reflect the creeping presence of advertising into the more intimate parts of people's lives but also changed people's relationships with one another as they became increasingly bound up with and defined by advanced capitalism.

Businessmen had been actively incorporating advertising specialties into their promotional strategies decades before Henry Bunting published *Specialty Advertising*, a book dedicated to the subject, in 1910. By then, Bunting was able to name over fifty "personal specialty advertisements," from celluloid game counters, monogrammed pigskin purses, and pressed-tin thermometers to embossed wooden rulers, brass letter openers, chromolithograph calendars, enameled stickpins, and glass paperweights. "Isn't it true," Bunting mused, "that these little unexpected gifts . . . somehow warm up one's heart to the giver and make it a pleasure to patronize him? Of course it is."[4] This kind of free was intended to perform a different sort of alchemy than premiums, kindling warm feelings that would manifest in business transactions.

Bunting recognized the power of free, but what he described was something much more sophisticated than the random handkerchiefs offered with bars of soap, the cheap dinnerware sets that came with stove polish, and the decoder rings hidden in cereal boxes.[5] Rather than rewarding consumers for buying the right kinds of goods, as premiums did, these "business intimacies," carrying the names, addresses, and logos of commercial entities, inserted themselves into people's lives even more effectively. Were they gifts or bribes? Special presents or crappy advertising vehicles? They often were both, suggesting the degree to which, by the early twentieth century, the commercial world understood how to leverage personal sentiment for financial gain. Giving away advertising specialties was a way for businesses to evoke the spirit of commercial transactions within an earlier "moral economy," when people were not anonymous consumers in a commodity-driven marketplace but, rather, *customers* who were known among and gave regular *custom* to particular shopkeepers, face to face.

But the logic of advertising specialties informed more than merely how people shopped. In the eighteenth and nineteenth centuries, the material artifacts used for sentiment were quite different from those used for business. Diaries, albums, and other intimate items belonged to the feminized domestic sphere. Women, more than men, created, personalized, and sentimentalized these things in ways that forged, maintained, and marked emotional bonds and defined relationships. Gifts were the ultimate embodiment of this sentimentality and emotion, particularly in the Victorian era, and they were used as part of "a social system for the transfer of affection and the establishment and maintenance of social ties."[6] Highly personal and highly personalized, they included objects such as hairwork jewelry, hand-stitched needlework, and hand-drawn valentines—"parts of thyself," in the words of Ralph Waldo Emerson.[7] Even purpose-made gifts, such as mass-market "gift books," were personalized in some way, typically embellished with heartfelt inscriptions that made them unique. Of primary importance was the object's ability to effectively embody and communicate one individual's personal feelings to another.

Much different, the material artifacts of commerce, such as ledger books, blotters, and business forms, were intentionally impersonal.[8] They were simply tools to help maintain an efficient, well-ordered, profit-maximizing machine: books were filled not with flowery verse but with the straight lines on which to memorialize profits and losses. When seen at all, personalization came in the form of crude, practical branding; for instance, company names stamped and stenciled on barrels and crates to identify goods and track inventory.[9] Over time, cultural advisors, from authors of etiquette manuals to popular tastemakers—influenced in part by the early twentieth-century turn away from the Victorian aesthetic toward the more streamlined Arts and Crafts sensibility—called for gifts to be more useful and practical, too. This shift was not, however, simply a result of changing aesthetic considerations but a sign of the market's continued encroachment into the domestic realm, blurring, if not erasing, the boundaries between commerce and sentiment. Gifts became yet another viable commodity form, "a reliable market to exploit and expand."[10] Retailers began offering gift certificates and touting the appropriateness of various merchandise to mark weddings, birthdays, and Christmas.[11]

At the same time, the world of commerce began to influence gift exchange in another important way, by creating a need for items *specifically for* gifting in a business context. This led to the manufacture of and trade in a new kind of thing. The very existence of the advertising specialty, to say nothing of its veritable overnight success, was a testament to the ability of marketing professionals to seamlessly commingle the public and private spheres, the commercial and the personal. That is why, at the dawn of the twentieth century, admen like Lewellyn Pratt could crow about their ability to appropriate the conventions of emotion and sentiment for business interests. In the process, these men, through their crappy things, transformed the public's ideas about gifting in general. So-called business gifts were material evidence of all that had changed, as sentiment became commodified and exploited. This, then, was yet another cost of free.

Bunting's seminal *Specialty Advertising* did not spur a new marketing technique so much as acknowledge and expound upon a strategy that promoters had been using for decades but had not yet fully taken advantage of. As early as the 1860s businessmen traded in tokens of lead and brass stamped with their names and addresses, and by the late 1870s the country was littered with chromolithograph trade cards and calendars embellished with colorful images imprinted alongside commercial information. By the late 1880s companies were touting advertising specialties, such as letter openers, rolling blotters, pincushions, and yardsticks (fig. 6.1a–b). These were gifts given by men to other men at trade shows and sales meetings, exchanged within the realm of commerce for the express purpose of solidifying business relationships. Businesses also gave these things to

Figure 6.1a–b. By the late nineteenth century, businesses were able to offer an array of advertising specialties. *Nearly Three Hundred Ways to Dress Show Windows*, 1889, advertising section. Hagley Museum and Library.

individual consumers as prompts to remember their names, addresses, and services. In the late 1880s Baltimorean J. H. Wilson Marriott, for instance, claimed that his Advertising Tape Measures were an effective and affordable form of promotion because "every lady wants one": "There is nothing you can give so cheap that will be prized as much as this," he claimed. What was more, even if recipients tossed it on the ground as trash, it would be "taken up and saved" by anyone who noticed its bright red lettering.[12] In this way, any number of trifling goods could serve as three-dimensional calling cards.

While he listed several different terms for the technique, Bunting preferred "personal advertising" for these items, characterizing them as "gift articles which effect individual advertising or *personal appeal.*" He explained that it was a form of advertising "designed for the cultivation of friendly relations with individual customers and prospects." "Nearly all persons," he continued, "are reached through their feelings easier than by way of their mental processes. Personal, gift or souvenir advertising captures the gates of sentiment."[13] Unlocking these "gates of sentiment" meant opening up hearts and homes to advertisers' overtures. And that could best be accomplished through crappy things.

Over time, the variety of advertising specialties became more expansive. Bunting listed several examples: a glass paperweight with a photo of a Burroughs Adding Machine; celluloid buttons embossed with "Mennen's Borated Talcum" (worn by "your office boy"); an imitation framed oil painting made of a stamped piece of steel with a "wonderfully refreshing picture" advertising Clysmic Water; state and county maps carrying the imprint of International Harvester; a bronze ashtray embossed with the emblem of the Long-Critchfield Corporation; and many more.[14] They could each be personalized by the advertiser, and different specialties could be presented as merchandise choices to consumers as well. In 1931 alone *Novelty News* carried display ads for some 230 advertising specialty manufacturers and over 900 classified advertisements for firms seeking "novel" advertising specialties.[15]

Advertising specialties created new contexts and opportunities for gift exchange. As one later marketing expert explained, "The clue here is not only to find the wants and needs of people and serve them through your inexpensive gifts, but also to serve these needs *as they coincide with the need for your product.*"[16] Thus, they were gifts and not gifts. They were generous yet created obligation.[17] They forged relationships on social terms that were, in fact, economic. They were cheap and trifling yet possessed emotional power: even something as insignificant as a small pocket mirror "warms the heart" and "gets the business" by appealing "strongly to the prospects' pleasure motives."[18] This kind of free crap, in short, perfectly embodied the new kinds of relationships forged by an ever-expanding market. Each met the immediate needs of the advertiser rather than the anticipated desires of the recipient. "The carrying of the advertisement," one trade professional noted bluntly, "is the **primary** purpose of the medium."[19]

The professionals were onto something.[20] Advertising specialties became frequent topics in professional trade literature and at advertising and marketing conventions. The National Association of Advertising Specialty Manufacturers was established around 1904 to promote the interests of makers and distributors of all of this "good-will, reminder, or novelty advertising."[21] The specialized trade journal *Novelty News* was established in 1905. By 1912 the advertising industry was recognizing advertising specialties as a distinct promotional category, just as important as billboards and print advertisements. And unlike these forms of "general publicity," specialty advertising was "personal, social and friendly," appealing "not to reason, but to the heart, to the emotions, to sentiment, to good will on the basis of implied acquaintanceship between advertiser and potential customer." Customers responded positively if singled out for such "implied acquaintanceship." A cynical thing, the advertising specialty was "a mild form of flattery."[22] As Ralph Waldo Emerson observed in the 1840s, "We love flattery, even though we are not deceived by it, because it shows we are important enough to be courted."[23] This need for connection and flattery was innate, part of being a human being in a functioning society. It was only a matter of time before admen and manufacturers figured out how to monetize it.

Promiscuous Distribution

The rise of "personal" advertising in the early twentieth century coincided with the expansion of national markets and the rise of product branding intended to help distant and faceless producers establish a sense of personal intimacy with consumers.[24] Brands at once acknowledged and attempted to mitigate increasingly attenuated commercial relationships. The advertising specialty, not surprisingly, found a place within this larger commercial environment. Neither true gift nor true commodity—and yet plausibly both—it was a novel kind of thing that not only embodied these new, fraught, and complicated market dynamics but also capitalized on them.[25]

The purpose of advertising specialties was to evoke, within the depersonalized market, the traditional but quickly fading relationships of patronage and custom, using flattery and attention to make people feel unique, valued, and worthy of special consideration. The particular gifts businesses chose to distribute helped distinguish recipients, based on their worth as customers. At conventions, for instance, exhibitors often kept on hand "a good supply of inexpensive advertising novelties" to give "promiscuously to all who ask for them." More important customers received special articles, "which are kept under cover and only handed out in a personal way to the individuals who count." Things like "fine" pocketbooks and gold watch chains and cuff links were "too valuable for promiscuous distribution."[26] By touting various "classes" of goods, manufacturers of advertising specialties acknowledged and catered to categorical distinctions among customers. Referring to embossed metalwork plaques, for example, one trade journal remarked, "High-class advertising specialties are being used more and more every year," while a "cheap" leather key fob, too, could be "yet valuable" in courting prospects.[27]

Advertising specialties were gifts with ulterior motives. Gifting creates power dynamics that, ironically, establish the giver, the benefactor, as the "creditor" who is owed something: all gifts trigger an obligation on the part of the recipient, and no gift is purely beneficent.[28] One contemporary explained that at their most effective these items not only generated some abstract and unquantifiable form of goodwill but, in addition to loyal patronage, provided something even more valuable: namely, a list of prospects. "Sometimes the specialty is given as a consideration for the name and address, sometimes it is given to induce a call at the place of business or a sample purchase." If specialties were not "used for the purpose of increasing the prospect list and the list of customers," then they were being wasted.[29] "What We Are Doing to Make Advertising Specialties Pay the Advertiser," a keynote talk at the national convention of Associated

Advertising Clubs in 1921, focused on "the extent to which the advertising specialty manufacturers are going to make specialties produce results."[30]

Advertising specialties, like "regular" gifts, created debts and obligations, but repayment was to be made through commercial mechanisms. Sales agents used them not only to generate new business but, just as important, to engender loyalty in existing customers.[31] Specialties solidified and concretely marked these ongoing relationships and hence fabricated a sense of personal (rather than commercial) obligation, a process the business world euphemistically called "generating goodwill." They were an investment in the future. A 1923 advertising circular for the Newton Manufacturing Company, for example, exclaimed, "YOUR SALES TOMORROW DEPEND UPON THE GOOD WILL WHICH YOU CREATE TODAY."[32]

The gesture of goodwill made via modest and crappy merchandise could nevertheless have a powerful impact. The breasts of even savvy businessmen themselves might swell with the obfuscating warmth of goodwill when bestowed with an exclusive rather than "promiscuous" gift. As the Sanders Manufacturing Company put it, "No man or woman is too 'big' to accept your gift, and be influenced by it. The effect of a [business] souvenir remains long after the cost is forgotten."[33] The king of crap, F. W. Woolworth, especially understood the psychological power of business gifts, even the shoddy ones: he prohibited his own managers from accepting gratuities or gifts of any kind from suppliers or "anyone doing business" with the company, lest they be unduly influenced by them.[34] Even when people fully understood the insincerity of business gifts, they often could not help but be swayed by them.

Odd hybrids of things and ideas and intentions, advertising specialties were the turduckens of promotional crap—advertising vehicles in the form of commodities masquerading as gifts. Their status as gift-commodities created not just a first-order obligation toward future patronage but a loyalty to the gift-commodity itself. Marketers relied on the fact that worthy recipients would accept the gift, take care of it, and ideally keep it close—in a purse, a kitchen drawer, an automobile glovebox. Business gifts carried with them the expectation that rather than being discarded, they would instead be integrated into daily life, cementing the patronage of present customers and forever cultivating fresh ones (fig. 6.2).

As fundamentally promotional tools, advertising specialties were prominently inscribed. But as perverted gifts, they were not personalized

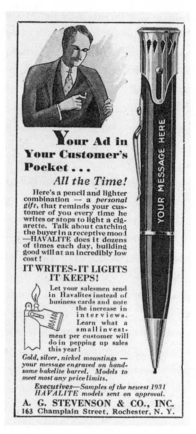

Figure 6.2. Advertising specialties would be most effective if they were kept close at hand or, better yet, close to the heart. "Your Ad in Your Customer's Pocket . . . *All the Time!*," advertisement for A. G. Stevenson & Co., *Novelty News*, April 1931.

with the names and mottos of the *recipients,* as were proper gifts such as monogrammed silverware and stitched linens. Instead, they carried the name, address, and logo of the providing company. In this way, advertising specialties took Emerson's dictum that the only true gift "is a portion of thyself" to its logical market-driven end: the gift bore the markings of the business's identity, while the recipient was just another interchangeable customer. And so business gifts could only ever convey a feigned, cynical sincerity.[35] The Newton Manufacturing Company pushed its line of fly swatter giveaways by pointing out that "Swat the fly" campaigns helped lower the death rate from insect-borne diseases: "Show your customers you are interested in their health by helping them destroy these disease carriers."[36]

The most promiscuous advertising specialties were distributed to housewives and their charges. Advertisers repeatedly stressed that the most successful examples, seamlessly incorporated into domestic life, would become familiar and reliable household objects—not just fly swatters but tape measures, pincushions, hand towels, yardsticks, dustpans, and other mundane necessities. Repeated exposure to advertising specialties would help users see the objects as company agents spreading goodwill. Ideally, through use and exposure, consumers would internalize those agreeable messages. That the majority of these gifts were utilitarian—humble yet practical everyday household items—shifted previous conceptions of gifts and their importance from communicators and repositories of sentimental value to communicators and repositories of use and economic value.[37] Regarding such pedestrian things as aprons and can openers, one trade journal noted, "Advertising specialties with a utility-value in household affairs gain the greatest response. . . . The environment of the home is an exceptionally fertile field for instructive advertising."[38]

Further, as Henry Bunting observed, the recipient of advertising specialty gifts was "pleased out of all proportion to the intrinsic value of the article."[39] Indeed, these typically cheap things, fashioned of celluloid, paper, wood, and base metals, were not unique but produced by the gross.[40] Novelty maker Whitehead & Hoag, for instance, routinely produced commemorative plaques in the tens of thousands, blotters in the hundreds of thousands, and buttons on a far greater scale. For one campaign, American Tobacco Company produced a million buttons a day for a hundred days.[41] Most specialty suppliers were, in fact, in the printing rather than the manufacturing business, simply customizing premade blanks with company names and logos, generating crap upon crap: woodenware toys, rulers, and tokens; celluloid buttons, shirt cuffs, pins, and card counters; leather coin purses, key fobs, and wallets; metalwork plaques, desk sets, and name plates.[42] The St. Louis Button Company, for instance, produced round and oval buttons combining various colors and typefaces and even photographs. The lines offered by the Sanders Manufacturing Company in Nashville included pencils, fans, whistles, and pocket mirrors. In addition to enumerating its products, the company's 1931 sales catalog offered suggestions about designs sourced from generic stock illustrations and provided advice

Page 15

Figure 6.3. Premade blanks to be customized. Sanders Manufacturing Company, *Price List and Catalogue No. 30 Illustrating a Few of Our Advertising Specialties*, 1931. Hagley Museum and Library.

about what kinds of gifts were most appropriate in various circumstances, making them thoughtless gifts in the most literal sense (fig. 6.3).

Advertising in the House

The American public accepted these cheap, commercially personalized false gifts with open arms. Advertising specialties' very triviality, in fact, enabled them to insinuate themselves even more deeply into people's home and lives. As one marketing professional explained, "Some occupy the consumer's wall (thermometer), some stay on the desk or table (ash tray), some stay in sewing closets (yardsticks), some are carried on the consumer's person (pens)."[43] They became "perpetual," "powerful," and "silent reminders" of a company's past and future largesse, and helped buyers feel rather more like patrons than consumers—who now owed a debt.[44]

Whether ashtray or bookmark, its commercial imperative remained the same. For instance, the J. B. Carroll Co. of Chicago offered a celluloid tape measure that promised to "maintain the good-will of the housewife," adding, "The Tape Measure is, of course, a very necessary household appliance, and the cleanliness of polished celluloid covers and rim make it especially attractive." Their Paper Weight with Mirror was a "greatly appreciated" gift, "immediately placed on the desk where for many years it constantly repeats the advertiser's message." The copy continued, "A 15-cent cigar tendered a good customer or prospect may soon be forgotten, but this can never be said about the most durable and permanent of advertising desk pieces."[45]

Advertising specialties were but one part of "the endless chain of salesmanship" that saturated consumers' experiences; as if by hypnosis, messages were intended to "condition the reflexes of the individual and group mind favorably" toward certain products and services.[46] People allowed advertisements in even the most private recesses of homes and offices. As specialty manufacturer Carroll & Co. noted, quite rightly, "The housewife would spurn any sum you might offer for permission to paint your advertisement on the wall of the home, but the same result can be accomplished . . . thru the use of the handsome, washable Dial Thermometer Plaque."[47] Printed calendars worked as "indoor billboards," and recipients would be grateful for the "opportunity" to display them in their homes. Any small item carrying an enterprise's imprint and contact information—whether a key case in a pocket, a nail file in a purse, or a broom holder in a closet—was considered a "pocket business card" whose efficacy relied on people seeing it or carrying it with them every day (fig. 6.4).[48]

While businesses' use of the unbranded kinds of retail premiums actually surged during the Depression (since consumers responded more positively to giveaways than to cash discounts), the trade in advertising specialties flagged. No longer a novelty by the 1930s, branded crap lost its ability to "get over the advertising message."[49] What was more, in previous decades manufacturers of specialty lines had overpurchased inventories of blanks, which left them stuck with languishing supplies of metal ashtrays, leather key cases, and wooden rulers; this limited their ability to offer new varieties of freebies during lean times. Finally, there was, simply, less general goodwill to go around during the Depression. Whether justified or not, consumers blamed the adman's puffery itself for contributing

No. 542
"Handy" Broom and Utensil Holder

Figure 6.4. Ideally, advertising specialties would find their way into the most intimate parts of the home and be used in the course of daily life. Standard Advertising and Printing Co.'s "Handy" Broom and Utensil Holder, from *Catalog No. 40M: Sales Stimulators*, 1940.

Back made of heavy steel, finished in White. Spring made of cold-rolled spring steel, assuring strong tension and rust-proof cadmium finish assures everlasting service

Spring formed with two slots permitting hanging of whiskbrooms, dusters, etc., in addition to the broom or mop. No other broom holder affords this feature

PACKED IN INDIVIDUAL CARTONS WITH SCREW FOR ATTACHING
No. 542—White finish with Blue imprint
No. 543—White finish with Red imprint
No. 544—White finish with Black imprint

Quantity	Price	Quantity	Price
100 to 149	10c	350 to 499	7¾c
150 to 199	9c	500 to 749	7½c
200 to 249	8½c	750 to 999	7¼c
250 to 349	8c	1000	7c

to overconsumption and, as a result, helping to precipitate the economic downturn. The advertising industry as a whole shrank markedly in revenues, salaries, and staffing. A $2 billion industry in 1929, advertising was worth about half that just four years later.[50]

Depression-era Americans were much less receptive to this version of free, which came with obligations and strings attached; they were already indebted enough. More practically, countless Americans found themselves homeless and on the move; more stuff was the last thing they needed. For those trying to get by in such straitened times, the intimacies conjured in the back rooms of ad agencies and then parceled out by their wheedling representatives seemed false and insincere. Advertising professional James Rorty confessed at the time that "the ad-man treats love pragmatically, using every device to extract pecuniary gain [from it]."[51] A writer for the *American Mercury* in 1937 captured the spirit of the general public, saying, "Make-believe good feeling is as useless as it is dishonest, and an era of sincere good feeling can not be handed off a shelf ready-made, nor can it be improvised out of any old shoddy stuff that happens to be at hand."[52] Try as they might, advertisers could no longer so facilely buy people's custom.

In the postwar era, insincerity was on the rise again. Americans once again welcomed advertising specialties, along with countless other consumer goods, as they set about accumulating stuff and more stuff to put in their spiffy suburban homes. Many midcentury producers were small, flexible, often family-run establishments able to engrave, imprint, paint, and etch corporate names, addresses, and logos onto any number of objects, from desk sets and coin purses to rain hats and rulers. One of hundreds was Brown & Bigelow's Remembrance Advertising, established in 1920 and headquartered in St. Paul. By the 1950s its sales force—some 1,100 agents equipped with over a million samples—was hawking over a thousand different lines each year, including beer scrapers, religious calendars, bookends, bottle openers, coasters, whistles, and playing cards; sales in 1950 alone were over $38.5 million.[53] The company gave out its own Remembrance-branded remembrances—some two hundred thousand calendars and fifty thousand pen and pencil sets in one year. And it was not alone. Experts estimated that businesses in the United States were giving away between $500 million and $700 million worth of advertising specialties annually ($3.34–$4.65 for every man, woman, and child in the United States at the time).[54]

Manufacturers and sales agents even established their own trade groups—the Advertising Specialty National Association and the Advertising Specialty Guild of America. New trade organs arose, like *Premium Practice*, the *Counselor*, *Specialty Salesman*, and *Premium Merchandising*, all to serve the interests of thousands of manufacturers, importers, distributors, and jobbers who profited from the "repeat exposure" that business enjoyed through free stuff.[55] Like any other effective form of propaganda, sales messages insinuated themselves into consumers' minds "hourly, daily, weekly, monthly . . . the year-round," with every promotional calendar, pen, and desk set.[56] Freebies didn't have to be functional to be effective: salesmen repurposed their old samples of advertising specialties, like outdated calendars and obsolete models, to use as gifts, generating goodwill even "when they are through with them as selling tools."[57] Appearing to be innocuous gifts and trivial things, advertising specialties were, at heart, powerful promotional vehicles.

The Business of Business Gifts

Not all gifts with an agenda were shoddy throwaway things made of paper and plastic, but they were crap just the same. Typically of better quality, "executive gifts" and "business gifts" also leveraged emotion in the service of commerce; they were sentimentally, if not materially, inferior. Fancy pens and pencils, engraved brass desk sets, veneered boxes, and other kindred objects were given to superiors and exchanged among high-powered business associates rather than being "promiscuously" distributed to ordinary customers. These items acknowledged relationships among members of upper management, whose advancement relied in large part on their ability to cultivate pseudo-personal intimacies. Relationships between subordinate and boss, between supplier and customer, or among fellow executives existed, and only continued to exist, because of commercial dependencies; they were understood and measured in terms of status, competition, and success.[58] The Dur-O-Lite company, for instance, explained that its products, "specially tailored to promote the interests of Business," were meant to "remind customers of What and Where to Buy" and to "develop better Customer Relations through friendly Giving." Among other business gifts, Duro-O-Lite offered Promotional and Gift pen and pencil lines, Friendship Gift sets, and the Fidelity set (fig. 6.5).[59]

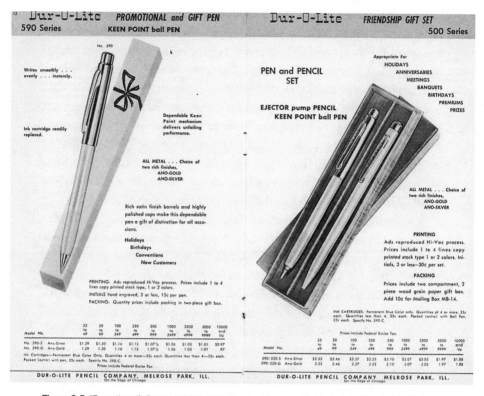

Figure 6.5. "Executive gifts" occupied their own category of commodified insincerity. *Duro-O-Lite Business Gifts Catalog No. 54*, [1954].

Unlike advertising specialties, executive gifts did not prominently display the gifting company's name, brand, or logo but employed earlier gift-giving conventions. One guide recommended, "If possible, find a way to personalize your gift—perhaps to have the recipient's name or initials imprinted on it; and to include your company name *in a way that is unobtrusive or removable*—never put a prominent advertising imprint on a business-gift item."[60] The most effective were individuated versions of generic things commonly found in executive suites—the desk set, the cocktail set, the paperweight. In their very explicit gestures to *seem* to be rejecting crass materialism, business gifts were also, and equally explicitly, the very embodiments of it.

Executive gifts would not have existed were it not for the attenuated and forced collegiality the executive classes enacted at company dinners and during rounds of golf. As marketing professionals explained, these gift-commodities were "given in appreciation of past business and in antic-ipation of future business." They were "reminders of a seller's thoughtful-ness" and had little to do with the preferences or wishes of the recipient.[61] And although everyone knew that business gifts were insincere, everyone continued to exchange them. According to the American Management As-sociation, more than half of the sales managers it surveyed gave business gifts to customers, "because it is customary *and expected*."[62] In 1955 Boeing issued a directive forbidding employees from accepting business gifts, stat-ing emphatically, "The company selects its suppliers solely on the basis of merit." *Business Week* translated this as "no bribes, please."[63]

Quasi-gifts, these myriad objects were, if not outright crap, then crap-adjacent. Givers and recipients alike were conscious of the obligatory and performative nature of these things.[64] They had become "taken for granted," notable only in their absence.[65] What was more, they had to be particular *kinds* of objects that were both gifts and decidedly *not-gifts*. Advisors recommended not only that business gifts be "modest in cost" but that everyone's gift should cost about the same, thus erasing the indi-viduality of each gift and recipient and also acknowledging that recipients would try to discern their place in the commercial hierarchy based on what others received.[66]

The crappiness of business gifts was measured not by their material quality—many were actually quite nice—but by their impoverished sen-timents and false sincerity. Due to their obligatory nature, perfunctory personalization, and self-conscious context of exchange, they were never truly gifts but only shoddy simulacra. Etiquette advisors actually recom-mended choosing from a fairly limited range of items and making them "impersonal in nature." Things that were "of a distinctly proprietary na-ture," such as cigarette lighters, fountain pens, and knives, were deemed the most appropriate.[67] This level of surface personalization made the otherwise generic object a unique gift and therefore signified, however emptily, thoughtfulness and sincerity. At the same time, personalization made these items all the more worthless, since they could not be passed along to someone who might find that (now-monogrammed) pewter tan-kard or cigarette lighter useful. People were stuck with this stuff, making the gift not only an obligation to give but also a burden to receive. Finally, professionals were advised, "Since the gift is from you as a representative of your company, and not from you personally, **make sure the enclosed gift card leaves no doubt that this is primarily a company gift, and that your personal role is secondary.**"[68]

The personal aspect of gift giving became even more attenuated and commercially driven when companies hired the services of "gift-purchasing agents" who picked out and purchased business gifts and sent them off to designated client lists. They chose gifts from categories determined by price and matched them to "classes" of recipients.[69] They even signed the cards. Eventually, individual agents became thriving enterprises whose very names—Premium Service Company, Selective Gift Institute—suggested the crass work they did. In 1962 *Business Week* told its readers, "You may want to remember your business contacts and customers with bigger-than-usual gifts this Christmas. Right now there is no Internal Revenue Service dollar limit for such items. The new $25-per-person limit takes effect Jan. 1." So people could be even more generous with their insincere largesse, as long as they remembered to obtain "some proof of a 'business purpose' to the gift."[70] The people who gave business gifts ultimately cared about the bottom line.

By the last quarter of the twentieth century, free had been thoroughly integrated into Americans' lives. All those anticipated giveaways, from prizes inside cereal packages to branded merchandise, however, came at a price. They dissolved what were at one time fairly clear distinctions between gifts and commodities, both as material things and as conveyors of sentiment. With consumers' often eager consent, gifts and the very practice of gift exchange were co-opted by business, which replaced objects embodying sincere sentiment with mass-produced crap meant for anyone who might make a purchase, and as a way to encourage those purchases. "Gifts" that built "goodwill" became a currency in themselves, and in the process defined consumers as subjects of the market whose emotions could be bought and sold just like anything else. What was more, free influenced the way individuals approached gift exchange outside of the market sphere, since they more often opted to buy rather than to make, and chose one from any number of infinite interchangeable commodities personalized only by company name or corporate logo: their signatures, and not those of the recipients, came to mark material possessions, even gifts.[71] Further, because they were free, crappy branded items were able to easily enter even the most intimate personal spaces: as can openers in kitchens, ashtrays in cars, shirt hangers in bedroom closets, and soap dishes in bathrooms. People sacrificed the inviolability of their private lives for someone else's profit.[72] In exchange, they got free stuff that was both materially and symbolically impoverished and crappy.

The fullness and profundity of intimate relationships, as marked and expressed through such practices as gift exchange, had been flattened by commodity culture and its cheap things. Cheap gift commodities not only hastened but reflected what had become, due to advanced capitalism, Americans' increasingly alienated relationships with one another. And that very alienation, in turn, generated an entirely new kind of (profit-driven and fittingly contradictory) object, the business gift. Ralph Waldo Emerson's essay on gift exchange remained relevant even though it was published in 1844: "The only gift is a portion of thyself. . . . Therefore, the poet brings his poem; the shepherd, his lamb; the farmer, corn; the miner, a gem."[73] People of the modern era would have to add to that list the businessman, who brought his advertising specialties, his "business intimacies," his "goodwill ambassadors." If it's the thought of a gift that counts, as the saying goes, then the thought behind free stuff was informed more by accumulation, materialism, and profit than by the bonds of deep and abiding relationships between individuals. This new form of the gift, born of commercial imperatives, seemed to particularly befit consumers now deeply immersed in commodity capitalism.

PART 4

(No) Accounting for Taste

7

THE BUSINESS OF HERITAGE

Curios. Knickknacks. Bric-a-brac. What-nots. Dust catchers. Tchotchkes. Gewgaws. Call them what you will, all of this stuff is crap, too. Items such as dried flower wreaths, distressed wood candleholders, and Amish dolls with no faces are not simply home decorations; they also carry subtle and not-so-subtle messages about self and other, savage and civilized, past and present. Curios are curious. They seem, at first, to be trivial things. But they can be insidious. Heritage pieces, in particular, often celebrate white identity but masquerade as harmless objects, with an alibi. They often pretend to be something they are not—and less than they truly are—which makes these objects seem innocuous and disarming. And crappy.

As with other kinds of crap, understanding the crappiness of knickknacks means tracing origins and unpacking histories. This story takes us back to the later decades of the nineteenth century, when there was a thriving market for commodities that did not seem like commodities—individuated items within a world flooded with mass-produced homogeneity. People found such goods in specialty gift shops, which were themselves distinguished from ordinary commercial spaces. Within, customers could purchase wares purporting to be unique in some way—foreign, different, handmade. Their value came from the extent to which they seemed to embody the labor of their makers and displayed the taste of their buyers. Distilled into objects, that labor was then put in the service of expressing status. Purporting to acknowledge the richness of diverse cultures, the purchase and display of these kinds of things was actually an act of appropriation.

Looking Backward in the Face of Progress

A new kind of retail space answered the revving up of mass production and rise of dime store chains in the later decades of the nineteenth century. The independently owned specialty shop, which began appearing in smaller towns and larger cities alike, catered to customers looking for unique things among the common, pedestrian, and mass-produced. One of these specialty shops was the tea room, the precursor of the modern gift shop. Located in urban and suburban areas, tea rooms were run almost exclusively by women, often out of their own homes, and provided passersby with simple refreshments served in quaint environments. Patrons found the charming décor and eclectic bric-a-brac as enticing as the food itself, a nostalgic pause amid the roar of the machine age. Often these respites were in historic taverns and mills, which capitalized on claims of history and tradition, much as antiques shops did.[1]

Tea room visitors often became so enamored of their surroundings that they wanted to buy them: the items hanging on the surrounding walls, the knickknacks arranged on tables, and even the rugs under foot. An 1882 issue of the *Decorator and Furnisher* explained that it was not enough for modern tea rooms to serve palatable food; they also had to create the right kind of atmosphere. In the article, a young woman opening a tea room chides her grandmother, Mrs. Ouldtimson, for not understanding that formerly useful articles were now prized only for their decorative value. The older woman had observed, a bit bewildered, "The pieces of china that appeared on my mother's table, or on the kitchen dresser, are suspended from the walls of this room as if to apprise visitors that appetites are out of date, and they have been transformed from the useful to the ornamental." Mrs. Middleman, a professional decorator, tries to explain the current taste to Mrs. Ouldtimson by relating the biographies of several items in and around a curio cabinet, speaking the in-the-know language of connoisseurship:

> The picture surrounding it is exquisite. The figures are painted on cloth, and the raised work is applique and Kensington. On the shelf above is a real Sevre tea set, and other articles are Japanese bric-a-brac. That curious looking affair on the next shelf is a genuine Chinese tea caddy—

Mrs. Ouldtimson still doesn't get it, seeing "an incongruous collection of modern and antique furniture"—in other words, a bunch of crap. She's told she doesn't understand "art culture" and is not properly "cultivating a taste for the beautiful."[2] Items in the tea room, like their gift shop progeny, were de- and then recontextualized, turned into saleable commodities that emphasized surface over substance, demanding the correct cultural pose of both buyers and sellers. "Atmosphere" helped imbue objects with a sense of importance even though—actually, *because*—any were devoid of provenance, historical import, and economic value. It was humbuggery, as Mrs. Ouldtimson had surmised.

Perhaps more than any other force, automobile travel greatly expanded the number of tea rooms, which began as modest retail outlets and would eventually give way to modern gift shops and souvenir stores.[3] When supping in tea rooms, automotive travelers embracing the modern age were taking in, ironically, the distillation of nostalgia, both in culinary specialties—delectables like English crumpets and freshly made cottage cheese—and the interiors' "distinctive décor." The Bottle Hill Tea Shop, for instance, highlighted its "quaint" hooked rugs and "delightful" wing chairs. A corner cupboard was "filled with interesting bits of china and old pewter," which "lent charm." "Quaint" prints, watercolors, and mirrors hung on the walls, and lampshades of orange sateen trimmed with blue and white silk braid repeated the room's "keynote" colors. The proprietrix's intention was to duplicate the eclectic and overstuffed trappings of the old Victorian parlor, making travelers feel at once at home and away in both space and time. Even better, this backward-looking stuff was all for sale; the power and importance of these objects as both free-floating cultural signifiers and profitable commodities were repeatedly highlighted in promotions. "Our furnishings," one brochure noted, "instead of costing us money, actually *made* money for us. As fast as one article was sold, it was replaced with another."[4]

Mementos and gifts sold the best. Sometimes they were modest postcards and the like, but more often "unique" and "exotic" articles appealed to customers' desires for keepsakes that demonstrated their own well-heeled entitlement. These souvenirs, it should be noted, rarely had any connections to the places where they were purchased, save the occasional jar of locally made jam or handcrafted potholder. For instance, the Bottle Hill Tea Shop sold "Only the Choicest Gifts":

Java brass, elephant bells; quaint Italian linens; and old pottery were among the choicest. We originated many gifts such as wrought iron candlesticks, made from our own design. . . . Frequently the customer ate her muffins and bought the plate; drank her tea, and ordered a tea set. The napkins—dainty squares of Japanese crêpe with hand-rolled edges done in Wellesley blue, and a tiny tassel at each corner—became a fad, and to date over one hundred dozen have been ordered and sold, and never an order solicited. Everything which the tea room had to offer was for sale, except the cook.[5]

Tea rooms eventually solved the problem of the cook by getting rid of the food entirely and offering merchandise alone, thus becoming full-fledged gift shops.

Just what were people purchasing when they bought Java brass pieces and elephant bells? A kind of cultural knowingness, however vaguely articulated. Gift shop items were both like and unlike souvenirs. Souvenirs evoke memories of places and experiences (which is why souvenirs can become important memory objects even if they were not made where they were bought). Gift shop items, however, represented cultures far and wide. They did not connect a customer's personal experience directly with the place of purchase but instead connected her to a constellation of associations conjured through the object and origin story. As status rather than memory objects, gift shop articles embodied the labor of others, and the more exotic, the better. Claims of uniqueness—indicative of fine taste and connoisseurship in a commodified world—gave these things value.[6] The speciousness of those claims made them crap.

While the practice of collecting decorative tchotchkes was not new—people of all classes had been purchasing plasterware figurines since the 1820s—the "automobiling" craze in the first decades of the twentieth century turned the commodification and consumption of nostalgia into a more elite activity. At first, the few hundred automobiles manufactured each year were prohibitively expensive (to say nothing of their upkeep), but by the 1920s recreational auto touring was increasingly taken up by the middle classes.[7] Travelers enjoyed motoring more than, say, steamship or railroad travel because it afforded more freedom, invited spontaneity, and promised serendipitous occasions to meet new people and see new things. It also gave them the chance to purchase stuff at the new roadside tea rooms and gift shops. During her cross-country tour in 1915, Effie Price Gladding made sure to stop at "The Sign of the Green Tea-Pot," in the Shenandoah Valley, "a charming little place" "kept by a woman of taste" and appointed, "home-like," with "simple, dainty furniture." Having reached Monterey on the other coast, Gladding "browsed about the curio and gift shops."[8] Such quaintness had become a necessary and expected part of elite auto touring. In her travel account, Beatrice Larned Massey complained about the lack of "charming places" on her itinerary and the countless hotels unfortunately decorated with "heavy, hideous furniture" and "impossible" wallpaper. The town of Bedford, though, was "charming" because of its Victorian trappings—a past of precisely the sort, "not soiled by modern gewgaws," being packaged and sold within gift shops."[9]

Retrograde Tastes in the Jazz Age

When Grace Knudson published her book *Gift and Art Shop Merchandising* in 1926, gift shops had migrated from rural roadsides to main streets, their merchandise appealing to consumers with pretensions. As Knudson explained, "Human nature always has loved and, in all probability, always will love to feel that it is being 'let in' on something choice which only a favored number have the faculty to appreciate." She continued, "We all like to feel that we have instincts a bit apart in individuality from those of the herd. And we are all quite certain to flock to the business or institution whose manager is keen enough to embody . . . this subtle implication of good advertising: It is choice, but you, my customer, can appreciate it!"[10] She provided vivid examples of the mass-produced objects that flattered the growing hordes of gift shop patrons. In addition to lamps, tea towels, candlesticks, and dishware, the Gift Nook in Courtland, New York, carried "fine rugs." The Happiness Gift Shop in Bridgeport, Connecticut, offered a "home setting of occasional furniture." Everything about them—from the "light and roomy" atmosphere to the retail stock—signaled good taste according to traditional, conservative, and older visions of domesticity and propriety (fig. 7.1).

In the era of the gift-commodity, it was not enough to choose a *nice* gift. The thought alone did not count. Givers were now expected to choose the *right* gift—that is, the one that most effectively conveyed something of the

In the shop of which views are reproduced here and on pages 16, 23 and 35, note how perfectly the illusion of a distinguished home is achieved, through taste and discretion in the arrangement of the stock

Figure 7.1. The first gift shops incorporated an eclectic assortment of homey goods that evoked conservative ideas of domesticity. The view of this shop shows "how perfectly the illusion of a distinguished home is achieved, through taste and discretion in the arrangement of the stock." Grace Knudson, *Through the Gift Shop Door*, 1923. Hagley Museum and Library.

status and cultural leanings of both the giver and recipient.[11] But a fundamental problem remained: these seemingly unique goods were still mass-produced objects and very much situated within market contexts and market relations.[12] And so they needed stories to make them seem more personal, personalized, and individuated, especially since showing discernment became more important than conveying sincere feelings. Gifts and gift shop bric-a-brac more often became outward representations of taste than symbols of personal affection. The narratives created around market-generated giftware, then, attempted to erase the commodity status of these objects and turn them into something more, or different—anti-commodities—by giving them individual biographies.

This process, ironically, merely reinforced their status as commodities. Many of the "distinctive" goods sold in gift shops, such as retro-style lighting fixtures, dainty china cups and saucers, tatted doilies, and homemade preserves, evoked the genteel and sedate home well removed from brash commercialism and the excitement of Jazz Age modernity. The shops' pretend domesticity helped customers imagine how merchandise might look in their own homes, decommodifying it. This specialized retail environment attracted female shoppers who had tired of the overwhelming grandiosity of department stores and the base cheapness of the five-and-dimes. Not only could patrons experience a personal intimacy within gift shops, feeling that their tastes aligned with the saleswomen and goods alike, but they also had the opportunity to purchase merchandise that they believed could not be found in more market-driven, crass shopping venues. Gift shops were so popular in this time, in fact, that department stores themselves opened up gift-shop-like rooms, such as the Davanzatti Room in Gump's in San Francisco, which "carrie[d] the atmosphere of the treasure palace," in a way that, presumably, the rest of Gump's did not.[13]

These retail curio cabinets strived to arouse all of their customers' senses, especially touch. Being able to handle goods both created another layer of intimacy and also helped distinguish gift shop displays from the sleek yet sterile glass showcases lining the floors of department stores. Gift shop proprietors preferred arranging goods on open shelves and tables (often themselves for sale) rather than storing merchandise in locked cabinets. "The liberty to touch, handle, taste, and smell in the gift shop," Grace Knudson noted, "has contributed largely to its popularity. This very factor unconsciously 'sold' the gift shop idea to a home and beauty-loving feminine public."[14] Allowing women to handle goods such as hooked wool rugs and velveteen pillow shams intensified their need to acquire and, at the same time, removed them even further from the bloodless realm of commerce. Customers were rather "guests," the shop was the proprietor's "home," and physical contact with such rarefied items was a "privilege." Knudson recognized that buyers psychically took possession of goods before they actually purchased them; simply through handling items and imagining future use, they became "theirs."[15] The gift shop was a (commercial) home away from home that seamlessly commodified taste and connoisseurship.

The environment of the gift shop helped promulgate the idea that its items held cultural, if not material, value. The home, especially of the refined, was the locus for sincere relationships and emotional transactions. And the home, rather than the market, was also the space in which family members were enculturated and values inculcated. Knudson and other retail advisors urged gift shop owners above all else to maintain their wares' "sincerity" and "integrity."[16]

The Commodification of Quaintness

Gift shops' attempts to offer "sincerity" and "integrity" meant treating those marks of character just like any other commodities—manufactured, packaged, bought, and sold. Only products that seemed to sit outside the market were imbued with these evanescent and human qualities. "Sincerity" and "integrity" only existed, ironically, because gift shop owners sought to make money by selling decorative wares, increasing desirability by placing them in faux-domestic commercial settings. Proprietors could accomplish this in a few key ways, including by carrying regional products and, as Knudson suggested, "capitalizing [on] local talent" that was, presumably, relatively unschooled and regionally primitive. "For instance," she wrote, "in sections of the South whole towns in the mountains are now making a certain style of basket and sending them throughout the country." Likewise, a town in New Hampshire was "happily employed in hooking rugs." Even these supposedly local craft traditions, bathed with the glowing aura of primitive handcraftedness, were products of the modern market. In the case of the hooked rugs, for example, a female "mastermind" created the designs and oversaw the workforce. "When they are finished, the same woman finds ready market for them in high class city shops."[17] Plucked from the backcountry and placed in "high class city shops," such items became firmly positioned in the distance both temporally and geographically, transforming them into seemingly hard-to-find products to be consumed by those in the know. Gift shop denizens only seemed to be purchasing gingham-covered jars of jam or pine-scented wreaths. What they were really buying was their own elitism.

To this end, merchandisers created enchanting biographies for gift shop items to mark them as handcrafted rather than industrial products fashioned by people far removed from one's own world. The labor of others was looked upon favorably, as something that could and should be exploited by customers and shop owners alike; in the process, these vaguely odd and foreign pieces of merchandise also represented, reinforced, and made apparent prevailing divisions between us and them, self and other, present and past, superior and inferior, subject and object. More expensive merchandise, whether handcrafted by locals or imported from remote sources, "require[d] explanations," or backstories, to create authenticity and make them seem like unique and individuated things.[18] Hand-painted woodware gifts ("Unusual–Distinctive–Fragrant") sold by the Pohlson Galleries in Pawtucket had "some of the most interesting personalities" that would "respectfully and funnily demand your attention." They had names, like Nancy the Twine Lady (a string holder), Laura (a thimble holder), and Bill, the Bell Boy (a bell with a handle hand-painted to resemble an African American). "We are sure," the proprietors noted, "they will be happy with your companionship."[19] Similarly, White's Quaint Shop offered "Original Gifts for Thoughtful, Discriminating People," including articles with hand-painted motifs, to be "just a little different," "enhance their value," and make them "very desirable as gifts to friends."[20]

Here, women could buy status in ways they could not in other retail venues. Gift shop merchandise commanded higher prices than, say, pedestrian variety store stock, not necessarily because it was better quality but because it came with symbolic trappings and personalized narratives, from the fancy display contexts to intriguing biographies. A higher price, too, seemed to confirm an object's cultural value. Unlike the dime store's hodgepodge bins of miscellanies, which were physical manifestations of big bargains, gift shops sold only a fraction of the number of things and displayed them in quasi-domestic settings that aestheticized them. Relatedly, gift shops were able to stock larger and more fragile merchandise because, since they were selling fewer items, they could display them with more care. All of these factors contributed to the creation of "classes" of merchandise, enabling gift shops in particular to shape—and monetize— ideas about the "distinctiveness" and "uniqueness" and "prestige" of their goods. In his manual *How to Run a Gift Shop*, Arthur Peel stressed the importance of locating one's store in a high-rent district since it would be "a neighborhood frequented by the class of people who are favorable prospects for gift shop merchandise," clarifying, "where the class of business generally is above mediocre." He added, "It is the difference between a gift shop and a novelty shop, a store of merchandise and an art salon or showroom." The different outlets, he emphasized, "do not appeal to the same market."[21]

Despite their pretensions, gift store items were often sourced from the very same manufacturers as variety store stock. And it was often the very same merchandise, simply elevated by its context. For both kinds of stores, crepe paper items and ceramic plates and figurines were imported from Japan, painted wooden figures and toys from Switzerland, greeting cards and dolls from Germany, and metal ashtrays, flasks, and desk sets were manufactured in the United States. Different grades of hand-painted Japanese chinaware, for instance, were exported by the crateful, bound for American jewelry stores, drugstores, general merchandisers, and retail gift shops. Sellers described these lines as providing "wonderful decoration for the money" and "awfully cheap for such artistic work." Even better, they cost a fraction of finer examples from England and France.[22]

Because the biographies of items sold in gift shops were essential to their commercial attractiveness—backstories made them stand apart from variety goods—gift shop suppliers and proprietors did not obscure but rather emphasized the origins of these things, especially the manual labor required to make them, a distorted form of commodity fetishism. Even items made "unique" through hand-painting, monogramming, or other manual craft processes were the products of mechanized and often exploited labor. Customers valued gift shop merchandise to the extent that it was clearly the result of labor-intensive handwork not just ably but happily performed by foreign and lowly workers. Linen doilies, hand-embroidered, came from Ireland and Czecho-Slovakia. Pretty handkerchiefs were "made in Porto Rico," hemstitched and hand-embroidered. "Villiager [*sic*] Handiwork" could be purchased at "moderate prices." All kinds of bric-a-brac and what-nots were hand-embellished, including metal bookends, wastepaper baskets, and serving trays; celluloid toothbrush holders and baby toys; and wooden egg cups, purse handles, and trump tellers for bridge games.

Often, the very hand-wrought decorative flourishes that increased goods' gift shop prestige erased their utilitarian value. Conspicuous uselessness (or waste, in Thorstein Veblen's terms) was part of the point: things like hand-painted trash cans and hand-embroidered handkerchiefs were, after all, too nice to actually use for garbage and snot.[23] Proprietors of gift shops acknowledged this, noting that even the purportedly "practical" goods they carried were not really meant to have "utility value."[24] They carried fancy slippers but not shoes, fragile hand-painted porcelain pieces to hang on the wall and not dinner dishes, flower vases that were not to be filled with water (fig. 7.2).

1368—.50
Gretchen the Hair Pin Lady.

1429—.50
Brer Rabbit. Full of Twine.

1359—.75
Hungry Hans Bank. Eats Pennies

1377—.85
"Bill," The Bell Boy.

1450—.75
Butterfly
"The Loiterer.'

1436—1.00
Bayberry Candlestick
For Bayberry Dips

1383—1.00
Soldier and Sal.
"A Corking Pair."

1373—1.25
"The Tall One."
With Hat Pins. Two of Sterling Silver.

1382—1.00
Cook and Count.
"Both indispensable."

Figure 7.2. Like most of its competitors, the Pohlson Galleries in Pawtucket offered a range of "unique" and "quaint" hand-painted items, "done by clever artists." From a catalog ca. 1925.

So value was generated by status. Compelling origin stories, real and fake, traveled with pieces of Spode china, Mexican pottery, Swiss figurines, and Swedish embroidery, underscoring their distinctiveness, creating demand, and justifying their higher prices. Merchandise rose in price the more clearly it evinced the labor of others. Creating narratives, though, was not an innocuous process but a deeply political one. Consumers were placing themselves, as buyers of the labor of others (especially as it was embodied—fetishized—in such unnecessary and useless pieces of crap), in positions of power. "Quaintness" added even more value, for it signaled the extent to which producers were further diminished—imagined as remote, premodern, and therefore racially, economically, and/or culturally inferior. Ideally, all of the above. The process that created that quaintness and then commodified it reinforced these racial, cultural, and economic hierarchies in the process. It was romantic to believe that people in faraway lands were toiling away *just for* gift shop customers' buying pleasure. For instance, Arthur Peel advised gift shop proprietors, "If you have a pair of Finnish curtains into which the tragedy of some Finnish peasant has been woven, this merchandise may fire the imagination of a customer to the point where the price you are asking, however high it may seem, will be paid willingly by a woman of means."[25]

Peddling cultural and material appropriation, gift shops sold *ideas* first, things second. Gift shop wares were often quite crappy, but the alchemy of high price and romantic narrative both decrappified them and transmuted them into "art merchandise of quality and distinction."[26] Peel explained how this worked:

A young man visiting a gift shop in the White Mountains, New Hampshire, was idly examining some Swiss hand-carved figurines, some of them quite grotesque in character. The girl assisting in the shop . . . asked the young man if he knew where many of the peasant craftsmen got their ideas for their whittled characters. Admitting he didn't know, she told him that most of this hand carving was done by shepherds on the Alpine and Jura Mountain slopes during the long hours they were away from home looking after the cattle, and that many of the pieces were real caricatures of people in the Swiss shepherd's own village. The young man's interest visibly increased. The set of Swiss figurines *had taken on for him a new value*. The wood carver of the

Swiss Alps had become an interesting human character with a strong sense of humor; the little figures were real people. He bought the whole set.[27]

The story gave these "grotesque" figures a "new value," which "visibly increased" the customer's interest, because now they could be understood as the work of "peasant craftsmen" who were neatly distilled, aestheticized, and commodified into "local color," rendered by the hands of "an interesting human character," and available for purchase. Customers experienced a pleasant synchrony between the men in Alpine villages—laboring twice, whittling their "grotesque" characters for export to American gift shops while working their day jobs, tending cattle—and themselves, laboring by making educated purchases in their own similarly romantic milieus.

Conveyors of these narratives, gift shop owners, like museum curators, became important arbiters of taste and distinction among the rising middle classes of the early twentieth century who had pretensions to upper-class society but not the money to back it up, or the money but not the cultural connections. They were attempting to buy cultural capital—evanescent "distinction," "quality," "uniqueness"—to which they might not otherwise have access. "Fine gifts" afforded them that opportunity. People possessing high-class cultural capital were, after all, "patrons" who purchased their decorative wares from fine art galleries, auction houses, jewelry shops, antiques shops, and boutiques that sold truly authentic things.

Gift shops carried the trappings of genteel life in image if not reality. Housewives bedecked in carved coral jewelry ("one of the season's richest conceptions") could use German sandwich tongs to place crustless sandwiches onto hand-painted porcelain plates that they served to their knowing friends on tolework metal trays. They could then discreetly dab the corners of their mouths with delicately hand-embroidered linen napkins while playing rounds of bridge—scored on custom-monogrammed bridge tallies using pens lodged in holders shaped like hearts, spades, diamonds, and clubs. Writing from Atlanta in 1932, Mrs. F. C. McClure marveled at the selection of "the gift unusual" offered by Robert W. Kellogg Inc. "Your orders are always dreams come true," she cooed. "Your establishment must look like Fairyland to people who love unusual and fine things." Even though she lived in rural Georgia, satisfied customer Miss M. T. Wilson was able to impress her urbane friends with Kellogg gifts. "I sent one of your little etched brass India bowls to a friend in New York, and she was charmed with it—wrote me that she had not seen anything so lovely even on Fifth Avenue. And I told her that of course she hadn't—it came from Kellogg's, where they knew how to select gifts."[28] Kellogg's good taste and cultural currency became her own. More important, she was able to impart that good taste and culture, in the form of the bowl and its defining characteristics (brass, etched, from India), to her sophisticated, citified friend.

Selling Time

With the exception of higher-end outlets, which specialized in avant-garde "art" pieces (discussed in the next chapter), most gift shops offered their customers items that were both geographically and temporally distant. Far-away pasts appealed as much as far-away places, and it was no coincidence that gift shops often traded in goods that replicated easily recognizable bygone aesthetic styles, like Gothic Revival doorknockers and reproduction portraits of Marie Antoinette set in Baroque-style picture frames. In this way, a historical object's aesthetic nuance and meaning was encrappified: distilled and flattened into a stylistic shorthand—a "look"—ripe for appropriation, commodification, and popular consumption.

The most redolent era for gift shop appropriation has long been Colonial America. The Colonial Revival style first became popular during the Centennial Exposition in 1876, appealing to Americans' rising embrace of nationalism, heritage, and anti-immigrant sentiment. The decades that followed saw the establishment of living history museums such as Colonial Williamsburg and Old Sturbridge Village and the formation of heritage-related groups like the Daughters of the American Revolution and the Colonial Dames. Anglo-Americans embraced their idealized colonial past in various ways, from reading romantic accounts of life in "ye olden days" penned by authors such as Mary Livermore and Alice Morse Earle to hanging Wallace Nutting's nostalgic hand-colored photographs of made-up Colonial interiors on their walls.

Embodying the seemingly idyllic domesticity of bygone times, models of open-hearth colonial kitchens came to exemplify the Colonial Revival ethos. First re-created for Civil War "sanitary fairs"—fund-raising bazaars organized to support the Union cause—they later became focal points of historic houses and public museums; the elite sometimes installed them in their own homes.[29] During the World's Columbian Exposition in 1893, colonial kitchens served as both public museum spaces and restaurants, promoting to fairgoers a particular vision of life in the past, a history uncomplicated by racial diversity, civil discontent, or women's rights efforts. They accomplished this by serving colonial-ish food against a backdrop of symbolically pregnant artifacts. Brown bread, apple pie, and pork and beans consumed amid spinning wheels, copper kettles, and bolts of calico cloth created a seemingly authentic, and properly quaint, "old-fashioned" American experience.[30]

The Colonial Revival presented an American past distilled and idealized by the present. A style of retreat, it was a response to the seismic shifts at the turn of the century: the country was recovering from a series of economic depressions, dealing with waves of immigration and attendant xenophobic reactions, adjusting to demographic changes that brought many African Americans north after Reconstruction, coping with scientific and technological innovations that rapidly and significantly changed how people lived, enduring spasms of labor unrest that threatened owner-worker power relations, and more. Those who had the means, therefore, took comfort in the apparent simplicity and virtue of the colonial past, a safe haven in an unpredictable, chaotic world. Valuing style over substance, nostalgia over reality, an Anglo vision of the past over the present polyglot, the material trappings of the Colonial Revival made perfect gift shop merchandise.[31] The kitchen, for instance—the thing that most embodied the movement as a whole—suggested work but never showed it actually being performed. Hanging artfully on the wall or placed just right in curio corners, tongs and ladles and kettles and distaffs were no longer tools of production but the stuff of consumption—mere relics of a longed-for but untenable way of life, a mourned-for past that never existed (fig. 7.3).[32]

In the Colonial Revival's version of the past, people heeded clear gender roles, and women found satisfying work as helpmeets to their husbands. In the present, the country was witnessing women's suffrage, the rise of the flapper, and the championing of "the new woman."[33] Women in the past relied on spinning wheels, candle molds, kettles, and the like to produce food and goods for their families and for the market, unlike the modern woman, who had reduced these tools to mere decoration: they carried symbolic value for the very reason that their use value was spent. While Colonial Revivalists venerated a specific past for its seeming simplicity, integrity,

Figure 7.3. The back of this postcard, showing a re-creation of a colonial kitchen with its open hearth and traditional tools of domesticity, reads, "In the Colonies for two hundred years the Kitchen was the center of life. An exhibit at the Newark Museum, 1926–1927." Courtesy of the Newark Museum Archives.

and handcraft tradition, their nostalgic material trappings could only be made possible due to advances in manufacturing capabilities that could produce so many door knockers, andirons, candle molds, and cast-pewter plates. This same expansion also resulted in a consumer base willing and able to buy these fraught faux objects.

By catering to the auto-touring crowd, creating carefully staged shopping milieus (echoing the diorama approach of the colonial kitchen), and stocking their stores with real antiques and modern reproductions that evoked nostalgic, if fictional, pasts, gift shops were able to capitalize on the anti-modern sentiments that pervaded American culture by the 1920s and 1930s. New firms such as Gifthouse and Art Colony Industries enjoyed great success making reproductions of candlesticks, wall sconces, stagecoach lamps, coffeepots, trivets, andirons, firewood holders, bed warmers, water pitchers, chamber pots, spice boxes, and other objects that "create atmosphere and lend that Color to a house which few things could replace."[34] Many of these colonial American items, Art Colony noted unironically, were imported from "around the globe"; even menorahs could be refashioned in the Colonial Revival style (figs. 7.4, 7.5). What was more, the style enabled savvy companies to disguise shoddiness as authenticity. For instance, the Albany Foundry Company, specialists in cast-iron bric-a-brac, explained that its replica door knockers, bookends, and andirons were intentionally crappy: "Hard and sharp lines and strong details are purposely omitted to imitate the antique," noted a promotional booklet.[35] In gift shops across the country, including places like Croston's in Boston,

"**The Most Distinctive and Fairest Priced Line in America**"

THE ART COLONY INDUSTRIES, pioneer craftsmen in brass and copper, have the pleasure to introduce to the readers of the SURVEY their products of solid cast brass candle-sticks, candelabras, door-knockers, wall sconces, etc., and vases, fruit bowls, trays, comports, fern dishes, etc., etc., in hand-hammered brass and copper.

These may be found at your favorite art shop, gift shop or store, but if you cannot get them, we will be pleased to sell you direct.

To insure quality and craftsmanship insist on ART COLONY PRODUCTS in Brass and Copper.

Price-list and catalogue on request

ART COLONY INDUSTRIES
326 Church Street   New York City

Figure 7.4. Gift shop merchandise tended to look backward in the face of modernity. Advertisement for Art Colony Industries, *Survey*, June 11, 1921.

Figure 7.5. As much as the items themselves, culture was on offer in gift shops. Croston's ca. 1920 catalog, for instance, offered merchandise that "attracted a most discriminating clientele."

customers could choose from these and other Colonial Revival product lines, from antiqued oak tavern tables (useful and artistic) to wrought-iron plant stands.[36]

The Colonial Revival style, as all retro styles tend to do, reflected how people felt about the present rather than the past. Colonial Revival homes were linked to "the founding of the country, ancestral homes, and a strong family life—what many middle-class members regarded as the foundations of American life, values, and institutions."[37] Proclaiming and perpetuating these conservative political and social attitudes, Colonial Revival pieces were marketed to those who not only believed in the supremacy of Anglo-American heritage but were actively trying to promulgate it. These same people also preferred to live in Colonial Revival housing developments. An advertisement for one of them, Wilmot Woods in New Rochelle, New York, stated with authority that "neither people nor houses of radically different traditions make congenial neighbors." The developer assured prospective buyers that home sales would be "restricted" to "American families of refinement," implementing what one historian referred to as "the spatial strategies of white supremacy." Meaning, privileged whites only.[38] The very concrete political dimensions of this persistent nostalgia explain both the appeal of Colonial Revival objects and also why they were crap. By turning conservatism, and even racism, into an aesthetic pose, "a look," the makers and sellers of these objects enabled their buyers not only to feel superior but to openly display this sensibility to others.

The stylistic pose also provided plausible deniability for the overtly racist items for sale in these quaint gift shops. These, too, conjured another particular American past and place, that of the plantation South during slavery and Reconstruction. Whether lawn jockeys or minstrel-themed ashtrays, these objects froze African Americans "in their place" as domestic workers under the control of the whites who might very well live in "refined" and "restricted" neighborhoods. Slavery was over, but in the era of Jim Crow, African Americans still struggled for work, for equality, for justice. Privileged whites continued to summon African Americans through material surrogates bought and sold on the open market, performing political work if not actual labor. Among the many objects "worthy of consideration" in Prince's Gift Shop in Lowell, Massachusetts, for example, was a brass letter opener with a handle shaped like the visage of a black boy wearing a ratty straw hat.[39] Fireside Gifts offered many "Objets des

Arts," including the Three Black Crows Bean-Bag Game, featuring Sambo, Rastus, and Isaac as targets: "They will obligingly fly back on their hinges and let the bean-bags through!" There was also the Ash Tray "Aunt Jemima from Jamaica," a "jolly" and "genial" figure that would "add spice to the after dinner smoke" because smokers could extinguish their cigarettes on her head. Spools of sewing thread constituted the topknots for The Topsy Spool Holder, an "amusing chocolate maiden." The Aunt Miranda Clothes Bag, in gingham, "greets the soiled clothes with outstretched arms and a beaming smile." Pliant things, these objects were perpetually ready to do the bidding of their owners, with "a beaming smile."[40] It was their reason for being. Despite their various forms, Colonial Revival gift items and racist bric-a-brac performed the same work. As "gifts of distinction," they normalized and reified divisions of class and race (figs. 7.6–7.8).[41]

Literal objectifications, tchotchkes incorporating racist stereotypes of African Americans were welcomed into cultured white homes to become

Figure 7.6. Racist merchandise, like the Sambo letter opener and other examples shown here, was integral to gift shop offerings. Detail of a page from a catalog for Prince's Gift Shop, ca. 1915.

388—Dinah, the Darning Set that's the newest thing out. Consists of needle case, brightly enameled with hand painted thimble for cap. Beautifully enameled darning ball in black, red, white and blue, with handle measuring 5½ inches. 75 yds. of Heminway darning silk, 25 yds. tan, 25 yds. white, 25 yds. gray, on spools arranged on special holder with head of Dinah at top and bottom. A very splendid gift.............**$1.75**

Figure 7.7. Dinah the Darning Set, sold by White's Quaint Shop in the 1920s, made for "A very splendid gift."

TOPSY SPOOL HOLDER
No. 741 (above)
This amusing chocolate maiden smiles proudly because she wears spools for pigtails. Hgt 5"; shpg wgt 6 oz; list price $1.00.

Figure 7.8. The Topsy Spool Holder was described as an "amusing chocolate maiden." Fireside Gifts, *Objets des Arts*, [1931].

domesticated and disciplined. Like the others, the Aunt Dinah blackface darning set made for "a very splendid gift."[42] The darning sets, spool holders, and rag bags functioned both as practical items for domestic work and amusing objects whose levity came from casual racism. Small, infantilized, and static versions of black people doing work, these gift items were material surrogates for the lost labor of the plantation South. The people who bought this coarse merchandise were turning slavery's legacy into just another crappy indicator of their "taste."

Handmade in a Plastic World

"Quaintness" provided a convenient cover story for the bric-a-brac of white supremacy, whether those things were handmade by "primitives," looked like relics of colonial times, or took the form of racist caricatures. Although they might have seemed innocuous enough, all of this stuff that sat on mantelpieces, hung on walls, and cluttered side tables reinforced their owners' worldviews and helped champion the supposedly civilizing forces of white Anglo-Americans. Being aestheticized and turned into "conversation pieces" thus made these political statements seem quite anodyne, which was part of their allure and also part of their dishonesty.[43]

By the middle decades of the twentieth century, there were even more ways that the market met sophisticated consumers' needs for the decorative crap of primitive producers. Mail order outfits enabled shoppers to purchase "authentic" items direct from the source. Tesori d'Italia Ltd., for instance, offered "magnifici" gifts mailed from Italy, "where for centuries the making of beautiful things in hand-blown glass, in precious metals, in carved wood and gold-tooled leather has been a tradition and an art." Buyers could order nut dishes that were "faithful reproductions of Verrocchio's fabulous fountain" (of pressed glass and cast pot metal); strings of hand-blown Venetian glass beads (factory seconds?); costume jewelry created by "a fine Italian hand" (silver plate and faux mosaic, $1.98/set); and dolls with hand-painted faces representing ten different Italian cities (likely made in Germany and indistinguishable from the "authentic" dolls made in other countries).[44] Receiving packages with "genuine" Italian postmarks and customs stamps only confirmed the items' authenticity and added to their value.[45] It also enabled people to be mail order tourists (fig. 7.9). Likewise, "lovers of handcrafts" could order objects capturing the "particularism" of Quebec; said objects "have been able, with the passage of time and taking into account the evolution of techniques and of taste, to forge a typically French Canadian style."[46] Another company, Shannon International, representing "the handiwork of craftsmen from more than a dozen lands," carried lines produced in Ireland and beyond.

To Delight A Little Girl
A Handbag That's A Foreign Doll

These vivid felt dolls have expressive hand-painted faces, soft pretty "hair" and skirts that are really zippered hand-bags! All are made in Florence, Italy, but each wears the gay peasant costume of a different country.
There's Maria, a black-haired, blue-eyed Italian beauty (a) left; flashing brunette Carmen of Spain (b) center; Ilsa from Austria, in Tirolean dress (c) right; and Dutch Gretchen with long yellow pigtails (d), not illustrated. Each is approximately 7½" long, with matching 5½" handle. Specify your choice by the letter.

#22 _____ Each, $1.98 ppd.

Your Selections Shipped Direct from Abroad

Figure 7.9. Mail order tourists could purchase "foreign" dolls that were also purses. Tesori d'Italia Ltd., *1954–1955 Gift Catalog: Magnifici Gifts Mailed Direct from Italy*, 1954.

Its "Gifts-Souvenirs" included miniature Blarney Castle music boxes and bookends in the shape of Irish monks (fig. 7.10). The company also sold subscriptions to its "high class" magazine, *Ireland of the Welcomes*, which featured "the varied and colourful facets of Irish history, culture, folklore and present-day life and arts."[47]

The work of gift shop proprietors in locating and selecting "exclusive" merchandise, too, became much easier thanks to the increasing number of middlemen funneling goods from remote production facilities to Main Street retail outlets. They often displayed their wares at trade fairs like the New York Gift Show, whose exhibitors included, among many others, Ancient & Modern Oriental Imports, Canastas Mexicanas, the Danish Candle House, Eur-Asian Imports, Sam Hilu's Odyssey Imports, and House of Jordan. Even when trading among fellow crap purveyors, these outfits adopted names that helped perpetuate the biographical fictions of the things they sold, since many items did not, indeed, come from their claimed points of origin. Names such as the Pan-American Barter Co. and African Wood Carvings Inc. also helped hide the taint of commodification.

Offering an assortment of eccentric goods helped gift shop purveyors in a practical way: (supposedly) unique goods thwarted comparison shopping. One proprietor noted, simply, "An item must be blind, so that it can't be compared."[48] Within the retail space, such eclecticism also helped create otherworldly, noncommercial milieus that enhanced the desirability of singular goods by making them part of a more interesting collective. Take, for instance, Shopping International, in Hanover, New Hampshire. The store created not one distinct retail environment but several. The space was divided into nine rooms, each decorated "in the style of a particular country." Folk music appropriate to each locale played in the background, thus "heighten[ing] the impression of 'being there.'" Customers were lured

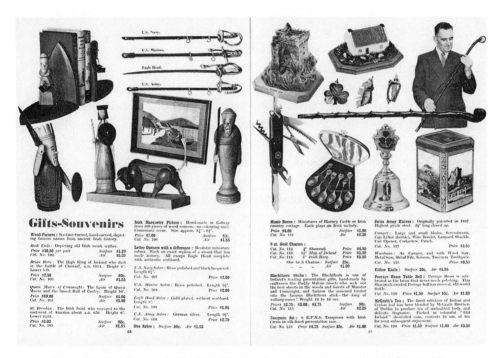

Figure 7.10. "Gifts-Souvenirs" of the kind sold to American customers by enterprises like Shannon International intentionally conflated handmade and mass-produced items. Shannon International, *Shopping and Mail Order Guide, 1962–3.*

by the siren song of the "mystery, whimsy, romance, and surprise," externalized via "products of time and tradition which are priceless because they cannot be duplicated elsewhere."[49] Here, the tourist/visitor/shopper could buy burro muzzles from Mexico, hand-woven door mats from the Philippines, hand-punched tin harem lamps from Iran, and carved teak cigarette boxes from Japan, curiously situated in the Africa section between a man's wallet from Morocco ("made to fit American currency") and the Casablanca Hassock.

The Past Is Present

The qualities celebrated by aficionados of the Colonial Revival and related styles remained popular throughout the twentieth century. Nostalgia, it seemed, was timeless. Merchandise in the style of "rustic," "primitive," "country," "lodge," and "shabby chic"—nostalgia by another name—was carried in gift shops with names like Village Peddler, Country Cupboard, and Yankee Trader. "The Country look is still popular around here," said the owner of The Wheel House in Bristol, New Hampshire, in the late 1980s.[50] Set in an old Victorian house with an incongruously "elegantly homespun feel," the Country Goose, in Washington State, "allows product to be displayed in its natural habitat," a weird yet apt phrase that captured the semiotic and semantic tricks that made commodities seem unique and yet enabled "product" to be priced. Style stereotypes valued surface over substance, as they always had.[51]

These handcrafted "looks," it bears stressing, were productions of industrial manufacturing techniques that created imperfections to imitate handmade work—even texture itself was a sham. An article touting the popularity of folk art in the late 1980s, for example, stated matter-of-factly that pieces prized for their unique eccentricities were "not necessarily handcrafted."[52] "Verdigris products are current fast-sellers," a gift shop owner reported in the late 1980s. She was referring to objects with a green surface that simulated the mellowed patina of aged copper.[53] People were drawn to the "earthy feel" of salt glaze, splatterware, and redware pottery, which enabled buyers to "envision the potter hard at work at his wheel!"[54] People who bought or displayed these surface-driven knickknacks were then able to articulate their own cultural politics, in subtle and not-so-subtle ways, through seemingly innocuous decorative choices.

Ironically, there really *was* something behind all those surface treatments and appropriations, from the ersatz crackle-glazed pottery and distressed-wood dressers with peeling paint to the faded flour bags made into tote bags and antiqued tea tins filled with dried lavender. The aggressive quaintness of objects in neo-nostalgic styles belied an inescapable and intentional conservatism. "The genre gives Americans a sense of heritage," noted one trade journal in the early 1990s.[55] Dish towels made of gingham, jars containing the preserved and the pickled, artisanal soaps, wooden plaques "hand-painted" with inspirational mottos (modern samplers), things with flags, and, distressingly, pickaninny figures (fashioned out of carved wood, clothespins, and cookie cutters) signified an idealized past that never existed and could only be conjured through these utterly false objects, deeply immersed as they were in the home, the heartland, and the country's mythos itself. As Michael Kammen famously observed, although "nostalgia tends to be history without guilt . . . this elusive thing called 'heritage' is the past with two scoops of pride and no bitter aftertaste."[56]

Indeed, trends in country-style giftware stressed "comfort" and "feeling good," its adherents having "revived their interest and caring for family, religion, and lots of basic values that many thought were long past and gone forever." Published in a merchandise trade magazine, that statement rested on the assumption that a segment of the population was actually *against* comfort, feeling good, and an interest in family and "basic values," whatever those were. An inherently reactionary style separating an Us from a Them, neo-nostalgic country chic loudly and proudly proclaimed American exceptionalism, conflating past and present, conservatism and aesthetics. One trade observer from the 1990s summed it up: "It's a time when More, apple pie, and the American flag are not to be sneered at, when nostalgia for the good old days is replacing the reality of what they were. Even the Depression is having a décor revival."[57]

Related to but different from heritage, nostalgia can be understood as a "yearning for a different time" and the "ache of temporal distance and displacement."[58] (The word "nostalgia" comes from the roots *nostos*, meaning homecoming, and *algia*, meaning pain.) People's desire to be more closely connected to distant pasts and rightful "homes"—whether places or times or both—helps explain why things like faux finishes and worn surfaces became such important aesthetic elements in late twentieth-century gift shop merchandise, whether intentionally distressed "shabby chic" or pieces made of wicker, twigs, and animal horns, known more generally as the "Lodge style."[59] Ironically, tactility was also nothing more than a surface gesture, used not only to create artificial imperfections but also to help sell the merchandise by making it more immediately attractive to consumers, just like the hooked rugs and velveteen pillow shams of earlier gift shops. "Display is vital," noted one proprietor, who continued, "Make the product accessible to the customer's touch and easy to buy!"[60]

The Smell of Heritage

The ability to touch was but one aspect of the multisensory experience conjured in gift shops specializing in the neo-nostalgic country style. Another was sound, like the harp recordings played in the Country Goose, which served as an "audible testimonial to soothe shoppers."[61] Most lucrative, however, were the many scent innovations adding olfactory redolence to shopping. The history of gift shops is not complete without discussion of their most popular and quintessential item: the scented candle. Considered a manifestly authentic and nostalgic item, the scented candle was, too, the product of modern times. Ever since the incorporation of electricity into urban and rural homes, candle production in the United States had been on the decline. Lower oil prices and high import duties revived the domestic manufacture of candles in the 1970s. Enabled by technological improvements allowing the production of better-quality candles embellished with decorative motifs, retailers were able to place a huge markup on candles and realize great profits.[62] By the late 1980s candles and candle accessories (!) accounted for upwards of 20 percent of the total market of gift shop items. Estimated total sales in the late 1990s for candles alone were between $968 million and $2.3 billion.[63]

Many factors contributed to the growing popularity of scented candles. Smell, perhaps even more than appearance or texture, evoked nostalgia. Candles, according to a giftware trade journal, "can scent an entire room with the heavy throw of their true-to-life fragrances." Echoing this, a company representative for Yankee Candle claimed that scented candles "have the ability to transport people to appealing places and times they remember."[64] Simply as objects, candles have been, over time, important and easily recognizable symbols, especially when used for decorative rather than utilitarian purposes. Their luminosity offers a warmer and more pleasant light source that seems a rebuke to newer and harsher fluorescent lighting. What is more, they harken back to the days before electrified and gas lighting, when candlelight was the only option for a predominantly rural population living by the cycles of the sun: The Good Old Days. They also could be seen as modern and much more compact iterations of the Colonial kitchen, symbolic shorthand for ye olde tyme warmth of the open hearth without the requisite fire pokers, copper kettles, cast-iron trivets, or even the hearth itself. Candles filled (and continue to fill) these symbolic and emotional needs.[65] "It's no secret," remarked *Giftware News*, "that customers are drawn to products that make them feel warm inside. . . . Customers are looking for a quick escape."[66] WoodWick Candles even integrated an aural component, since "their natural wooden wicks . . . create a soothing sound reminiscent of a crackling fire." One of WoodWick's most popular "unique" fragrances was Evening Bonfire.[67]

The appealing places and times conjured by scented candles were not, of course, places and times that people *actually* remembered. Rather, they existed as vague, romanticized, imagined pasts: a time without electricity and running water; a time when white men dominated; a time when women tended only to a family's domestic needs; a time when the country seemed less complicated and more wholesome—pasts that were pleasant, uncomplicated, palatable. In other words, pasts that never existed at all. These were fantasies redolent with pine, cranberry, and vanilla. In the scented candle world, "Rustic Chic" became a "fragrance zeitgeist" described as "the emotional counterpart of the natural trend. Focused on the authentic rather than the ideal," according to one marketing expert, "these scents are all about realistic recreation, vegetable, earth, and wheat inspirations."[68] The 1803 Candle Company's top-selling scents included Shoofly Pie, Grandma's Kitchen, Perfect Morning, Orange Caramel Scone, and Perfect Evening.[69]

Whether scented Yankee candles made in New England or leatherwork purses imported from Morocco, gift shop crap traded on consumers' intense investments in nostalgic outlooks and celebrations of heritage. Since well before the dawn of the twentieth century, shoppers had avidly and consistently bought and bought into sanitized pasts and the aestheticized labor of outsiders. When they purchased this crap, buyers were actively appropriating and commodifying others' work and, in the process, declaring themselves members of a certain economic, social, and racial elite. And by displaying these ersatz objects in their homes (and giving them away as gifts), they became evangelizers, champions, and normalizers of colonizing impulses and neo-nostalgia. This was the cultural politics of crap.

8

CONNOISSEURSHIP FOR SALE

As we have seen, gift shop customers were able to purchase sanitized versions of the past. They found a lot of decorative crap attractive because it offered up vague but easily understood and conveniently packaged notions of heritage and nostalgia. But heritage-seekers weren't the only ones who bought giftware. Others selected their bric-a-brac not because it romanticized the retrograde but just the opposite: they believed their tchotchkes and knickknacks projected sophisticated discernment. People tried to buy connoisseurship and classiness in gift shops as well.

Fashioning Pedigrees

Commodified connoisseurship could be found in retail spaces that fashioned themselves as elite places of refinement and gentility, and the cultural process of commercializing class was already in full swing by the final decades of the nineteenth century. Shops like Hartford's T. Steele & Son issued prescriptive "manuals" like *What Shall I Buy for a Present* (1877), which was dedicated to "Our Customers . . . who by their patronage and aesthetic taste have helped to cultivate . . . a love for all that is refining and beautiful in artistic wares." Like those that would follow, the firm was no simple middleman but an arbiter of taste, hiring "commissionnaires" to scour the Continent, using their "taste in selection" to procure "fine artistic articles of the various European manufacturers at the lowest prices." The interior of the store also appealed to the pretensions of the clientele it sought, flattering them with "choice decorations," showcases inlaid with ebony imported from France, and "elaborately carved ornaments as a finish" (fig. 8.1).[1]

People acquired connoisseurship and taste in gift shops as much as

Figure 8.1. Interior cut of T. Steele & Son's Hartford establishment, showing off its good taste to potential customers. T. Steele & Son, *What Shall I Buy for a Present: A Manual*, [1877]. Hagley Museum and Library.

merchandise itself; proprietors' cultivated eyes helped guide and inform their choices. Steele's office, for instance, featured a large window onto the store; customers who peered in could view, among other things, a well-stocked library of books on mythology, treatises on gemstones, and even classical dictionaries—advertising his desire to see to the "education of the taste of the public." Steele's son, who eventually took over the business, was also a man of culture, making not only "literary contributions to the Press" but also paintings of trout, "which have received well merited praise in Hartford and elsewhere."[2]

Trout paintings, reference books on mythology, and lessons about classical art might seem only remotely connected to the act of shopping for gifts. However, because gifts were becoming commodified things—objects of the market that were produced, sold, and purchased *as gifts*—consumers needed reassurance about their authenticity, legitimacy, and efficacy as markers of high culture, whether they were, like Steele's stock, "of the finest quality," priced at $1,000, or "much less expensive" at fifty cents.[3]

Whether merchandise was choice or pedestrian, what mattered was its pedigree and claim to authenticity, rarity, and distinction. An 1875 advertising puff credited T. Steele's fifty-year longevity to "the cultivation of the best taste in all matters connected with it." In other words, it was the degree to which, in a world of mass industrialization and the rise of chain stores, the business could offer its patrons out-of-the-ordinary wares—things that *seemed* like they were something other than commodities. High-end gift shops held themselves out as institutions more closely aligned with museums than retail stores: "The entire stock of goods is exceedingly rare, both as to its extent and quality," reflecting the "excellent judgment and exquisite taste" of the proprietors.[4] By displaying these objects in their homes, customers, too, would reveal their good taste. And by giving away these objects as gifts, they not only signaled their own good taste but also flattered friends and loved ones by acknowledging their discernment as well.

Like antiquers, who at this same time were in search of objects possessing historical authenticity ("real" artifacts), gift shop customers were looking for items with cultural authenticity, if such a thing could exist.[5] While antiques experts could point to concrete features of objects, such as workmanship and provenance, to mark pedigree and confirm legitimacy (and hence confer value), gift shop customers could look only to the representations and judgment of gift shop proprietors. What was more, while antiquers had to occasionally deal with objects that were fake, all gift shop gifts were fake in that they were inherently not pristine artifacts but mediated—contaminated, if you will—by commercial imperatives. No matter how much they claimed or implied otherwise, items in gift shops were firmly embedded in the market.[6]

And so, like heritage objects and antiques, "finer" gift shop items needed origin stories that would imbue them with positive associations and a sense of uniqueness, which would increase both their purchase price and their symbolic value as vehicles through which to convey discernment. Promotional literature for gifts and gift shops emphasized the uniqueness and irreplaceability of these mass-produced things, deploying vague terms like "distinction" quite liberally, just as gift shops sold heritage and nostalgia by creating "quaintness."

Successors to businesses like T. Steele similarly touted the idea of fine gifts among people who aspired to be, but were not quite, members of the elite. High-end gift shops offered patrons accessible versions of high-class stuff in styles that possessed cultural currency among the refined and knowing, and therefore supposedly meant something. For instance, by the turn of the century many shops were offering merchandise in the Oriental style as a way to capitalize on a collecting craze among wealthy American women, who had become enamored of Chinese and Japanese goods displayed at international expositions. Like heritage goods, these things were political as well as decorative and "represented and celebrated America's expansionist and imperialistic power." Trailed by a popular press that reported widely on their exploits traveling through "the Orient," elite women bought up pieces of jade, chinaware bric-a-brac, elaborately embroidered silk robes, bronzes, tapestries, prints, and dishware.[7] By the 1920s many had amassed sizable and important collections. Although these rarefied pieces were within reach of only the one-percenters, aspirational middle-class women in turn developed a taste in Asian and Asian-inspired goods, for which gift shops, department stores, "novelty," and "curio" shops were happy to oblige.

Middle-class women hoped to accrue prestige and show refinement by purchasing objects similar to those favored by the elite. Tastemakers deemed such studied foreignness and staged eclecticism, all the rage in the early decades of the twentieth century, to be "daringly artistic" and essential to creating "cosmopolitan décor." By owning and displaying these items, middle-class women seemed to be embracing nonconformist and perhaps decadent lifestyles fashioned after the fine artists and actresses who appeared on the pages of popular magazines. Women who never left their hometowns could yet pretend to be cultivated world-class travelers by purchasing exotic goods from the local gift shop.[8] What was more, by favoring objects with a particular stylistic gloss—a "look"—women could parrot the sophisticated aesthetic language of the cultured and in-the-know. Items offered in bric-a-brac boutiques both relied on and created women's preconceptions of what those stylistic elements might be, based on stereotypes concentrated and distilled into specific material forms.

These vaguely exotic styles, though, were neither culturally nor historically authentic. Nor were they tied to specific countries or regions. Nor did it seem to matter. "Oriental" treatments incorporated aesthetic elements from Japan, China, India, and the Middle East. This "Aesthetic Orientalism" was about imperialism and colonialism, casting the West as modern and forward-looking (consumers) and the East as premodern, traditional, and simple (producers).[9]

Gift shop Orientalism was crappy in much the same way as the Colonial Revival. Although faux colonial chandeliers did not at all resemble Asian-inflected goods, they were similarly emptied of meaning and authenticity: Japanese Jardinieres were merely "of Japanese inspiration"; Chinois Lamps were rendered "after the Chinese fashion"; and Kashmiri Flower Holders were "not imported."[10] As Edward Said observed, "A white middle-class Westerner believes it his human prerogative not only to manage the nonwhite world but also to own it, just because by definition 'it' is not quite as human as 'we' are."[11] Thus, in their miniaturization and pseudo-practicality, these curious items literally objectified—turned into objects—the "them" to be consumed by "us": a Cossack cigarette holder, whose body contained cigarettes retrieved by pulling off its head; a Chin Chin window shade pull, whose entire body one would grasp to raise and lower shades; and a "dark" Gypsy head that doubled as a decorative pin holder.[12]

More broadly, gift shops created the impression that all their goods were "chic," "distinctive," and "unique" by leveraging foreignness of all sorts, whether referencing distant times and faraway places or conflating them. The stylistic shorthand of the "look" created positive associations between a thing and its owner.[13] Women valued goods identified as "imported" because of their ready-made connection to high culture. Consider just two pages from a lengthy merchandise catalog for Fireside Studios, which offered these items, among others: Quimper pitchers ("Imported"; "worthy of a collector's enthusiasm"); Del Mare wall pockets ("Imported"); La Neige vases ("Created by the famous artist and worker in glass—Sabino!"); Delft Blue flowerpots ("Not Imported"); Rouen vases ("A piece of genuine

Rouen pottery, a faithful reproduction of a museum antique"); Le Dauphin pitchers ("from the hand of a master potter, Delacourt of France!"); Sevilla jardinieres ("imported from Spain"); vases from "Czecho Slovakia" ("Imported"); Andalusian vases (with "lines of a Moorish olive oil cruse"); and so on. Foreignness in this context meant something exotic, hard to find, unique. The word "imported," for instance, appeared on these two pages alone thirteen times. Even the Imported Vase was further identified as "Imported" (fig. 8.2).[14] Labels linked items to their places of origin while at the same time obscuring the realities of production, hence creating "chic" and "charm."

As with other forms of crap, qualities such as "charm" were ultimately meaningless. Likewise, merchandise was not more valuable simply because it was handmade or imported; it just seemed that way. Producing "chic" and "charm" required using human labor as if it were machinery and often implementing "extreme specialization." Overseas workers did not fashion entire pieces from beginning to end, because they did not know how to. Different processes were performed by different people, and different phases of manufacturing often took place in separate facilities altogether. The cups in a "foreign style" tea set of the sort offered by Fireside Studios might be made in one place, while the matching teapot was made in another and the saucers in yet another. Similarly, slip-casting the chinaware forms might take place in one facility, while glazing occurred in another, and decorative embellishments were added across town; one painter might be assigned to flower petals, while another did leaves and stems.

The impression that imported goods in this realm were well made was itself often a mirage. Japan, for instance, manufactured better lines of merchandise for its own domestic market. Popular exports, in "the foreign style," tended to be produced by less sophisticated makers. Ironically, they were, according to one report, "generally of lower quality than those turned out by the large factories," since the unskilled hands working as fast as possible in these smaller manufactories could not match the consistency and quality of merchandise mass-produced by machine.[15] Observing a "flood" of items in the early 1930s, advertising professional James Rorty likened foreign sweatshops to America's own highly systematized manufacturing system, noting, for instance, "the neo-Mayan design in pottery and textiles which results when the primitive social-economic pattern of a Mexican village is shattered and the native craftsmen are Taylorized by a capitalist

Beautiful Imported Vases and ---

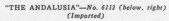

"THE SEVILLA"
No. 6111 (above)
(Imported)

A practical flower jardiniere showing real beauty and distinction. This piece was imported from Spain especially for Fireside Members. The ivory background is almost covered with quaint designs in blue, yellow and green. Tiny lines of black outline the figures and give snap to the composition. Tremendously smart and will give your home that desired touch of chic! Height 4¾"; diameter of top is 5¼"; shpg wgt 4 lbs; list price $5.00.

"THE ANDALUSIA"—*No. 6112 (below, right)*
(Imported)

How fascinating to own a piece of pottery from sunny Spain, especially an endearing bit like the dainty pitcher to the right below. It has the lines of a Moorish olive oil cruse, and bears a tricky design of dark blue on white.

Here is genuine distinction and good practical use, for this pitcher is just the right size for cream, syrup and sauces. A rare bargain! Hgt 5⅝"; shpg wgt 2 lbs; list price $4.00.

"LE DAUPHIN" PITCHER
No. 8027 (above, left) (Imported)

When your friends see this they will exclaim, "Where did you get that **darling** pitcher?" It's perfectly adorable, and is from the hand of a master potter, **Delcourt** of France! The amusing dolphin design is orange, blue and black on tan. Dolphins and seaweed inspired the motif. Hgt 5½"; shpg wgt 3 lbs; list price $1.50.

ROUEN VASE
No. 6024 (above)
(Imported)

A piece of genuine Rouen pottery, a faithful reproduction of a museum antique. The delightful flower motif is done by hand in soft yet vivid hues, and clever antiquing makes the vase look centuries old. It has genuine beauty, charm and decorative value. Hgt 7¾"; shpg wgt 4 lbs; list price $7.50.

GLINT O'GREEN FLOWER POT
No. 1719 (left)

Made of porous pottery, glazed on outside with a delightful shade of green. Height 4¼"; diameter of saucer 4¾"; shpg wgt 3 lbs; list price $1.50. (Not Imported.)

KASHMIRI FLOWER HOLDER
No. 288 (below)

A distinguished flower arrangement is possible with this clever flower holder. The top is perforated for flower stems. May be finished with enamels or left just as it is. Hgt 4½"; diameter 5¼"; shpg wgt 4 lbs; list price $1.25. (Not Imported.)

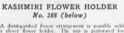

"LA NEIGE" *(The Snow)*
No. 6022 (above) (Imported)

Created by the famous artist and worker in glass—Sabino! Like the wondrous patterns created by Jack Frost on the winter window pane. There is the ice-clear and frost-white effect, glimmering, crystalline, spotless and pure. The raised parts of the designs have been polished off and catch the light in gently twinkling points, giving a delicate sparkle that emphasizes the exquisite snowy effect. Hgt 4"; shpg wgt 2 lbs; list price $5.00.

DELFT BLUE FLOWER POT
No. 1718
(right)

Here is a really beautiful flower pot, glazed with Delft blue and porous within! Height over all 4"; diameter of saucer 3½"; shpg wgt 2 lbs; list price $1.25. (Not Imported.)

"THE BIT O'BLUE" GLAZED VASE
No. 1794 (right)

The glorious blended hues of this beautiful vase run all the way from pale sapphire blue to darkest violet. It is a gorgeous piece of color. Hgt 6¾"; shpg wgt 4 lbs; list price $1.75. (Not Imported.)

Figure 8.2. Highlighting the "chic and charm" of imported goods helped obscure the fact that they were mass-produced commodities. Fireside Studios, *Fireside Gifts*, [ca. 1931].

entrepreneur."[16] The specialty goods found in gift shops were made possible by mass production, giving women, primarily, steady access to what purported to be rare, unique, and special decorative items but were, in fact, no different from other mass merchandise.[17]

Call It Giftware

These peculiar goods came to have their own generic name, which both acknowledged and obscured their commodity status. In the early twentieth century, advertisers had struggled to accurately characterize all this merchandise. A 1913 advertisement for the D. H. Holmes gift shop in New Orleans, for instance, described its merchandise as "Imported Gift Things," "Glass Wares," "China Wares," "Novelties," "Ornamental Wares," and "Gift Wares." In the world of commodities, this ambiguity wouldn't do, and soon tchotchke traders coined the neologism "giftware" to uniformly characterize all this stuff. By the 1920s, the word frequently appeared in the popular press as the accepted term for faux-sophisticated crap.[18] At the same time, the National Gift and Art Association (NGAA) was established to support and promote the trade in it. Thirty-three manufacturers and vendors attended the first trade show, held in New York City in the spring of 1928. Members' common goal was to capitalize on the "rapidly growing interest in gift merchandise."[19] The NGAA published a trade journal as well, *Gift and Art Shop*.[20] By 1930 New York hosted two competing trade shows—the Art-in-Trade exhibit and the NGAA show—right across the street from each other; some nine hundred people attended.[21] Reports noted different aesthetic valences among giftware lines, not only "glass and novelty chinaware" but also "products of mountain industries, hand-wrought silver, and lamps and items for home decorations."[22]

Over time, members of the NGAA became pivotal tastemakers in the giftware industry, determining what items might make it onto gift shop shelves and eventually into women's homes. The most marketable lines of giftware had to be distinctive but not *too* distinctive. Uniqueness and novelty were acceptable within a limited range. Firms attending the 1928 trade show hawked merchandise lines that were "a modification of the modernistic note," backing away from "the extremes in this type of decoration."[23] Bric-a-brac with hackneyed distinctiveness continued to rise in popularity.

Cultural critics recognized the contradictions embodied in the commodified gift and decried modern American society for measuring sincerity of feeling in monetary terms. A 1931 article pointed out the "wantonness" of every price tag in gift and souvenir shops, since items were "marked double for just such sentimental suckers as yourself." But increasingly, cost *did* stand in for emotion. And that emotion could be the source of substantial economic returns for suppliers and sellers, hidden as it was under giftware's veneer of affection, affectation, and wrapping paper. People justified spending more on giftware crap with excuses such as, "She's our only sweetheart. . . . Christmas comes but once a year . . . we're engaged only once and married only once. . . . So WHY SHOULDN'T we shoot the moon—blow our all—about three times a week."[24]

Not only were commodified gifts bad bargains, but they overpromised and were insincere. One send-up recommended that the best gift should be one

> of distinction, delightfully different, demanded by fashion and priced within the reach of every purse, an heirloom of tomorrow yet intensely practical in the moderne mode and fraught with old-world craftsmanship and the swagger [of] sophistication . . . for it is to go to a home of social importance and must have rare beauty, the spirit of Christmas past, and the smartness of the Long Island set, and be at the same time always in good taste, unbelievably aristocratic, obtainable only in the better-class shops and more representative department stores and made of a secret, easily cleaned new substance which modern science has discovered.[25]

The perfect gift was impossible: it embodied the old and the new, was affordable yet looked expensive, seemed at once a long-lasting heirloom and a thing of the moment, and conveyed sincere heartfelt sentiment while also being "unbelievably aristocratic." It was hard to know what was worse: the crap itself, which either pretended to embody or ignored these absurd contradictions, or the people who literally bought into its various claims and affectations. Capturing this dynamic, a cartoon published in *Life* magazine in 1929 showed a haughty clerk in a bustling store chastising a befuddled man, "No, sir, you can't buy yourself a watch—this is a gift shop!"[26]

Despite the economic downturns during the Depression, the giftware industry, improbably, continued to thrive. The *New York Times* reported in 1932 that "jewelry and personal accessories, china, glass and pottery novelties, art metal goods, linens, pictures and frames, books and novelty gift items" continued to be strong sellers.[27] Thousands of retail buyers attended the shows in the spring of 1933, and sales volumes were up some 35 percent from the previous year. Anticipating holiday sales, buyers were "purchasing freely," noted one report.[28] In the wake of Prohibition's repeal at the end of 1933, cocktail sets "in glass, silver, copper, and other materials" were selling extremely well, as were beer steins and mugs.[29] Giftware sales continued to be brisk in the middle of the decade, with very high-end and very low-end wares selling the best. Exhibitors' most robust trade was in imported "novelty wares" retailing for $1.[30] Cheaper and more useful articles that people could better afford and justify were also popular— constituting some 75 percent of the featured goods at giftware trade shows during the heart of the Depression. Just a few years later, as one article reported, "luxury items of a type rarely encountered in recent years stood out as active sellers" at the New York Gift Show. Many of these were "considered impractical from a sales standpoint a few seasons ago."[31]

The trade expanded to include more stuff and more potential consumers. Distinctive crap was coming to the countryside. Trade show buyers were not only proprietors of urban shops but retailers from more rural areas who were purchasing higher-priced items, such as "Chinese lamps in semiprecious stones," selling for up to $100, "porcelain and pottery figurines in the better price ranges," and "better glassware of the Lalique type."[32] Larger department stores, too, were purchasing stocks of giftware in greater quantities, from antique Russian icons to English silver. Things were looking up, if cautiously.[33]

The fact that much giftware came from foreign countries bound the market to global politics and international events. World War II foreclosed some markets and opened others. Due to Japan's singular focus on producing for wartime needs, by 1940 giftware such as bamboo baskets, hand-painted chinaware figurines, and wirework articles had become scarce. Similarly, because Persia's railroads were being used almost exclusively to move military personnel, its exported giftware was also absent from the trade shows. China, in contrast, worked to keep its channels of commerce open "to facilitate exports." Domestic producers, too, saw an uptick in orders by the early 1940s because of the "curtailment of serious competition from abroad in the form of imports of such items as Czech glassware, French porcelains, Swedish furniture, and various Belgian and Dutch products."[34]

Rather than curtailing the giftware trade, the onset of World War II created new opportunities. It opened up new markets for domestic giftware entrepreneurs, especially "American artists and designers [who] have never taken kindly to what they regarded as the unfair competition of a 'romantic' foreign label."[35] Keeping the giftware industry well stocked, the five hundred or so exhibitors at the 1941 NGAA trade show were selling American and British goods almost exclusively, and at lower prices. Many products once made in Europe were now coming off the lines of American manufacturers, such as handbags (previously produced in France, but now at a fraction of the cost), costume jewelry (with a "strong trend toward the military motif"), and ceramics from Goldscheider, which had moved its production facility from Vienna to Trenton, New Jersey. Even the popular press, rather than acknowledging the limitations on giftware sourcing during the war, recast it as a change in tastes that indicated national pride: "Major trends at the show reflected patriotic, American Indian and fiesta types of Mexican and South American inspiration," according to reports.[36] In 1944 the Robert W. Kellogg Company of Springfield, Massachusetts, was still able to offer items from Staffordshire, Scotland, Switzerland, and Mexico. Scarves of "pure Scotch wool," for instance, came from "the Lake Country, north of Glasgow, the only place where scarves of this grade are attainable." A six-piece set of fluted china from Finland could be had for $4.95, sheepskin rugs from the Andes for $9.50, and small painted chairs for children, "Mexican Made," for $2.95.[37] Even global upheaval could not stop the circulation of crappy "exotic" giftware.

Dens of Distinction

Despite lack of access to a full complement of global giftware, the drive for distinctiveness and authenticity remained as insistent as ever, even though, or perhaps because, trade in foreign-made goods had slowed to a trickle. Thus, foreignness and exoticism had to be created by domestic manufacturers. French "opera song" plates ($1 each or eight for $7.85), for instance, were "exact reproductions of those famous plates made in France years ago." Mayflower dinnerware was hand-painted "fine American semi-porcelain," and its "shape was devised by Royal Hickman, the noted ceramic designer of Swedish ware."[38] Responding to increasing demand, potteries in Ohio and West Virginia boasted of mechanization that used machine-poured molds, automated glaze dippers, and even "liner machines" that applied decorations. Ironically, the rage for handcraftsmanship made humans obsolete. Craftsmen able to work through about 800 dozen pieces a day were replaced with machines manned by only three people churning out some 1,800 dozen pieces in an hour.[39]

After the war, Americans' disposable income soared; and the giftware industry, among others, flourished. Some ten thousand products were offered at the International Trade Fair in 1950, and vendors at a Chicago giftware show in 1951 saw their sales double from the year before.[40] US-sponsored postwar programs to revive the economies of "ex-enemy" countries started with export goods such as Japanese Christmas tree lights and German toys and porcelain. In one of the first such arrangements, the government purchased $1 million worth of clay from Czechoslovakia, which was then sent to Germany "for stepping up German output of ceramics. If this plan goes through," according to a report from late 1946, "Dresden china will return to world markets next year, along with other types of German chinaware, artware, and pottery."[41]

On the home front, American suburbanites were working, socializing, and buying. They embraced with gusto the material trappings that increasingly defined middle-class lifestyles. More women were getting married after the war than during the Depression, deciding on spouses at earlier ages, forgoing professional careers for occupations as homemakers, and giving birth to more children. This not only reinforced the centrality of the home in their lives but invited many more occasions for buying giftware for themselves and to give as presents at weddings, anniversaries, and baby showers.[42] The market continued to encroach: gift shops and department stores more often serviced bridal registries, so that gift givers could be let in on the bride's "secrets . . . tastes and interests." And proprietors of gift shops became "liaison agents" to ensure gift giving was done right.[43]

As wives and homemakers, women faced cultural pressures, through movies, prescriptive literature, magazines, and their own friends, to keep their men happy and satisfied. The home became the locus of consumer spending, and in the five years immediately following the war, money spent on household furnishings and appliances rose exponentially.[44] William

Whyte observed that because of the homogeneity of the new suburbs, otherwise imperceptible differences in house styles and home furnishings became critically important. "People have a sharp eye for interior amenities," he observed, "and the acquisition of an automatic dryer, or an unusually elaborate set, or any other divergence from the norm is always cause for notice." That is, "the marginal purchases become the key ones."[45]

Even for those "marginal purchases," women were expected to spend wisely, since men kept a close eye on how their wives were spending their money (or the household's earnings—if both worked). Therefore, decorative objects had to seem useful, too. Not surprisingly, men were increasingly considered viable consumers of giftware as well. In addition to feminine Brahmin scarves ("as colorful as a bazaar in Bombay") and hand-embroidered Tyrolean blouses ("strictly for the fashion connoisseur"), Boston giftware retailer Madison House, for example, offered lines specifically for men. Within the "enforced intimacies" of postwar suburban life, wives could show their husbands they were "pampered" and "loved" by giving them crap. The golfer might like the Putt-Trainer, which "simulates actual putting conditions." The outdoorsman might prefer the battery-operated Tri-Color Lantern ("A honey!").[46] Along with risqué items like beer steins concealing nude ladies, gift retailer Bancroft's offered fraternal emblem pens, sectional desk files, and pocket adding machines.[47]

By the early 1950s the market in giftware for middle managers had only expanded. Much of this merchandise reflected and normalized stereotypically masculine conformism and social behavior, including drinking, smoking, traveling, and leisure pursuits like fishing and playing golf. As such, this crap helped men show that they belonged to a social set, or "gang."[48] Quasi-utilitarian accessories, men's giftware also typified how work and leisure were intertwined, since entertaining, traveling, and even playing golf were activities meant to cement both personal and professional relationships.

This is why a lot of giftware intended for men ended up in the office—all those desk accessories, daily calendars, cigarette lighters, and cocktail sets. Like the postwar suburban home, the office was, in many ways, a stage on which the middle class could demonstrate a compliant kind of distinction—"the middle course," in Whyte's words. These material things made it easier for coworkers and rivals to "siz[e] up the relative rankings around the place" based on the smallest differentiations among the most seemingly insignificant, inconsequential possessions: "It is easy to joke about whether or not one has a thermoflask on his desk or whether the floor is rubber tiled or carpeted, but the joking is a bit nervous and a number of breakdowns have been triggered by what would seem a piddling matter." Whyte added, "Even a thermoflask is important if it can serve as a guidepost—another visible fix of where one is and where others are."[49] Although trifling and interchangeable, crappy giftware items nevertheless clearly marked one's membership and rank within status-seeking groups.

Men and women alike remained keenly aware of how home entertaining might affect both a middle manager's advancement at work and his ability to fit in with his neighbors at home. The pages of a 1956 catalog for the Washington, DC–based Game Room were filled with merchandise for socializing: twenty-two sets of drinking vessels (tumblers, cordials, whiskey glasses, brandy glasses, lowball glasses, steins); eight decanters; five bar sets and the same number of travel liquor cases, carriers, bottle openers, and pourers; three sets of bottle and drink labels, bar caddies, and tables; and sixteen miscellaneous items related to drinking (from ice cube trays and cocktail napkins to ice buckets, novelty ice cubes, coaster sets, and an Inebriation Computer). The Game Room also sold items related to more generalized forms of entertaining. In addition to wiener skewers and nut bowls, there were ashtrays shaped like compasses, buoys, and golf bags, plus barbecue paraphernalia, from aprons to meat-carving boards, all imprinted with recognizable motifs—anchors, fishing lures, sailing flags, pheasants, horseshoes—that marked one's place, however generically, among a class who participated in well-defined categories of the same leisure activities.[50]

In addition to signaling their membership in particular cohorts, the seemingly incidental trappings of an anxious middle class also mirrored the crap of their superiors. The upper-level executives, whom they aspired to be, owned the same basic, if slightly better-quality, stuff. Rather than snuffing their cigarettes in ceramic ashtrays shaped like revolvers, they might choose to wear sterling silver cuff links that looked like miniature medieval dueling pistols. Their travel liquor sets were encased in genuine pigskin instead of Naugahyde, and their monogrammed handkerchiefs were made of real Egyptian cotton. They might have paid more for ever more distinctive items but chose from among the same genres of goods:

bar sets and multi-tools, hobby-themed clocks and beer steins, novelty ashtrays and drink caddies on wheels.[51]

Gift shops were successful in part because they promised romantic domesticity for women by trading in exoticism, foreignness, and nostalgia; giftware for men could function in much the same way. Although men's goods did not typically reference particular pasts or invoke other cultures, they did serve as props for places of refuge. Unlike women, who escaped from the often suffocating confines of the domestic sphere by frequenting public retail environments like department stores and shopping malls, men created spaces of retreat—from the wife, from the kids, from the job—within their own homes, in dens, workshops, and garages. In 1943 *Better Homes and Gardens* profiled the basement hideaway of Harold Hahn, of Kansas City, Missouri, who was proud—"mighty chesty"—of his room. "On certain nights it's . . . strictly reserved for Dad's men friends, who refer to it fraternally as 'the gaming room.'"[52] Like their offices, men's dens were showcases for forms of cultural currency that, when shared with like-minded neighbors, coworkers, and bosses, reaffirmed their rightful status and memberships. This was true regardless of what crappy lifestyle accessories they chose: whether these gents were playing golf using their gold-plated tees or stubbing out their cigarettes in nautical-themed ashtrays while drinking from Kings of the Turf racehorse-themed highball glasses (fig. 8.3).

Personalized Commodities

Buoy ashtrays sure to "inspire a lot of sea talk" and football-shaped music boxes to "bring nostalgic memories to old grads and bright-eyed alumni" provided prepackaged personalities and backstories for their owners— "types" of people who would buy things with a particular "look."[53] Yet there were other ways that midcentury Americans could distinguish themselves— in the most conforming way possible—within a world of like consumers buying the very same distinctive things. Everyone sought to be an individual just like everyone else.

One way that giftware purveyors answered the need for individualized conformity was through personalization. Monogrammed items once belonged only to the very elite, since they required an added layer of handwork, of surplus adornment, that could not be achieved by machine or

Figure 8.3. By the mid-twentieth century, men's domains also became cluttered with crappy giftware, like "Gifts of Distinction" from the *Gentry Gift Guide*, [1952].

scaled up in production. By the 1930s monogrammed products were becoming more popular and slightly more affordable for middle-class consumers. Technological innovations such as stamping and etching machines and iron-on embroidery decoupled manual labor from the creation of individual things. *Business Week* noted that the "interest in 'personalizing'" in the mid-1930s was responsible for "boosting" sales for shirts, towels, linens, and even cheap pocketbooks. Even kitchenware and appliances were "going strong on the initialling craze."[54] "Women," the *American Home* remarked in 1933, "love to see their own monogram on their possessions," adding that "their own particular three-letter identification brings a feeling of pride to almost every feminine heart" and was a way for women to "boast a little quietly and yet in good taste."[55]

By the next decade, people could pay extra to have just about anything monogrammed. Now even personalization could be readily manufactured and machine-made items rendered unique and individuated—"distinct" in the giftware lingua franca. By 1944 giftware outfits like Kellogg Company were selling monogrammed money clips, key chains, pencil sharpeners, clothes brushes, coasters, necklaces, napkins, corsage pins, leather pencil pockets, pencils, and pencil cases. What was more, personalization was no longer tied simply to strings of initials or short names. Christmas cards carried family portraits, and cocktail napkins could be embellished with personal images of one's distinctive possessions ("your house," "your doorway," "your dog"). People could even order bars of soap for servicemen customized with specific military insignia along with a name and title, "up to 12 letters."[56] People went crazy for personalized things, stamping marks of ownership on everything from diapers to cigarettes.[57] More typically, companies like Madison House offered cigarette boxes embellished with personal signatures written on top in copper wire, magazine racks and trash cans embossed with Olde English letters of one's choosing ("sheer elegance and practical"), and even pocket printers for people to do it themselves (fig. 8.4).[58]

Lillian Vernon, the maven of mail order, built her entire multimillion-dollar business on personalizing mass-produced consumer goods. Married with a child in the early 1950s, she nevertheless had professional ambitions

There's No Beauty Like Copper . . . PERSONALIZED

TO BE DISTINCTIVELY YOURS

Four superb pieces from the famed Boston studios of G. Nelson Shaw! Embossing is hand wrought on antiqued sheet copper. Only the finest woods are used, finely finished in deep tone. Because each piece is made to your order, please allow several weeks for delivery. State desired personalization clearly when ordering.

Magazine Rack—Prize of the lot—large enough even for Life. Three Roman letters, or two if you wish.
DM115—Copper-faced Embossed Rack $22.50

Waste Basket—Sheer elegance in any setting! Big 14-inch height makes it practical, too. In solid antiqued copper, with 3 Olde English letters, surname initial last. Or order it with a single large Olde English letter. A beauty!
DM116—Copper Basket $15.00

Desk Letter Box—Adds both beauty and convenience to any desk. With single Olde English letter, or with signature copied from original you send with order. Five inches tall.
DM118—Desk Letter Box $12.50

Cigarette Box—The signature you send us is hand-duplicated on the copper top of the cedar-lined sectional box that holds 60 regular or king-size cigarettes. It just couldn't be more perfectly made!
DM117—Cigarette Box, $15.00

Figure 8.4. New manufacturing technologies enabled giftware companies to produce personalized multiples. Madison House, *Gift Digest*, 1953.

and chafed against the domestic roles foisted upon postwar wives. While brainstorming about what she could sell through the mail, Vernon, who had worked in her father's leather goods business as a young woman, settled on handbags and belts. "My handbags," she remarked, "would offer something special: each one would be personalized with the owner's initials. I knew with absolute certainty," she recalled, "that teenagers would go for items that made them feel unique—as long as their peers had them, too."[59] The same-only-different approach was a great success for Vernon and many other mass merchandisers.

Vernon applied her personal philosophy to the general merchandise she chose to stock her mail order catalog. "I try to be imaginative and find something a little offbeat—useful, perhaps, but still unusual."[60] The more things people owned, the more challenging it was for giftware purveyors to offer them something new. How could they continue to sell surplus goods to people who seemed "to have everything"? One answer was monogramming. "There is a school of gift donors," wrote one disapproving commentator in the 1960s, "who feel that just about anything, as long as it is gussied up with a monogram, will take the place of a really thoughtful gift, painstakingly sought for. But the recipient is rarely fooled—or pleased."[61] A giver's sincerity and a receiver's satisfaction, though, had become entirely beside the point. The giftware market only continued to expand; in the process, it promoted and normalized the idea that distinction and individuality were not, in fact, anathema to the process of commodification.

Putting the Crass in Class

The giftware industry existed—and thrived—because it could convincingly monetize vague ideas about refinement, taste, and uniqueness that customers came to value. This was in part because these abstract, symbolic associations helped obscure the mass-produced nature of this merchandise. When talking among themselves, of course, manufacturers and distributors of giftware were much more open about their aim to turn taste and culture into marketable commodities. Craft Potters, for example, who attended the 1953 NGAA New York Gift Show, exhibited "smart new additions to our profit-making lines of matched Boudoir Accessories and other Giftware."[62] At the 1959 show, the Philadelphia Manufacturing Company noted that its cast-iron mallard bookends were a "Traditional sales leader" in the medium-priced category. Like its other lines of "new low price gifts for volume producing sales," the company assured, "you will find them priced for a highly profitable business." Some outfits tried to class themselves up by employing euphemisms for profitability, like Charles Martine Imports, which presented "a selection of Limoges giftware beyond compare, and at prices so interesting your customers will empty your shelves the moment you refill them."[63] Others described themselves as "Creators of the unusual—as usual," who were offering "idea items" rather than massmarketed products; they even distanced themselves from the profit motive, claiming what they did was "a privilege, not a job."[64] (fig. 8.5).

Customers might have found giftware attractive because it was an expedient way to express "taste" and "distinction." But for people in the giftware business, "taste" and "distinction" mattered only to the extent those qualities could generate a profit, turning "interesting" prices into dollars and cents. So all of those "authentic" and "exotic" imported goods were the results of suppliers' attempts to suit the tastes and meet the economic needs of the domestic trade. Crappy things had to be made, and then made crappier still. For instance, one American wholesaler persuaded a British factory of export ceramics to "apply the same motifs it was using for lamp bases to cookware," because he thought that would appeal to American women. He convinced a French cutlery company to substitute the wooden

Room 716

BEVERLY HILLS ACCESSORIES

Box 202, Jenkintown, Penna.

Oldfield 9-0549

★   ★

Beautiful Decorated Items

for the better shops

Figure 8.5. In the hands of outfits like Beverly Hills Accessories (based in a suburb of Philadelphia), trash cans were tarted up as "Beautiful Decorated Items for the better shops." National Gift and Art Association, *44th Semi-Annual New York Gift Show*, [1953].

handles of their knives with plastic so they would not split apart in the dish-washer, since American housewives could not be bothered to wash their fine French pieces by hand. And he "spent some lucrative hours arguing" with the owner of an Italian factory about the way to produce copper- and brass-covered pepper mills, finally convincing the manufacturer "that the metal-covered mills could be produced more cheaply if the metal 'skins' were cut out in advance instead of fitted individually to the wooden bases." Six months later, "New Yorkers could buy the same design, with a bet-ter tarnish-resistant finish than before, for $5, or one-third the previous cost."[65] Gift shop owners who wanted to dispense with international trade altogether could source their exotic goods from domestic manufacturers like the Portland, Oregon, pottery company Norcrest, which offered lines of painted ceramic figurines including Balinese and Spanish dancers, Chi-nese boys and girls, Hakata dolls, and Cannibal figures (fig. 8.6).[66]

Despite the imperative to provide new, unique, and distinctive items for the giftware industry, manufacturers continued to come up with what was basically the same stuff they had been peddling for decades. Year after year, vendors at the New York Gift Show offered hackneyed wares simply gussied up, surplus on surplus: so many new and novel monogrammed playing cards and compact sewing kits, cute pieces of chinaware bric-a-brac, and stamped brass trash cans in the Early American style. A few enterprising companies created truly novel items—what one account de-scribed as "gayer oddities"—that tried to be somehow distinctive and prac-tical, leading to truly improbable things: The Tiny Hors D'Oeuvres Cart, a "hand-decorated miniature peddler's wagon that does a Charles-of-the-Ritz job of serving a drink and a canape"; The Johnny Clock, a clock shaped like a toilet lid; Hee-Haw the Donkey, the ceramic burro tape dispenser, paper clip holder, and pencil sharpener (plate 6).[67]

There seemed to be no limit to how crappy giftware could be. Even industry insiders thought so. Lillian Myers, owner of the Gift Shop in Cal-ifornia, for one, complained to the editor of *Souvenirs & Novelties* in 1966 that gift shops were becoming notorious for selling "Junk." Suppliers "do make cheap things and the wholesalers jack up the price too high. . . . We the retailer are on the short end," she wrote. "I'm always trying to buy the best I can to sell as cheap as I can. It has to be half way decent merchandise that won't turn green or break before they get home." She implored, "Make decent merchandise for a fair price."[68]

Plate 1. Sellers of "flummeries" and "quirks" often puffed their petty and cheap goods. [Edward Clay?], *Or Fair samples of MILKY DUMPLINGS offered for CORNBREAD* (Philadelphia, ca. 1830s). Library Company of Philadelphia.

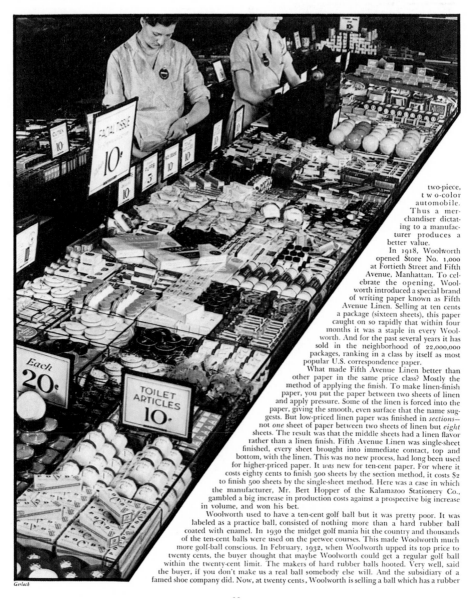

Gerlach

two-piece, two-color automobile. Thus a merchandiser dictating to a manufacturer produces a better value.

In 1918, Woolworth opened Store No. 1,000 at Fortieth Street and Fifth Avenue, Manhattan. To celebrate the opening, Woolworth introduced a special brand of writing paper known as Fifth Avenue Linen. Selling at ten cents a package (sixteen sheets), this paper caught on so rapidly that within four months it was a staple in every Woolworth. And for the past several years it has sold in the neighborhood of 22,000,000 packages, ranking in a class by itself as most popular U.S. correspondence paper.

What made Fifth Avenue Linen better than other paper in the same price class? Mostly the method of applying the finish. To make linen-finish paper, you put the paper between two sheets of linen and apply pressure. Some of the linen is forced into the paper, giving the smooth, even surface that the name suggests. But low-priced linen paper was finished in *sections*—not *one* sheet of paper between two sheets of linen but *eight* sheets. The result was that the middle sheets had a linen flavor rather than a linen finish. Fifth Avenue Linen was single-sheet finished, every sheet brought into immediate contact, top and bottom, with the linen. This was no new process, had long been used for higher-priced paper. It *was* new for ten-cent paper. For where it costs eighty cents to finish 500 sheets by the section method, it costs $2 to finish 500 sheets by the single-sheet method. Here was a case in which the manufacturer, Mr. Bert Hopper of the Kalamazoo Stationery Co., gambled a big increase in production costs against a prospective big increase in volume, and won his bet.

Woolworth used to have a ten-cent golf ball but it was pretty poor. It was labeled as a practice ball, consisted of nothing more than a hard rubber ball coated with enamel. In 1930 the midget golf mania hit the country and thousands of the ten-cent balls were used on the peewee courses. This made Woolworth much more golf-ball conscious. In February, 1932, when Woolworth upped its top price to twenty cents, the buyer thought that maybe Woolworth could get a regular golf ball within the twenty-cent limit. The makers of hard rubber balls hooted. Very well, said the buyer, if you don't make us a real ball somebody else will. And the subsidiary of a famed shoe company did. Now, at twenty cents, Woolworth is selling a ball which has a rubber

Plate 2. By the Depression years, variety store proprietors arranged merchandise more rationally and made sure prices were clearly marked. Photograph by Arthur Gerlach, from "Woolworth's $250,000,000 Trick," *Fortune*, November 1933.

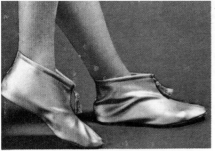

OH! THEM GOLDEN SLIPPERS! . . . A glittering fashion touch for sleek silk & velvet pants, floor-length skirts, all your "at home" outfits. Marshmallow-soft golden vinyl; sparkling tassles. Suedine foam sole; faille-lined.

☐ **Golden Slippers**
Small 4½-6 (H-32623D) **$2.95**
Medium 6-7½ (H-32631D) **$2.95**
Large 7½-9 (H-32649D) **$2.95**

GENUINE HUMMEL NOTES . . . The original, world-famous Berta Hummel designs reproduced on quality note paper. Adorable collector pictures in soft rich colors—trimmed in gold. A pleasure to receive! Ass't prints. Fine quality paper stock—single fold—4½" x 3½". 15 with envelopes.

☐ **Hummel Notes** (H-31948D) **$1**

LIVE THE LIFE OF A MERMAID . . . Dive, swim, shower and keep expensive hairdo in. Specially fabricated rubber strip fits comfortably under bathing cap. Absorbs no water. Adjustable. Velcro closing. Seals at the touch. Protects bleaches & tints.

☐Mermaid Band (H-35683D) **$1**

SHUFFLE CARDS AUTOMATICALLY!! . . . 1, 2, even 3 decks at one time . . . Card Shuffler does a thorough job automatically!! Never a shadow of a doubt! Fast, easy . . . just place cards on tray and revolve! Presto; a "square deal" every time! Use bottom side as a Canasta tray! Sturdy plastic, ass't colors.

☐ **Shuffler** (H-51177D) **$1**

34 PRESIDENT STATUES . . . COMPLETE FROM WASHINGTON TO KENNEDY . . . A magnificent collection . . . your own museum display of miniature carved statues of every president of the United States. Each authentically detailed from head to toe—from the lifelike, familiar faces to the typical gestures & dress of each president. Each poses on a gilded pedestal printed with name & dates of office. An impressive display for den, office, living room, hobby room! Comes in "picture frame" box to hang on wall. Plastic statues, 1¾" high.

☐ **34 President Statues** (H-49825D) **$3.98**

49

Plate 3. The vinyl slippers, rubber bathing cap, automatic card shuffler, and presidential statuettes offered on this page from the 1964 Spencer Gifts catalog were likely all made in Japan.

Plate 4. Companies often went out of their way to create an aura of prestige around their cheap free premiums, like presenting them in full color. Page from the Lee Manufacturing Company's *Wonderful Catalogue of Easy Selling Goods and Premiums*, 1924.

Plate 5. Thanks in large part to full-color action-packed ads, millions of children sent away for free stuff offered by cereal companies. General Mills newspaper advertisement for the Kix Atomic Bomb Ring, *New York Sunday News*, February 9, 1947.

Plate 6. Giftware was often weird and inexplicable. Helen Gallagher–Foster House catalog, Fall/Winter 1964–5.

Plate 7a–b. The First Christmas Eve plate, a.k.a. the Official Bethlehem Christmas Plate, issued in 1977 by Calhoun's Collector's Society. The back of the plate shows some of the collectible's authenticating hallmarks, such as plate number, signature, and "official marks." Tim Tiebout Photography, www.timtiebout.com.

Plate 8. Old things could also be crappy, like this early nineteenth-century Staffordshire figurine of Benjamin Franklin identified as George Washington. Collection of the Museum of the Shenandoah Valley, Julian Wood Glass Jr. Collection. Photo by Ron Blunt.

Plate 9. To their collectors, mass-produced figurines like Precious Moments seemed, nevertheless, to possess uniqueness. Tim Tiebout Photography, www.timtiebout.com.

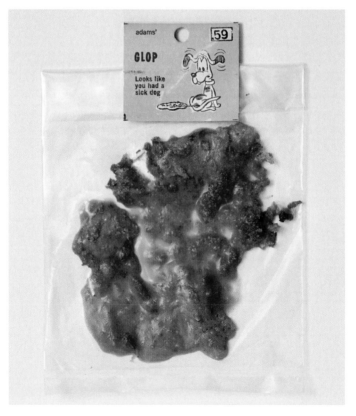

Plate 10. Spin-offs of novelty excrescence included fake bird poop, fake turds, and fake dog vomit. Tim Tiebout Photography, www.timtiebout.com.

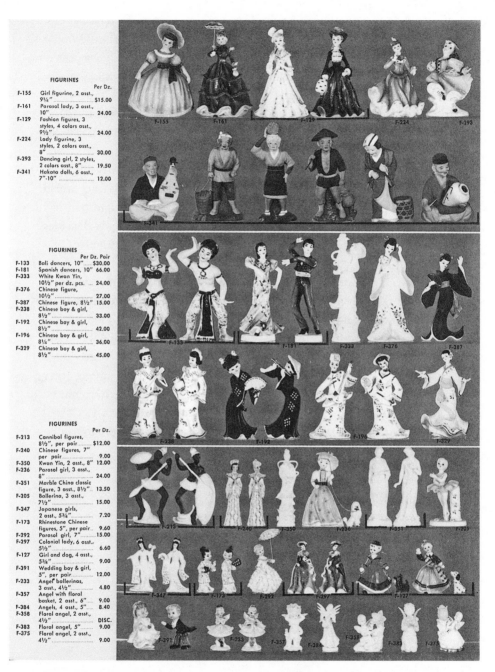

Figure 8.6. Domestic producers also manufactured exotic figurines in Oriental styles for the home market, like these, made by the Norcrest China Company in Portland, Oregon. Norcrest China Co., *Fine China and Gifts for 1959–60*), [1959].

Higher, more prestigious prices often misled giftware purchasers into thinking they were buying something of better quality. "Recently," noted one gift shop proprietor, "there has been a trend toward 'higher priced' merchandise. This can be a fallacy if it means only larger and gaudier 'junk.' New ideas, new designs, and better quality is needed." Appearing opposite her column was an advertisement for the Primitive Artisan, offering "New imports from Haiti at low, low price [*sic*]," such as necklaces made out of seeds, straw Go-Go Hats, and crepe paper flowers.[69]

At the end of the century, the trade in giftware continued to be big business. Despite everything that had changed over the years—from world wars and profound cultural shifts to economic upheavals and political controversies—consumers still identified with, and defined their identities through, articles that were by turns "quaint," "primitive," "distinctive," and "refined," the same basic "looks" that had been popular for decades. Like other forms of crap, giftware embodied many contradictions. It purported to be thoughtfully made but was mass-produced. It conferred status on its owners and showed their refined taste but looked pretty much the same as everyone else's stuff. It promised to express aspects of individual uniqueness but came as the result of mass production. Finally, giftware items claimed to exist outside of the market but were commodities among other commodities. Many people who bought giftware, caught up in romantic and colorful narratives, were blind to their many contradictions. Others simply did not care, since giftware merchandise enabled them to be distinctive just like everybody else.

PART 5

Value
Propositions

9

COLLECTING COMMEMORATION

A truism of the collecting world is that something is worth as much as someone else is willing to pay for it. That's the market, in a nutshell. That collectors more highly prize objects of particular age, beauty, workmanship, and material finery is self-evident and self-explanatory. It makes sense that some people might dedicate their lives to collecting, say, pieces of Chippendale furniture, Fabergé eggs, or the paintings of modern masters. Those things—old, beautiful, scarce, and possessing a kind of ineffable integrity—speak for themselves. People invest their financial resources, emotional energy, social capital, intellectual curiosity, and personal space in their collections. Acquiring a particularly sought-after piece can establish a person's reputation as a serious connoisseur and cement membership in a particular social clique. These beautiful things can accrue cultural capital in the present and, as sound investments, promise monetary benefits through appreciation in the future.

Such value systems are reinforced in popular culture, with television shows like *Antiques Roadshow* appealing to and inspiring a public of nascent collectors. People enjoy seeing professional curators and dealers explain the history of objects, their provenance, how they were made, and, perhaps most important, what they might sell for. Fortunes, it seems, are simply waiting to be discovered in junk piles and dusty closets. It is romantic to think that even seemingly pedestrian things might be not simply precious but priceless.

One category of goods decidedly absent from *Antiques Roadshow* is that of objects produced *specifically to be collected*.[1] We might call them deliberate collectibles, born collectibles, or intentional collectibles. Their absence—not just from the *Roadshow* but from premier auction houses and exclusive galleries—belies their popularity. They have been valued, certainly, but not by the right people, and not for the right reasons. For arbiters of elite tastes and appraisers making economic determinations, they are in no way special. Their provenances trace not through the lineages of great families but to factories in Asia. Because they emerge from automated assembly lines rather than workshops, their craftsmanship is not remarkable or unique or even very good. Because they are made of plastics, resins, alloys, cheap porcelain, and gilt, there is nothing praiseworthy about their materiality. And because they are created from the imaginations of profiteers, they lack the sincerity of older objects that have not only withstood the test of time but in many cases were actually used by their owners.

But like other genres of crap, intentional collectibles are more complicated than they might appear and raise fundamental and intriguing issues about the nature of value itself. Intentional collectibles are a reminder that value is often subjective, elusive, and arbitrary; people can prize, often intensely, things that by some measure might not deserve such devotion. On the surface, it may seem inexplicable that so many people have placed their faith in mass-produced collectibles, since they are not scarce, well made, or even, often, aesthetically pleasing. Nevertheless, for myriad reasons collectors have chosen these to be *their* objects over all others. In the process, they have fashioned identities as connoisseurs who hope to enjoy, just like the elite, the material and social benefits that such status confers. The fleetingness of these material and social benefits, however, illustrates the limitations of intentional collectibles, the perils of placing faith in things that are not considered to have worth by the right kinds of people, and the ultimate triumph of economic regimes over all other measures of value. This is a story of how manufacturers, retailers, and collectors have created value out of nothing, and how all too often that value has vanished into thin air.

Making the Modern Collector

Simply calling something a "collectible" or "collector's item" or "something collected" is enough to set it apart from other material artifacts (useful things like tools and appliances, and things to be literally consumed, like food). By dint of occupying its own category and special space, the "collectible" is distinguished from and elevated above other pedestrian objects. Collectibles also draw from histories of collecting practices that are linked to the surplus time, money, and knowledge possessed by the elite. During the Enlightenment, for instance, royalty kept *Wunderkammern*, or "wonder rooms"—cabinets of curiosity containing weird and fascinating objects from remote places and people. *Wunderkammern* were status symbols that showed their owners' mastery and possession of the physical realm.[2]

Most humans collect things, though it isn't clear why. Some scholars believe the impulse comes from our need to order the world as a way to understand it—the "material embodiment" of classification. Collecting shows "how human beings have striven to accommodate, to appropriate, and to extend the taxonomies and systems of knowledge they have inherited."[3] Others see collecting as a kind of psychological malady born of neurosis and maladaptation. From this perspective, collectors are not just "dedicated" and "serious" but "infatuated" and "beset" by an "all-consuming drive." They are compelled by an insatiable "hunger" toward the next acquisition, their "habit" pursued with a "chronic restiveness."[4] And still others see collecting as an intimate practice of self-fashioning and memory-making. For them, collections create highly personal material worlds within which people can feel comforted and comfortable. The individual items in those collections serve as memory objects that invite recollections about their acquisition and perhaps offer a connection to the past.[5]

Motivated by these factors or a combination of them, Americans in the mid-nineteenth century began building collections. "Stamp mania" was one of the first collecting crazes, initially pursued by middle- and upper-class women who had leisure time, were attracted to the aesthetic qualities of stamps, and considered activities such as indexing and arranging to be productive and educational pursuits. Increasingly, though, the pastime became segregated by gender. After the Civil War, men took up the hobby in greater numbers, likely driven by a number of factors: stamps' increasing ubiquity, their association with the adventure and discovery of foreign places, and their direct connection to monetary systems (since in many cases stamps could be used in lieu of currency). Women continued to collect "junk stamps by the millions," according to one historian, "but a woman philatelist was rare, indeed."[6] The masculinization of stamp collecting cast it as a serious and cerebral activity. In a sense, stamp collecting was another way, like *Wunderkammern*, for collectors to colonize and then impose order on foreign lands and people. Stamp brokers, practicing a new occupation, offered bounties on used stamps and helped collectors think of their hobby as an outgrowth of intellectual pursuits rather than collecting passions.[7]

Collecting coins and commemorative medals, too, was a male province. Aficionados of the allied hobbies of philately and numismatics often branched out into other masculine collecting fields such as antiquities and natural history specimens. Popular literature only reinforced these gender distinctions. The authors, advertisers, and correspondents in late-century publications such as the *Collectors' Monthly*—which published ads for stamp dealers, articles on taxidermy, and offers to trade artifacts and specimens—were all men. Publications like the *Philatelic West and Collectors' World* carried a separate "Woman Collectors' Department," suggesting the hobby was an entirely different practice depending on whether collectors were men or women.[8]

Women collected historical and commemorative things, too, but tended to specialize in objects associated with the domestic sphere, and especially artifacts tied to particular pasts. By the end of the nineteenth century, spurred by the Colonial Revival movement, the elite were putting collecting in the service of heritage and preservation, rescuing even humble American artifacts—spinning wheels, copper kettles, iron candlesticks, and the like. As long as they were American and old, fine and pedestrian pieces alike warranted reclamation. Even crappy commemorative plates were valued for their decorative qualities and because they depicted "memorials of early America which they perpetuate."[9] As we have seen, these efforts, extensions of anti-immigration sentiment, sought to instill and reassert "traditional" values of order, simplicity, and integrity by discriminating against newer objects with "mongrel" shapes.[10] White upper-class women from the Northeast, mostly, worked with "enthusiastic zeal" to, in the words of *Harper's Weekly*, "[strip] so many country homes of antique clocks, chairs, bureaus, and pieces of china" (fig. 9.1).[11]

The people involved in these heritage reclamation efforts also believed that their quarry would rise in monetary value—the market was never very far from their minds. In fact, objects discovered to be American- rather than foreign-made suddenly appreciated in value, brushed as they now were with the patina of patriotism. Books such as Irving Lyon's *Colonial*

1. The Sale of a Veritable Antique.   2. How "Veritable Antiques" are made.   3. A Bargain.   4. "There, Mr. Mortice, I think if you put new Legs, a Back, and Seat on that, it will make a lovely Chair."
5. "Lor', no, Mister! I ain't seen no old Clocks hereabouts."   6. A Prize.   7. A Rumor having spread that an old Lady on Long Island has some old Chelsea China, a few Collectors go down to see it.

THE RAGE FOR OLD FURNITURE.—DRAWN BY A. B. FROST.—[SEE PAGE 718.]

Figure 9.1. Fighting against "mongrel" shapes and people, New Englanders would often go to great lengths in pursuit of antiques. "The Rage for Old Furniture," *Harper's Weekly*, 1878. Library Company of Philadelphia.

Furniture of New England (1891) inspired people to reexamine their old possessions through the lens of reawakened nationalism.[12]

By the final decades of the nineteenth century, Americans of all persuasions, "suffer[ing] from the mania of collecting," were amassing everything they possibly could. People uninterested in or unable to afford artifacts like Chippendale chairs were gathering up old patent medicine bottles, luggage labels, buttons, cigar bands, and even, apparently, things like streetcar transfers. Industrialization made it still easier to indulge collecting passions. Things like brightly colored chromolithographic cigarette cards, postcards, and trade cards often came free with product purchases. Issued in themed series (holidays, famous actors, birds, landmarks), they were quite pretty and could be pasted in purpose-made scrapbooks, "fat albums with slotted pages," to form individual and personalized worlds through selection, curation, and arrangement.[13] People had become, according to one observer, "slaves" to the material world, lacking in the "powers of self-restraint." "Not to be a collector," he declared, "is a distinction."[14]

The Political Economy of Commemoration

Recognizing a commercial opportunity when they saw one, late-century entrepreneurs capitalized on the collecting craze. In the 1880s American cutlery companies began issuing souvenir spoons, one of the earliest intentional collectibles. Not only were they relatively easy to acquire and fairly affordable, but souvenir spoons were issued serially and created pleasing displays. They commemorated historically important people and events that helped build a grand American narrative: likenesses of George and Martha Washington, the Salem witch trials, Detroit's founding, the completion of the Erie Canal (fig. 9.2a–b). They became quite popular, and by the early 1890s well over two thousand different kinds of souvenir spoons had been produced domestically. Like other commemoratives, these intentional collectibles commodified history by rendering it uncomplicated and easily consumable. The very act of commemoration-commodification, too, gave significance to otherwise forgettable people and events. In a contemporary book dedicated to spoons, George B. James Jr. wrote, "Many a legend which has long been forgotten in the town of its origin, many a beautiful story which has long since been lost, have

The "Hutchinson, Kan.," Spoon.

The "Warsaw, N. Y.," Spoon.

The Hutchinson, Kan., Souvenir Spoon represents the salt industry of the town. Underneath the town, at a depth of about four hundred feet, is a vein of salt nearly four hundred feet thick. On the tip of the handle of the souvenir spoon is represented three barrels of salt, while on the shank of the spoon is the word "Hutchinson," and in the bowl is the word "Kansas."

This spoon is designed to represent the thriving salt manufacturing town of Warsaw, N. Y., located in the beautiful Warsaw Valley, forty-eight miles from Buffalo, and forty-two miles from Rochester. Warsaw salt is known as the purest salt in the world, and has taken first premium whenever exhibited. Warsaw also has the only salt baths and sanitarium in America. The sanitarium is a magnificent new building with all the modern improvements, and is considered a great health resort.

"*Hutchinson*" *Spoon.*
Sent to any address on receipt of price, $3.50, in ten sizes. Made in sterling silver only.

J. S. DUNN,
Hutchinson, Kan.

"*Warsaw, N. Y.,*" *Spoon.*
Made in sterling silver only, in both coffee and tea sizes. The coffee differs from the tea in one respect, leaving only the words, "Fine Dairy Salt,"
Tea Spoon . . . $2 50
Coffee Spoon . . 1 50
For sale by
JAS. A. MAIN, Jeweler,
Warsaw, N. Y.

140

The "Chauncey M. Depew" Spoon.
PEEKSKILL, N. Y.

Chauncey M. Depew was born in the beautiful village of Peekskill, N. Y., the picturesque valley of the Hudson, and is to-day in the vigor of his manhood, the noblest type of an American citizen.

Chauncey M. Depew, known to the world as the "silver-tongued orator of America," is no less the man of business than a philosopher. Man, woman and child claim him as "Our Chauncey," not the least as fearing his greatness, nor the greatest as exempt from his influence.

Chauncey M. Depew's life and character are without a parallel in this generation; the wisdom of his thoughtful mind, the value of his sense, the beauty of his speech, are the pride of the nation; and to every boy in America's common schools, and to every young man in America's colleges, are an inspiration to vie with each other to approach the acme of his perfection. Peekskill pre-eminently claims Chauncey M. Depew as her own; every hill, every valley, every wood and dale which gives picturesqueness to its situation, is the source of affection, and of reminiscence, and of story, in which "Our Chauncey" delights.

Peekskill, the birthplace of Chauncey M. Depew, sends forth its souvenir spoon, conscious that the designer and silversmith have gracefully wrought in silver and gold a correct and exquisite likeness of her illustrious son.

Bonbon Spoon.

PRICE LIST.

Tea Spoon, Silver Bowl	$3 00
Tea Spoon, Gold Bowl	3 50
Coffee Spoon, Silver Bowl	3 00
Coffee Spoon, Gold Bowl	3 50
Orange Spoon, Silver Bowl	3 75
Orange Spoon, Gold Bowl	4 25
Bonbon Spoon, Gold Bowl	3 50
Paper Knife	2 50

Sent to any address on receipt of price. The above engraving is given simply to show the outlines, and cannot reproduce the exquisite workmanship of the portrait on this elegant spoon. Manufactured only by

ARTHUR J. BIRDSEY,
Jeweler,
PEEKSKILL, N. Y.

84

Figure 9.2a–b. Any number of personages and historical events became subjects of collectible souvenir spoons, as illustrated in George B. James Jr., *Souvenir Spoons*, 1891. Courtesy American Antiquarian Society.

been happily brought to mind, and tradition honored by its incorporation in the souvenir spoon."[15]

Silverware and cutlery manufacturers shrewdly encouraged and then capitalized on the souvenir spoon craze, offering the world of commemoratives to the general public at various price points. Some pieces were creations of extraordinary workmanship, issued by the likes of Gorham and featuring finely detailed designs rendered in sterling silver and embellished with gold plating and enamel. Others were much more pedestrian. Made of sand-cast pot metal with few details and only the most ephemeral surface decorations, they were prone to discoloration and corrosion.

Women collected them all: spoons were not only ornamental but conso-
nant with their prescribed roles as domestic caretakers and nurturers.
Collecting commemoratives was not an "unruly passion" but rather a form
of curating—providing caretaking for one's family and, by extension, the
nation's heritage.

Men consumed different forms of commodified history. Along with
stamps, they also collected medals and coins. Major events celebrating
American progress like Philadelphia's Centennial Exhibition in 1876 and
Chicago's World's Columbian Exposition in 1893 helped fuel popular interest in commemoration.[16] The fairs themselves generated commemorative
items including books, albums, medals, pins, ribbons, pamphlets, trade
cards, postcards, and other paraphernalia. Many of these were cheap and
intended to be distributed in great numbers. Others, though, such as commemorative medals, were finely engraved productions struck in bronze or
solid silver in limited quantities, sought after by a more exclusive group of
aficionados.[17]

While commemorative "medalets" had been produced in the United
States since the late eighteenth century, the first commemorative coins
were issued at the World's Columbian Exposition to mark the four hundredth anniversary of the discovery of America. One coin featured Columbus and the other Isabella. The exposition's organizers intended the
coins to "bring considerable revenue," since buyers had to pay double
their face value and could keep them as souvenirs or use them for ad-
mission. But the aims of commerce and commemoration, profit and remembrance, remained at odds—the public blanched at "such expedients
to make money." One wag thought organizers should "go a step further and
make the pieces of base metal," rather than bothering to strike them in a
quality material, since they would never amount to anything as collectors'
items anyway. It did not help that people judged the designs substandard.
The image of Columbus was particularly inartful, and some thought the
rendering looked more like Daniel Webster, Henry Ward Beecher, or Sitting Bull.[18]

Whether designed to appeal to men or women, commemoratives, their
producers understood, were powerful objects that simplified and distilled
history. These objects were lifelines of sorts, connecting their owners to
the past in tangible ways. The other side of the coin, as it were, was that

commemoratives' symbolic value was easy for cynical and savvy producers to exploit and turn into dollar signs

Not surprisingly, some of the first and most popular American commemoratives were dedicated to George Washington. Although "totally imaginary," many early likenesses still satisfied a public eager to own busts, medallions, pitchers, tankards, mugs, and plates embellished with the visage of "the father of the country."[19] Washington's death in 1799 only intensified the market for objects, no matter how cheaply made and ephemeral, into which his image had been woven, etched, engraved, cut, or carved.

Washington's demise provided opportunity and economic incentive for producers to supply even more. Savvy entrepreneurs like the early nineteenth-century author and bookseller Mason Locke Weems could make a mint. "Primed and cocked" to capitalize on the myth of Washington, Weems recognized his death as a unique commercial moment. He wrote to his Philadelphia publisher, Mathew Carey, about producing a biography of the man who personified "God, patriotism, sobriety, industry, justice, &c. &c." "Millions," he argued, "are gaping to read something about him." Weems was prepared to write a book extolling Washington's "Great Virtues" as a war hero and president, for he understood that glorifying Washington would make a lot of money, perhaps even a threefold return on investment. He underscored this when, frustrated by the book's delay, he wrote to Carey, *"You have a great deal of money lying in the bones of old George, if you will but exert yourself to extract it."*[20] Similarly, commemoratives—however artificial or opportunistic—became powerful and lucrative tools that helped solidify what were often hagiographic narratives of the past, which the public could then claim for themselves.

The early market in commemoratives included all manner of things that championed easily understood patriotic and nationalistic themes: important personages such as military officers and government officials; historical events such as naval battles and mass tragedies; notable locales like the new nation's bustling harbors; and the United States writ large—symbolized by eagles, figures of Columbia, liberty poles, and flags (fig. 9.3).[21] Because these early patriotic subjects were so revered, it didn't much matter that they were affixed to objects made not in America but overseas. In fact, many commemorative pieces, like china embellished with transferware designs, came from British potteries whose own political and economic interests ran

Figure 9.3. Entrance of the Erie Canal into the Hudson, 1825, Wood Enoch & Sons, pearlware transfer print plate. Fenimore Art Museum, Cooperstown, New York, N0009.1996.

counter to the messaging on the merchandise.[22] For instance, British manufacturers profited greatly from American demand for patriotic-themed chinaware celebrating naval heroes and triumphant battles during the War of 1812—American victories in a war whose embargo stifled the importation of these very things.[23]

As they are today, commemorative objects, though secular, nevertheless were sacralized, a process made easier because their subjects are frozen in time, depicted without imperfections or nuance, and tied directly to significant events—often origin myths—that have shaped Americans' conceptions of who they are as individuals and a nation.[24] Commemoratives became and remained popular because they were able to "place a part of the past in the service of conceptions and needs of the present."[25]

This was not lost on manufacturers, who understood Americans' needs to possess material expressions of simplified pasts. Endowing objects with commemorative associations transmuted those mythic pasts into profitable commodities.[26]

The Rise of Collectibles

And yet there was no sustained American market for commemoratives until the later half of the twentieth century. In the nineteenth century pieces were manufactured in relatively limited numbers and available for only brief periods. Too, early commemoratives were produced in conjunction with or shortly after significant events—whether a steamboat explosion, a visit by a foreign dignitary, the death of a celebrated figure, or the laying of a cornerstone for an important building—in order to maximally capitalize on their immediacy. People did not purchase these items serially or with the intention of building collections but to pay homage to things they found personally and contemporaneously resonant.[27]

With the exception of the Danish porcelain company Bing & Grøndahl, which began issuing annual Christmas plates in 1895, the rise of truly mass-market collectibles did not occur until the mid-twentieth century, when companies began exploiting commemoration much more deliberately and profitably. In the postwar era, consumers—particularly in the white middle classes—had more leisure time, more disposable income, and more places to display their possessions. By first selling the *idea* of collecting to this demographic, entrepreneurs could then sell them mass-produced merchandise, especially if it seemed to convey qualities that the elite prized in their own collections, such as distinction, exclusivity, rarity, craftsmanship, and provenance. This manifested both as articles that ornamented the home (still purchased primarily by women) and as "serious" things like coins, stamps, and medals, which tended to attract men. As we will see, they were all crappy in their own ways.

Such "collectibles" were new things and their own things, existing not between antiques and reproductions but beyond them entirely. One of the first intentional collectibles commemorated what to many was the most commemoration-worthy event of all time. In 1950 the Florida-based outfit Kilgore Antiques and Gifts introduced genteel readers of *Hobbies* magazine to The Lord's Last Supper Plate. The ten-inch plate presaged the defining characteristics of the modern collectibles—claims to material quality and fine workmanship, a distinguished pedigree, rarity, and resale value. The "distinctive collector's item" featured, according to the ad copy, "The World's Most Beloved Picture" rendered on porcelain in nine "glorious" colors and framed by a "Lacy border" of 23-karat gold (fig. 9.4).[28] In addition to putting the plate in its own class as a "collector's item," Kilgore used other strategies to create artificial value. The company reassured readers that "those marked 'First Edition' are still available," suggesting they would soon be hard to find. The plate's gold rim and full-color image attested to its artistry and quality. Finally, because the plate was a blue ribbon winner ("at the 1949 Indiana State Fair. Need we say more?"), it possessed a pedigree: not quite authentication but validation.

By the 1960s the market in collectibles really began to take off. In addition to mail order, retail outlets like gift shops and "galleries" offered

Figure 9.4. Creators of the Lord's Last Supper Plate, one of the first intentional collectibles, promoted it through this poorly printed advertisement, which appeared in 1950 issues of *Hobbies*, a magazine for serious collectors of antiques.

women, primarily, a seemingly infinite range of manufactured collectibles to purchase, from eclectic owl figurines and clattering bells to pewter thimbles and porcelain teacups. The promise of free display racks and cabinets often incentivized collectors to acquire an entire series, investments that could amount to hundreds if not thousands of dollars. Among these myriad mass-produced collectibles were porcelain plates decorated in a range of pleasing motifs—puppies, cats, angels, pastoral landscapes, English cottages, clowns, Native Americans, Elvis, scenes painted by Norman Rockwell, and more. Collecting these plates offered several benefits. Collectors found satisfaction in amassing a cohesive series of things, while they were able to both beautify their homes and display their connoisseurship by hanging them on the wall. And they seemed to be investing in something that would appreciate in value.

It might be easy to dismiss this final point as sheer folly, especially given the number of cheap vintage collectible plates lining the shelves of antiques emporiums and listed on eBay today. People cannot seem to give them away. But futures markets are, by their very nature, perpetually uncertain, and without the benefit of hindsight, collectors tended to believe what manufacturers and retailers told them. (Just as people today are placing their faith in collectibles they hope, but have no certainty, will appreciate one day.)

So it is important to understand the fuller context of how collectibles were produced and marketed to appreciate their allure. Recounting his days as a copywriter for Calhoun's Collector's Society in the late 1970s, the filmmaker turned marketing consultant Herschell Lewis explained, step by step, the process manufacturers used to construct value artificially. In those heady days, when everyone was "rockin' and rollin' with the plate craze," Calhoun's was trying to figure out how to get in on an already glutted market. "Sitting cynically around the conference table," according to Lewis, employees brainstormed for a theme that "hadn't been worked to death." They settled on The Official Bethlehem Christmas Plate, which they hoped to promote as "actually fired in the Holy Land and bearing the imprimatur of a major cleric." Commissioning the artwork was easy, since "plate artists breed like rabbits." The challenge was providing evidence of being "official." This meant it had been approved and verified by a seemingly authoritative source who resonated with collectors. They hired a "fixer" out of Tel Aviv, who found a manufacturer near the Lebanese border. Plate collectors, they knew, cared as much, if not more, about the marks appearing on the plate's back, which would affect the monetary value, as the design on the front, which held only aesthetic value. "Equally ridiculous to producers and equally significant to collectors," backstamps, however bogus, created ready-made provenance and validity, to be replicated and further validated on the certificate of authenticity. Calhoun's fixer found their man in Archimandrite Gregorios of the Greek Church in Bethlehem, who licensed the use of his name and image as the plate mark and in sales literature (fig. 9.5). It did not matter that the manufacturers didn't know what an archimandrite was, nor that Gregorios was not really an important person; they liked that his title sounded exotic and religious-y. In the pre-internet era, "if we couldn't find out what an archimandrite was, neither could any plate collectors." Gregorios's original publicity photo, which he used to

The Sacred and Ancient Town of Bethlehem

The view of Bethlehem which we see in this beautiful painting has portrayed the essence of Bethlehem's history and its nature.

The portrait is a delicate interweaving of the colors, textures, and personality of our ancient and sacred town. Superimposed on the rolling hills and mountains is the image of Mary, Joseph, and the Christ child. Thus the artist, Gerald Miller, has successfully captured the religious aura as well as the natural beauty of the site.

With high artistry, Mr. Miller has depicted the terraced hills of Bethlehem. The countryside is typical of that of the Judean hills. Each hillside has been painstakingly terraced to provide small areas for cultivation. The countless rocks which dotted the terrain have been laboriously arranged to retard soil erosion; the result is a picture of brown, honey-colored, and olive green hues. Scattered over the hillsides are small flocks of sheep tended by shepherds reminiscent of King David, whose birthplace was this ancient town. All these elements are captured in this painting.

Many are unaware of the meaning of the name of my city. The name Bethlehem comes from the Hebrew *Bet lehem*, "House of Bread", indicating the fertility of this area. The town is revered by both Christians and Jews because of the tradition that Rachel, Jacob's beloved wife, is buried nearby. It is not only the place where David was born; it was here that he was annointed king. Moreover, it is also associated with the events recounted in the book of Ruth.

Most significantly, however, Bethlehem is sacred to all Christians as the birthplace of Jesus. Of the many sacred sites located in and near Bethlehem, the most famous is the Church of the Nativity, located in Manger Square. This church was first built by the Roman Emperor Constantine after he converted to Christianity. Our church which now stands on the site was built in the Sixth Century, though it has been enlarged and restored many times over the past 1400 years. It is the oldest Christian church in use today.

I have spent thousands of hours in this church, and still, whenever I enter it, I am filled with a sense of awe and the memories of the centuries which fairly cling to the walls. Occasionally I think of how often Bethlehem has been the focus of inter-religious rivalry and conflict. But more often I am reminded of the many, many branches of Christianity which have been drawn to this sacred place and which have established churches, convents, monasteries, hospitals, and schools.

For centuries, Christian pilgrims have streamed to this holy site to worship at the birthplace of Jesus. They must bow low in order to enter the church through a narrow opening (the Christians continually re-

duced the size of the entrance for security reasons to prevent enemies on horseback from entering the church to kill the worshippers). This low doorway is appropriately named the "Door of Humility". Though its original purpose was one of physical security, today it serves to remind all who enter this famous church of Bethlehem that God is best worshipped when approached with a simple and humble spirit.

At Christmas, the holiness of my church is at its highest point. I hope that this spirit will transmit itself to all who enjoy, as I have enjoyed, Gerald R. Miller's painting.

—*Gregorios*
Archimandrite of Bethlehem

The Archimandrite Gregorios
Chief Abbot of Bethlehem

I hereby lend my hand and seal as authentication.

Bishop Maximus of
Nazareth and Galilee

Figure 9.5. In 1977, Calhoun's created a "patron" for its "official" plate, "Gregorios, Archimandrite of Bethlehem," providing an information sheet about his "native town" accompanied by his "portrait," "signature," and "seal."

promote his real gig as a greeter of tour buses, showed him wearing sunglasses, which the company painted out. Even his likeness was a mirage.

Creating a convincing aura of authenticity presented one challenge. Constructing ideas about rarity presented another. The company set an "edition limit" (Lewis's own sardonic quotes) of ten thousand plates "to assure collectors of scarcity." It was similar to the "excellent duplicity" of other companies, he noted, who defined limited editions by the number of "firing days," a convenient hedge allowing manufacturers to produce hundreds of thousands of plates. The box was stamped with an official-looking red seal, initialed, and marked APPROVED FOR EXPORT AS A WORK OF FINE ART—printed on the box during production rather than actually stamped after the fact by a neutral arbiter. The "Certificate of Origin and Authenticity" was similarly marked with "official" signatures and stamped with seals that were not handwritten by individuals but printed by machines and ultimately meaningful only to collectors. And on the back was a place to record the "Transfer of Ownership" (*Cession de propriété*) (plate 7a–b).

Companies also created a sense of heightened value by touting collectibles' supposedly superior material properties. Calhoun's, for instance, fired its plates on "genuine Royal Cornwall china," a distinction that, like limited firing days, had no meaning. Establishing the Royal Cornwall imprimatur had been the most challenging part of the process, Lewis explained, "because we submitted 32 names to the trademark office before those picky bureaucrats accepted that 33rd name." The fictitious name would nevertheless encourage collectors to associate it with the lineages of both monarchy and famous British potteries. Because it now owned the name Royal Cornwall, the company could embed it into the backs of subsequent lines of plates, which they could cheaply import from Japan.

Despite being a fabrication on every level, the plate "was an instant smash hit," and the company was able to trade on the "magical cachet" of the First Edition. In 2003 when Lewis's account was published, he noted that copies of the plate could still be found online. And he cheekily encouraged people to seek them out (as I did). "After all," he remarked, "these are the *only* plates carrying the rare and famous signature of the Archimandrite Gregorios." It was perhaps the only honest thing about them.[29]

Yet another purveyor of collectible plates and other mass-produced commemoratives, the Bradford Exchange (est. 1973), launched one of the most audacious efforts to prove its products were "verified" and "authentic."

Figure 9.6. Companies encouraged plate collectors to track their purchases like investments in the stock market, a comparison made all the more legitimate by the establishment of the Market BradEx, which had its own trading floor and was profiled in *Collectibles Illustrated* magazine in 1983.

The company devised its own market index to track the activity of some three thousand plates on the secondary market. The Market Bradex created at once a metamarket, a phantom market, a parallel market, and a false market. It used transactions occurring in the mainstream market to create its own index of value that, for a time, tracked the trade in false objects. Collectors considered the Market Bradex to be "the Dow Jones of plate collecting." It was described in its heyday of the early 1980s as a "half-million-dollar commodities trading floor" featuring state-of-the-art computer equipment installed behind glass walls. A team monitored trading activity, "Wall Street-like," on a "big board," using a proprietary "Instaquote Trading System" (fig. 9.6). The BradEx matched buyers and sellers of collectible plates: "A bid from anywhere in the world can be immediately matched with various asking prices phoned in by collectors wishing to sell their plates." Buyers were charged a $4 or 4 percent commission, and sellers were guaranteed their asking price minus a 30 percent "Exchange commission."[30]

The existence of the BradEx heightened and validated the market in collectible plates, by both facilitating trade in the secondary market and making it seem as legitimate as the actual stock market. And it seemed to work. By the early 1980s over a million people in the United States alone were collecting commemorative plates.[31] Because these plates were described in the terms used by high financiers, collectors assumed they behaved in the same way as other commodities. In this way, they seemed to be sound investments. At the time, BradEx officers pointed to a rapid increase in recent sales, predicted robust future trading, and noted that "22 brokers are now needed to handle the volume at the Exchange."[32]

The market value of commemorative plates was characterized in terms that would have been meaningful even to those—actually, *especially* to those—who might never set foot on Wall Street. Some plates remained "mainstays" on the exchange, while other issues were "volatile," could "cause excitement," and might "perform" well or poorly. After supplying the requisite caveats about collectible plates being "no guarantee of getting rich quickly," a BradEx representative nevertheless directly compared the index to the New York Stock Exchange: "Looking at the action through the windows that enclose the Bradford Exchange trading floor, it's easy to imagine that millions of dollars worth of AT&T or GTE stocks are changing hands. Closer scrutiny reveals that the excitement stems from the fact that the 1971 issue of 'Maria and Child' by Bjorn Wiinblad is up $100 to a high of $1620."[33] An author touting plate collecting in *Rarities* magazine said that with the advent of the Market BradEx, it was "difficult to understand the reluctance of plate manufacturers to portray their products as investments."[34] *Collectibles Illustrated* published a cover story called "The Great Plate Explosion" in 1983, featuring a woman who exclaimed, "In 1976, I bought this plate for $45. It has since appreciated nearly 600%." In less than twenty years, a Lalique plate issued for $25 in 1965 was worth $1,540. And another collectible plate had reportedly "skyrocketed" from $35 to $1,050 in just six years.[35] Given these success stories, people would be foolish not to invest in rapidly appreciating collectibles like plates.

Making a Mint

At the same time women were participating in the collectible plate craze, men, too, were buying into the commodification of commemoration. Getting in on the ground floor was the Franklin Mint, whose inspired name alone associated it with both a well-loved (and famously honest) historical figure and a producer of coinage. Established in 1964, the company started out as a purveyor of privately minted gold and silver coins and medallions and a producer of gaming tokens for Las Vegas casinos. Later, it produced a veritable pantheon of intentional collectibles, including dolls (Marilyn Monroe, Princess Diana, Liz Taylor), historical replicas (Colt revolvers, samurai swords, Coca-Cola machines, Apache helicopters), "precision" scale-model die-cast classic vehicles (Model T's, Edsels, Corvettes), commemorative plates (the usual subjects), fantasy-themed figurines (witches, warlocks, dragons, castles), Harley-Davidson merchandise, sculptures of wolves, eagles, and miniaturized *Star Trek* space ships. In 1979 the self-described "world's largest developer and marketer of fine collectibles" netted $64 million from its porcelain pieces alone.[36]

Recognizing the broad interest in coin collecting, founder Joseph M. Segel established the National Commemorative Society, which issued series of special medals. Each of its 5,200 members had the opportunity to purchase one coin per month. Announced in popular coin-collecting magazines but unavailable to most readers, new issues became quite desirable even though they were inferior in many ways: made of a lesser-grade silver (.925 sterling rather than .999), embellished with low relief rather than highly detailed images, and not for general circulation. Segel described his products as "coin-medals," a term that was, according to one critic, "distinctly contrived but starkly necessary."[37]

(Coins, it should be said, are particularly complicated things since they possess many values at once. Their "collectible" value is determined by several factors, including historical interest in why they were issued, their scarcity, and their condition. Each of these points can be highly variable and will significantly affect a coin's value. Condition, for instance, is determined by independent "grading" companies, and the same coin can have a wildly different rating depending on the company and due to sometimes seemingly imperceptible defects. At the same time, however, "errors"— imperfections during minting—can make a coin highly prized. Coins also have a "melt" value, based on the intrinsic value of the amount of gold, silver, or copper it contains, which is determined, simply, by the current precious-metals market. Coins also have face value as legal tender. So collectors might pay upwards of $600 or $700 for an Indian Head $10 gold piece issued in 1912. To a metal assayer, that same coin might be worth

$150 or so, depending on the day. In your pocket, it would buy a few cups of coffee.[38])

Most numismatists prized coins, which tended to appreciate in value, and they disdained medals, which typically did not. Creating the hybrid "coin-medal" category enabled Segel to capitalize on the éclat of fine-quality medals that had been issued over the decades by reputable groups such as the Circle of Friends of the Medallion (est. 1908) and the Society of Medalists (est. 1928). He was also paying close attention to the more recent successes of companies such as the Heraldic Art Company, which issued "limited edition" series to a closed subscriber list, and Presidential Art Medals, which produced bronze and fine silver President John F. Kennedy medals and other series honoring the Signers and the states of the Union.[39]

Segel commissioned the skills of Gilroy Roberts, chief engraver for the US Mint, thus attaching a highly regarded name to the coins. In order to sell to a mass market, he experimented with metals and packaging and adopted the standards and value categories accepted among serious collectors. For instance, Segel represented inferior metals as something newer, better, and proprietary—exclusive to Franklin Mint issues. Many early examples, made of alloys of various sorts, were given exotic names, such as "NICON," "Sterling Plus," and "Franklinium I." Helping to further enhance its initial reputation, the Franklin Mint forged ties with numismatic societies by attending trade shows, donating shares of stock to collectors' organizations, and hiring their experts to serve as in-house specialists, archivists, and dealers. By the 1970s the company had secured contracts from countries around the world to strike commemorative medals and proof coins.[40]

By the end of the decade, however, the Franklin Mint was drawing more and more nonexperts into its orbit and alienating mainstream coin collectors, who not only refused to pay the Mint's inflated issue prices but also saw how they depreciated on the secondary market. *60 Minutes* even aired a damning exposé about the Franklin Mint in 1978. When silver hit a record high at about the same time, smart collectors sent their collections to the refinery to recover melt value.

Nevertheless, that many collectors held on to their Franklin Mint coins and continued to purchase new releases was a testament to their belief in the company's savvy marketing efforts, which engaged more downscale collectors by making the hobby accessible to them while simultaneously characterizing it as an elite practice. As commemoratives, the coins embodied simplified and valorized versions of American history. By emphasizing that people could "own a piece of the past," promotional literature hinted that collectors might be motivated as much by intellectual curiosity as by economic interest. This pitch flattered those who wanted to envision themselves not as crass materialists but as erudite thinkers interested in ideas over money: just like antiques collectors, they could admire and covet an object for reasons beyond resale value. For example, the selling script for the Franklin Mint's Carson City coins prompted sales agents to encourage potential customers to "imagine holding in your hand a rare and valuable piece of American history—an heirloom minted in solid gold dating 1870 to 1893. Sound impossible? However, you can actually own a genuine Wild West $20 'Double Eagle' gold piece.... These are authentic, US Mint gold coins, struck at Carson City, and steeped in the rich history of the western frontier."[41] Likewise, buying sets of Morgan dollars was a chance to "own the silver dollar that won the west!!" and, romantically, was "the preferred coin of gamblers, pioneers . . . and bank robbers!"[42]

Marketing efforts also appealed to collectors' sense of their own discrimination and connoisseurship. Because coins were "exclusive," they seemed scarce and therefore more valuable. But exclusivity also helped collectors imagine themselves as members of a more sophisticated collecting group, "clients like you."[43] Agents often let buyers in on a "little-known secret" about new collecting programs, or stressed that recent issues were "the perfect complement" to sets collectors had already purchased. Using flattery, sales agents were able to persuade collectors not only to buy coins virtually identical to ones they already owned but to feel like it was a smart, informed decision.[44]

Sellers of intentional collectibles also invoked vague and often spurious categories of distinction, ascribing to their mass-produced merchandise an aura of rarity and uniqueness. The claim that "quantities are limited" only intensified the urgency to buy. And because coins by nature were inextricably tied to the market, their commodity status was not obscured, as with some other intentional collectibles, but instead emphasized. Historicity and presumed investment value worked in tandem. Take, for instance, the sales script for "very rare Washington 'Error' Dollars," which "most people aren't aware of" and yet curiously were also "one of the most sought after" coins: "Few collectors will ever have the chance to own this rare error coin due to the limited number of coins. We were fortunate to secure a very small quantity of these phenomenal error coins and we are offering them to you. Quantities are limited, and **once they're gone, they're gone for good!**"[45]

Marketers also spoke in superlatives that, like other strategies, added only illusory value to the coins. Indian Head Gold Pieces were **"widely believed to be the most beautiful coin ever designed and struck by the U.S. Mint."**[46] The US Silver Eagle Dollar coin was the **"largest, heaviest and purest of all Silver coins** produced by the U.S. Mint!"[47] The Ultimate Nickel Set contained versions of "one of America's most important coins in the 19th and 20th centuries," with designs that "are also among the most beautiful and enduring of all time."[48] And so on.

Appeals to material properties helped concretize the value of the Franklin Mint's collectible coins and were yet another way the company encouraged buyers to envision themselves as members of a coterie of high-end collectors. People enrolling in the Presidential Dollar Completion Program (for which they had to prepay) received "TWELVE **Brilliant Uncirculated and richly layered in 24-Karat Gold** coins and a FREE museum-quality cherrywood-finish display." Meaning, people would get coins dipped in a gold coating (an "enhanced" feature) and a faux-wood laminated box to put them in. Collectors who committed to purchasing the entire set were rewarded with two additional coins dipped in a platinum coating.[49]

The company constructed value in other ways as well. As with collectible plates, artificial credentialing appropriated the terms and value scales used in antiques and art collecting. Many coins, encapsulated in clear plastic packages secured with elaborate seals, came with "official" papers and were expertly graded. While some coins were, in fact, graded by reputable independent outfits, most authentication occurred internally, creating a closed feedback loop that proved highly profitable. Each Presidential Dollar, for example, "ha[d] been encapsulated with a hologram 'Mint Security Seal' by The Franklin Mint and signed by Jay W. Johnson, the 36th Director of the U.S. Mint," who by then was no longer acting in that capacity but was an employee of the company.[50] The Franklin Mint also never clarified what, exactly, was being authenticated and failed to point out that it alone was doing the "authenticating," including printing its own Certificates of Authenticity and making its own "official" hologram package seals (fig. 9.7).[51]

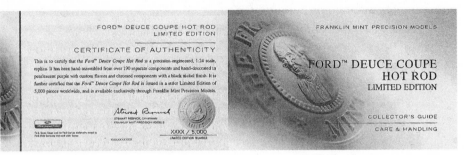

Figure 9.7. Printer's proof sheet for a certificate of authenticity generated by the Franklin Mint for its own Limited Edition Ford Deuce Coupe Hot Rod.

While legally the Franklin Mint could not claim its collectible coins would appreciate in value, its marketing materials encouraged collectors to draw that conclusion.[52] Sales scripts often mentioned the "intrinsic value" of the coins, a phrase that instilled confidence among collectors who did not understand its meaning and were seduced by the sound of "intrinsic" and "value."[53] Similarly, Washington Presidential Dollar Error Coins were "certified and graded," which "only enhances their collector value. Similar loose coins—ungraded—are selling for more than our graded coin."[54] The company reinforced these suggestively potent claims through repetitive promotions delivered over the airwaves, within the slick, full-color pages of product catalogs, and in personalized pitches over the phone. It was easy for people to become true believers (fig. 9.8).

Other companies supplying the market in commemorative collectibles used similar rhetorical strategies. They gave themselves important-sounding names that appealed to buyers' sense of authenticity, distinction, and cultivation. Those interested in commemorative items—from belt buckles marking the hundredth anniversary of the Gunfight at the O.K. Corral to George Washington Coins—might invest in the offerings of the Historic Providence Mint, the National Collectors Archives, American Heritage Art Products, Ltd., or the Westport Mint. Their very names helped imbue products with historical importance and often traded on the integrity of respected organizations. Since there were countless local and state historical societies across the country, an item bearing the imprimatur of, say, the United States Historical Society seemed to carry the validation of an august institution. People could not help but equate historical authenticity with economic value. Even casual familiarity with

Updated:
1/5/10
Author: WK

The Franklin Mint
The Morgan Mint

GOLD

Product: Indian Head Gold Piece $10 with Motto (1908-1933)

MOM# & NS#	B11F905	C001141
Offer	Vintage U.S. Gold coin (1/2 ounce)	
Packaging	TFM "Vault Collection" wood display	

PRODUCT HIGHLIGHTS

Unique Selling Proposition:
- Own a vintage *$10 Indian Head Gold Piece guaranteed to be at least 75 years old!*
- This is the *perfect time to start or add* to your collection of United States gold bullion coins.
- Struck from gold mined in the U.S. and *.900 fine Gold.*
- The last circulating gold coins were minted more than 75 years ago, and **over 90% of all U.S. gold coins were melted by the U.S. government in the 1930's.**
- As a result, *only a tiny fraction of the original coins remain today* – especially examples such as this in Extremely Fine condition, or better.
- Each coin is *sealed inside a clear acrylic case and certified by The Franklin Mint.*
- *Includes a handsome furniture finished "Vault Collection" display case*, plus a reference story card and a Certificate of Authenticity signed by Jay Johnson, 36th Director of the U.S. Mint.

Overview:
Gold has always been treasured for its beauty and lasting value, and classic United States gold coins are the most respected and sought-after in the world. But the last circulating gold coins were minted more than 70 years ago, and **over 90% of all U.S. gold coins were melted by the U.S. government in the 1930's.** Millions more were melted for quick profits when the price of gold bullion shot up. As a result, only a tiny fraction of the original coins remain – especially in Extremely Fine condition.

This is the perfect time to start – or add to – your collection of United States gold coins. However, you must order now, because the rising cost of gold means that today's low prices could soon be just a memory!

Figure 9.8. Internal sales sheets like this one for Carson City Gold Coins, used in 2010, enabled Franklin Mint representatives to easily access market information about the coins in addition to providing scripts for targeting specific customers.

real institutions was enough to lend credibility to the ersatz ones delivering crap.

The Reckoning

The principles governing today's collecting markets held true for those in the past as well: the long-term success of any collectible relies on the continued enthusiasm of collectors to maintain market demand in order to sustain price. This is especially true for mass-produced collectibles that do not possess the qualities that traditionally determine value, such as scarcity, artistry, or a true connection to the past. Enthusiasm and value have to derive from other factors.

Then as now, collectors of intentional collectibles found pleasure in the practice as a hobby. For them, collecting itself was an enjoyable pursuit, and the acts of amassing, arranging, and curating were quite satisfying. Collectors also gained prestige among their peers and found meaning and membership within collecting groups.

But, of course, collecting was (and is) never completely free of the market but deeply embedded in it, as examples from the BradEx, Franklin Mint coinage, and other phenomena from the world of intentional collectibles demonstrate. Most collectors hoped (and continue to hope) for a payoff from their hobby. For owners of commemorative plates, classic car replicas, limited edition medals, and other intentional collectibles, the thought of making a good investment proved a powerful animating force. Contents of dedicated magazines, modeled on established and reputable publications like *Antiques* and *Hobbies*, routinely intermingled articles about famous collectors with columns about the latest trends in mass-market collectibles. Publications like *Rarities* ("The Magazine of Collectibles"), *Collectibles Illustrated*, *Plate Collector Magazine*, *Plate World*, and *Collector's Mart Magazine* only reinforced the validity of what collectors of intentional collectibles were doing—they were specialists with their own magazines!— but also continued to suggest a thriving aftermarket for their acquisitions. According to *Rarities*, "Our basic criteria for any area of collecting will be to ask: 'Is it fun?' and, *equally important*, 'Can it be considered as a legitimate investment?'"[55] Likewise, *Collectibles Illustrated* described itself as a magazine for "savvy" and "smart collectors" who wanted to stay "on top of the marketplace," providing "up-to-the-minute information, in-depth articles celebrating the rich heritage of collecting, and a full array of ads and classified columns to put you in touch with other collectors who share your interests."[56] A contemporary trade magazine noted that collectibles "[offer] artistry, emotional appeal, home enhancement, nostalgia, tradition, companionship and investment value."[57] No matter how much their things might have been rooted in the past or pleasing to look at, collectors always considered their monetary future, placing trust in the available professional opinions, from those of Franklin Mint representatives to those of columnists for collectibles magazines.

And so, when collectors started disposing of their intentional collectibles in the late twentieth century, many were quite understandably surprised, if not shocked, to learn how little value their acquisitions actually possessed. Collectors of commemoratives—in particular those who had amassed sizeable collections of coins and medals—faced the greatest reckoning. Patrons of the Franklin Mint and other "exchanges" and "mints" had perhaps the most reason for assuming their collections would command robust resale values. After all, they hadn't been wasting their money on frivolous chinaware bric-a-brac, vinyl fashion dolls, or dog-themed plates but had invested in serious pieces with historical gravitas and meaning. Many of their specimens, as financial instruments, were *already* monetized.

But like other collectors of mass-produced crap, they figured wrong. One story among many was that of "DA" from Troy, Michigan. He wrote to financial expert Malcolm Berko in 2012 inquiring about the value of his collection of Franklin Mint coins:

During the past 25 years, I purchased over $47,000 in collectable silver coins and beautiful non silver coins from the Franklin Mint for my retirement because I thought the scarcity and limited edition minting of these coins would drive up their value over the years and because I believed the silver content in the silver coins would also increase in value. Now I'm 64 and decided to sell these coins to a coin dealer who offered me $2,500 for the whole lot. He told me most of the coins were worthless, and the only coins that had any value were those with silver in them. I was devastated because when I was buying all those coins, the people at the Franklin Mint told me these coins were minted in limited production and would be more valuable to collectors in the future. I called two coin dealers in Detroit . . . and both said they had no interest in Franklin Mint coins and said they don't know any dealers who would buy them from me. My son told me to write you because he said you might know of buyers for them and at this point I'd be very happy to get at least half of what I paid for them if possible. Please help me if you can. And if you cannot help me, do you think I can sue the Franklin Mint and recover my cost? And could you recommend a lawyer for me to sue them?[58]

Berko had little to offer. He explained that the products the Franklin Mint and others sold were not monetarily valuable and likely never would be. He confided that he himself paid $150 for a Franklin Mint die-cast model of a 1935 Mercedes 500K Roadster. "Although it was 'to scale,'" he said, "it was made in China, poorly and cheaply constructed; parts and pieces would fall off, and when the Mint wouldn't return my money, I tossed the Benz in the garbage." DA might as well do the same with his coins. Berko explained that countless collectors had overpaid "by orders of magnitude" for collectible coins, fooled by "clever buzz words" that made it more difficult for buyers to "make an intelligent buying decision." Even truly collectible silver and gold coins, which the Franklin Mint also offered, sold far above prevailing retail prices. One numismatist gave the example of a "collection" of five silver Morgan dollars, "portrayed as being nearly impossible to collect on your own," which the Mint was selling for a hefty $549. The set actually comprised quite common coins that could be acquired from coin dealers or on eBay for $30–$40 each, or $150–$200 for a set. Even truly rare coins like the $3 gold "Indian Princess" could be purchased on eBay, often in better condition, for about one-fourth of the Franklin Mint's asking price of $3,000.[59]

Berko did not even address the issue of DA's base-metal commemorative coins, likely worth nothing at all. Serious collectors had no interest in them, especially since they had very little precious-metals content. The "distinctive," "limited edition," and "collectible" surfaces of legal tender coins (electroplated, colorized, and embellished with stickers) would first have to be removed, using expensive processes, if they were to be smelted down or recirculated: in other words, they were worth *even less* than face value. One coin expert explained that The Color-Enhanced Collection was "an example of the Franklin Mint's business style because many people might be moved to buy these coins, believing them to be special and valuable. The coloring was added by the Franklin Mint itself," he explained, "and both the U.S. government and serious coin collectors view this action as a defacement of the coin which renders it worthless as legal tender and destroys most of its collectible value as well. Far from being special, valuable coins," he noted, "these are essentially ruined pieces of currency which are only valuable if you want a colorized set for your own enjoyment and don't mind paying many times the worth of the coins for it."[60]

Despite the Bradford Exchange's claims that it would "recommend only the best [plates]—those that combine artistic merit with strong potential to appreciate in value," its own BradEx market index reported very little activity. There were trading records for only 165 of the 3,000 plates listed in 1993, for instance; only 5 percent of all known collectible plates garnered any market interest. Of those that had been traded, the price for 22 percent of them had depreciated. The value for most others had remained stable but would also decrease in subsequent years. A broker who made his living finding desirable collectible plates for dealers described them as "a poor man's art collection."[61] The bubble for these particular mass collectibles had burst.

More than collectors themselves, companies trafficking in born collectibles realized that value was highly contingent and artificially created—which for a time worked to their great advantage. In this, the collectibles market *did* intersect with more high-end markets. Those, too, remained socially constructed, performative marketplaces that established value by tangible factors such as rarity and material quality and by intangibles such as artist's reputation, authenticity, and validation by important institutions and fellow collectors. To be sure, value in all market realms has always been highly contingent, arbitrary, illusive, and faith-based: while the Franklin Mint was misleading its customers about the worth of their collectible merchandise, the company was also inflating the value of its stock and eventually sued by investors.[62]

However, high-end objects do tend to be more truthful and less crappy than intentional collectibles, and in more salient ways. Pieces of fine art are rare if not unique and, carrying evidence of the hand of the artist, have been thus marked and sanctified.[63] Pieces are also authenticated and validated through an extensive art market, gallery system, and museum complex (however problematic those may be). Likewise, antiques inherently have *an actual connection to the past*; hence antiques collectors' concern with fakes, which, because they cannot claim that lineage, possess spurious pedigrees. Created decades if not centuries ago, true antiques were made and owned and used by people in the past. Having withstood the test of time and taste, they are rare survivors. Higher-end pieces, too, are works of art in their own right; beautiful and well made, they possess inherent validity. They are often made of the best materials and evince the finest craftsmanship of master artisans who have spent lifetimes perfecting their skills.

In contrast, born collectibles were (and continue to be) mass-produced commodities created simply to be marketed as things to be collected. Everything about them, save their ability to evoke personal sentiment and emotional attachment, had to be constructed out of puffed rhetoric, ersatz materials, gilt surfaces, and misleading marks. With very few exceptions, they could only ever be false things, true only in what collectors imputed to them—which, although based on false pretenses, was often sincere. Commemoratives might have pretended to lay claim to the past, to lineage and legitimacy, but were just mass-produced crap made by companies worlds away from the subjects they depicted. Focused on making money by creating false value, the companies that made them were never sincerely interested in projects of commemoration. Their products were neither rare nor scarce nor directly connected to people of bygone eras.

For a long while, producers of intentional collectibles were able to sell connoisseurship to a wide range of Americans. But when those collectors sought to realize the economic value of their dedication and appreciation in art and history, they came face to face with the inherent contradictions of investing in mass-produced merchandise. The authenticity, truthfulness, and worth of these things could only ever exist in the artificial world of quotes: "Franklinium" metal coins, "Royal Cornwall" porcelain plates, and "authentic" replicas were virtually (but not really) rare, virtually (but not really) historical, virtually (but not really) works of art. By finally seeing the quotes that had been so skillfully disguised, these collectors could at last understand the true value of their crap.

10

MANUFACTURING SCARCITY

Over time, makers of intentional collectibles have successfully monetized the nearly universal passion for collecting that is driven by a combination of enjoyment, emotional fulfillment, intellectual engagement, and collectors' hopes of making a good investment. But they have made their appeals in different ways. Commemorative coins and plates, as we have seen, were enticing because they had a connection to august figures and important events and therefore capitalized on collectors' interests in history and their place in it. What was more, since a lot of collectible merchandise was manufactured in the form of financial instruments of some sort (stamps, coins, medals, "medalets"), they seemed to be directly bound to and of the market and therefore safer, more natural investments.

But intentional collectibles have come in other forms as well, and companies have had to adopt alternate strategies to get people to buy their crap. While they played up the commodity status of commemoratives in order to artificially create and enhance their value, producers and distributors used the opposite strategy to do the same for collectible knickknacks and figurines, which were wholly untethered from the market, significant events, and meaningful import. By erasing rather than enhancing the commodity status of such merchandise, purveyors of intentional collectibles were even more successful at both creating value where it did not exist and selling that false value to countless collectors. As with commemorative coins and plates, explaining the mass appeal of things like figurines and beanbag animals is essential to understanding collectible crap—and how value is created (and destroyed). These things were different, but the same.

The First Collectibles

Ornamental collectibles, like other kinds of crap, did not simply appear out of nowhere but existed within a longer historical continuum. Beginning in the early nineteenth century, middle-class American women became passionate about ornamental Staffordshire figures that were quasi-intentional collectibles, products of mass production, and generally low-quality imitations of fine pieces fashioned by Meissen in Germany, Capo di Monte in Italy, Sèvres in France, and Minton, Spode, and Wedgwood in Great Britain.[1] Together, large and small manufacturers churned out cheap porcelain "chimney ornaments" by the thousands, helping to democratize gentility and respectability. As one observer wrote, "We can but deplore the loss to the wealthy and artistic, while congratulating the more ordinary citizen on his gaudy toilet ware or his cheap china tea-set."[2]

Designed to sit atop mantelpieces, chimney ornaments were often finished only on the front; their flat backs lacked detail and remained unpainted. Most of the figurines' painted embellishments, possessing what might euphemistically be described as "a certain unpretentious charm," merely suggested colors and patterns but were not applied precisely or artfully.[3] Made with little quality control, pieces with sometimes obvious production errors made their way to the market—for instance, statuettes of Benjamin Franklin mistakenly labeled "George Washington" (plate 8). Some, according to one observer, "show[ed] no recognizable likeness to their subjects."[4] This seemed to make little difference to buyers, who simply wanted to decorate their homes with objects that at least gestured toward middle-class respectability.[5] So relentless was consumer demand that by the time of the Civil War the county of Staffordshire alone was employing some twenty-three thousand people to produce common transferware table settings and tea sets for export.[6]

In order to realize a profit, suppliers sacrificed material quality and aesthetic finesse as they met consumers' increasing demands for decorative knickknacks.[7] Like other producers of crap, British potteries used exploited labor, whether the sprawling concerns employing thousands of people or the smallest backyard "potbanks" run by a few family members. Boys did much of the labor to make so many pieces that sold so cheap. In the 1840s each boy working in the larger Staffordshire factories, paid but two shillings a week, helped produce some 2,640 figurines a week. Two decades later, the 180 potteries in North Staffordshire employed some 30,000 people, 4,500 under the age of thirteen. Daily, they worked with clay slips and glazes impregnated with lead and arsenic.[8] Much like the home-based Japanese manufacturers that would succeed them, smaller potbanks typically employed family members who were paid nothing—all necessary sacrifices in order to satisfy America's "cheapening mania."

Staffordshire potters not only implemented the lowest production standards they could get away with but knew the pieces were intended for show rather than investment and therefore chose easily understood subjects for their wares, such as dogs and cows, ladies and gents, and members of the royal family. Even a century later, very few Staffordshire pieces had appreciated in value, and many survivals had not weathered the test of time very well. Even though deemed "solely of decorative value," a twentieth-century antiques repair manual provided instructions for how to fix broken and chipped pieces. Notably, it cautioned against overly skillful restorations: "In the early Staffordshire pottery," the book advised, "this may mean that the modeling will seem rather coarse, without much shape to it, but it is often difficult to achieve the original simplicity with conviction."[9] In other words, convincing repairs had to look as inartful as the originals.

Plasterware knock-offs of Staffordshire were crappier still, made of pedestrian materials and fashioned by untrained hands. Nevertheless, they, too, found a ready market, hawked by itinerant "image peddlers," often Italian immigrants, who sold their tchotchkes for less than fifty cents for the simplest figures and up to a dollar or more for larger and more intricate examples.[10] While visiting Italy in the late 1820s James Fenimore Cooper came across an exporter of plaster statuary (which "sent its goods principally to the English and American markets") who was selling insultingly poor imitations of the high art he had been enjoying in Florence galleries: "Grosser caricatures," he remarked, "were never fabricated: attenuated Nymphs and Venuses, clumsy Herculeses, hobbledehoy Apollos, and grinning Fauns."[11] But they sold because even the middling sorts aspired to possess the trappings of middle-class respectability, and Staffordshire-like plasterware offered myriad easily understood subjects from which to choose (fig. 10.1).

By the later decades of the nineteenth century, Americans began amassing even larger arsenals of stuff. Goods became their own language, and as they continued to flow into the market thanks to mass production and

Figure 10.1. The material markers of nineteenth-century middle-class respectability included chimney ornaments on the mantelpiece, shown in this image from a George Cruikshank series of prints. They are repossessed when the father drinks his way into insolvency. *The Bottle. In Eight Plates*, 1847. Library Company of Philadelphia.

expanding global commerce, people became more and more literate in the often complex and sophisticated semiotics of their material world. More than simply being decorative or useful, objects became even more important markers of ever-finer gradations of status, refinement, and cultivation. Even clutches of relatively worthless knickknacks showed the extent to which people were able and willing to spend their money on trivial things to show their economic prowess and social standing. Collections themselves became status objects; spaces purpose-built for their conspicuous display—"best rooms," "drawing rooms," and "best parlors"—were often incorporated into the very design of houses. While one critic described these spaces as one of "the follies prevalent in the middle classes," they enabled the genteel to fully participate in the "goods life," laying claim to cultural currency and economic worth. Displays of eclectic bric-a-brac sitting on mantelpieces and corner display stands were key, "arranged," ironically, "according to stiff, immutable law."[12]

By the early twentieth century, collecting had gone from a "mania" to part of Americans' daily lives. Frequent accounts in the popular press spotlighted the collecting practices of the rich and notable, cementing their status as cultural standard-bearers. For the elite, collecting remained a status contest reminiscent of the *Wunderkammer* keepers of long ago. It enabled people to practice connoisseurship, make investments, and compete for prized trophies. Elite collections often ended up in museums, where they could occupy entire wings under their owners' names. Henry Ford famously made it his mission to acquire "a complete series of every article ever used or made in America from the first settlers down to the present time" (an ironic quest, given his own contributions to mass production and modernity).[13] Others, such as the Morgans and Huntingtons, amassed nothing but the best. The elite's collecting preferences, much publicized, influenced even amateur collectors of more humble objects on more modest scales.

The Great Depression marked the end, for a long while, of the country's collective "collecting mania." The objects people had been bringing into their homes for decades—souvenir spoons, china figurines, printed ephemera, bottles—were now going out. They were sold to the antiques dealer, pawned at the pawnshop, traded at the junk market, or left at the curb, the aftermath of eviction (fig. 10.2). The collections of the very rich, of course, remained the exception. The elite's ability to continue to amass more possessions in such straitened times—material evidence of surplus time, space, mental energy, and money—simply reinforced collecting's character as an elite activity despite its previous widespread appeal.

The Birth of Collectible Figurines

After the Depression and Second World War, the collecting bug once again bit the American public. Served by antiques dealers in both large cities and small towns, newer generations continued to seek out relics from the past, and dedicated magazines like *Antiques* and *Hobbies* helped advise them. Shrewd businessmen, however, recognized that costly antiques—and the effort involved in acquiring them—did not appeal to everyone. Perhaps people simply weren't interested in things that were old or eccentric or expensive or hard to come by, even if they wanted to enjoy some of the associated prestige. But the countless postwar consumers might be *potential* collectors with a latent interest in collecting *something*, since they had disposable income and spacious suburban houses to fill up with stuff.

Intentional collectibles, fairly cheap, easy to acquire, and simple to un-

Figure 10.2. During the Depression, average Americans were getting rid of, rather than amassing, more stuff. Samuel Gottscho, "Junk Markets IV," 1933. Library of Congress.

derstand, became just their thing. Commemoratives held some appeal, but they weren't the only intentional collectibles that found purchase in the market. At the same time Kilgore's came out with The Lord's Last Supper Plate, Hummel figurines were also gaining popularity. They, too, featured highly accessible subject matter, benefited from an unassailable origin story, and offered expansive opportunities for collecting and connoisseurship. The first series of Hummels were three-dimensional ceramic incarnations of drawings and paintings of children rendered by Berta Hummel (1909–46), a formally trained artist who became a Franciscan nun. In the 1930s the German ceramics manufacturer Franz Goebel approached Hummel (by then Sister Maria Innocentia) about turning her sketches into porcelain figurines. Hummel's convent would approve each drawing for production and receive proceeds from sales. The first Hummels came off Goebel's line in 1935 and soon found ready space in shops all over Germany. Sales of Hummels, in fact, saved Goebel's from bankruptcy. Later, American soldiers returning from the Second World War brought the porcelain urchins home as souvenirs, sparking a domestic craze. By the early 1950s the factory was employing over seven hundred people and selling "to every country in the western world." The company "could scarcely turn out Hummels fast enough to meet the demand."[14]

Collectors were captivated by the figurines' innocuous, banal subjects: children singing, children carrying umbrellas, children huddled under umbrellas, children holding baskets, children sitting on fences, children sitting on fences holding baskets, children with dogs, children with cats, children with lambs, children with rabbits. They were innocence, purity, and nostalgia neatly distilled into material form. Preserved in porcelain, they could live out their days eternally young and chaste, denizens of an idyllic world forever uncomplicated and unchanging.

It might seem ironic that Hummels found traction in the postwar American consumer market, since most new products of the era embraced an optimistic future delivered by new technologies. The atomic age of chrome and Plexiglas and Technicolor was leaving the sepia-tinged pastoral days behind. But it was, in fact, Hummels' refreshing anachronisms and charming old-fashionedness that most endeared them to enthusiasts. Their subjects were not adults in a changing world but children in guileless stasis. Representing the past as a foreign country, too, Hummel's children were not even American but German. And they lived not in the present but in Germany's distant past, signaled by the figures' "traditional" costumes—lederhosen and dirndl skirts, handkerchiefs, Alpine hats, and plaited ponytails. These children were "uncontaminated" by the present.[15] All white, they were also, apparently, uncontaminated by ethnically diverse neighbors. Goebel did experiment with a darker clay base, but any such figurines that survive are considered outcasts—"experimental rarities" among collectors.[16]

Hummel promotional literature reinforced the value of these pristine worlds. For example, a booklet issued by Boston retailer Schmid Brothers in 1955 illustrated the various models of "original" figurines. The "official" Hummel story, approved by the convent, opened the catalog and showcased the unassailable life of Sister Innocentia and her life's work as an artist and woman of God.[17] "These charming, but simple figurines of little boys and girls capture the hearts of all who love children," for "in them we see, perhaps, our girl or boy, or even ourselves when we were racing along the path of happy childhood." (Wearing lederhosen?) Even transgressions were wrapped in the rosy gauze of innocence and purity: "These enduring figures will take you back to . . . the time when you, perhaps, stole your first apple from a tree in the neighbor's garden and were promptly set upon by his dog, as shown in the 'Apple Thief' figure."[18]

Contemporary Hummel literature decontaminated the figurines in another way as well, by downplaying their commodity status while emphasizing their potential as collectibles. The figurines' subjects themselves did not refer to the market (save Little Shopper, with her basket) but frolicked in eternally pastoral landscapes. March Winds and any Stormy Weather were greeted with Spring Cheer. The Sensitive Hunter, with a case of Puppy Love, became a Barnyard Hero. Going to Grandma's was a Retreat to Safety for the Happy-Go-Lucky and Joyful Weary Wanderers. If the advertising was to be believed, Hummels were not produced by a manufacturer but created by a higher power, sui generis. Berta Hummel went to art school, became a nun, and then used her creativity in the service of God. It was not so much that, by capitalizing on her artwork, the convent had made a wise business decision but that "the world became the recipient of her great works," as if they were a beneficent offering from heaven. "These little images were, after all, her childhood friends as she remembered them and one by one they appeared before her eyes until she had immortalized those who made her early life 'Heaven on Earth.'"[19]

The figurines, therefore, were not materializations of collectors' imagined versions of their own pasts but were imbued with the sentimental and spiritual power of Berta Hummel's own childhood memories. Her experiences became appropriated as theirs. What was more, these origin stories emphasized Hummel's reluctance to use her art for commercial purposes; she only did so to "give her beloved convent a telling financial assistance." Long after Berta's death in 1946, promotional copy was still reassuring collectors that "her royalties continue to support her Order and its principally charitable works," thus transforming their collecting activities from consuming goods into doing good works.[20] None of this early promotional material mentioned just *how* Hummel's "great works" took material form, thereby further distancing them from the realities of manufacturing and the fact that the pieces were mass-produced in a German porcelain factory. When collectors bought Hummel figurines, they were really buying, and buying into, convenient, feel-good fictions.

A seemingly simple thing, the *Story and Picture* booklet was actually a sophisticated marketing tool—part myth-maker, part merchandise catalog. Several pages of sepia-toned images of the figurines on offer followed the brief story of young Berta's life. The last pages listed model numbers and dimensions for easy reference. Word and text together gave consumers reasons to collect Hummels, helped them apprehend the totality of the Hummel universe, reassured them that the figures could be readily obtained, and enabled them to easily categorize their acquisitions into rational themes and series—in other words, to transmute purchases into collections. Goebel and his retailers were thus able to make collecting as easy and unchallenging as possible. It was a masquerade that turned the thrill of the hunt, the joy of historical curiosity and discovery, the pleasure of conceiving a coherent yet personal collection, and the satisfaction of being a collector and all that that conferred into a simple economic exchange. Or, to put it the other way, the simple economic exchange was a way to acquire everything else (fig. 10.3).

Literature aimed at Hummel collectors appealed to their budding sense of connoisseurship and helped them imagine themselves as elite collectors. By the 1970s rafts of Hummel-related literature were being produced, including regularly published histories of the Goebel Manufacturing Company, descriptions of "genuine," "authentic," and "original" Hummels, and collectors' catalogs and guides. "Hummels are a race unto themselves with their own terminology, symbols and markings," noted one such manual.[21] In addition to images and descriptions of the latest Hummel issues, guidebooks contained advice about buying, details on how to discern trademark variations (from the small bee and the full bee to the baby bee and the stylized bee), and tips on how to spot counterfeits.

The presence of fakes was both a bane to collectors and a validation of their collecting practices, since Hummels were "worth faking." There were the outright fakes, to stay away from. And there were the imitations, which collectors considered flattering more than duplicitous, because they were evidence of desirability: "Hummel fanciers" who could not afford the real things would have to make do with these inferior yet "closely akin"

Figure 10.3. Over time, a Hummel collector marked off her purchases of members of the Hummel Orchestra. Marie Lynch, ed., *The Original "Hummel" Figures in Story and Picture*, 1955.

versions.[22] Imitations helped establish quality and price hierarchies that situated Hummels at the top and also constructed a collecting universe that more closely resembled the parallel world of high-end collecting. An awareness of the market in "pseudo-Hummel," which supplied the "low budget Hummel fanciers," also underscored the cultural capital of collectors who were purchasing the "real" and "authentic" versions.

The M. I. Hummel Club, established in 1977, reinforced collectors' sense of connoisseurship and status. Members received a quarterly newsletter, a ceramic membership plaque, a "handsome" binder with a collector's log, and price lists. Members also received information and "facts about M. I. Hummel history and production," which was an important way to compensate for the products' lack of actual history. Since the figurines themselves were new off the assembly line, the company's story and Sister Maria Innocentia's biography helped enrich collectors' connection to the past. Membership also gave people the opportunity to purchase an exclusive figurine each year and special access to a Hummel "research department" and, similar to the BradEx, a "collectors' market to match buyers and sellers."[23]

Like other aspects of intentional collectibles, the Hummel Club was at once marketing boon and community builder. Local clubs forged bonds among like-minded collectors and funneled their business to retailers who carried the exclusive club figurines in addition to standard Hummel lines. Retailers surveyed in 1990 noted that servicing the collectibles market and collectors clubs increased their revenues by encouraging additional traffic and repeat customers, conferring on their shops "the exclusivity the category brings to them." Indeed, members typically spent twice as much as nonmembers, and often up to $1,000 each year.[24]

By gathering aficionados, the Hummel Club not only fueled the passion for collecting by introducing an element of competition but also enabled collectors to discuss and compare variations of figurines. Particularly zealous Hummel collectors became obsessed with these differences. Each piece, hand-cast and hand-painted, was merely a mildly variant doppelgänger. While the differences were nearly invisible, production inconsistencies became collecting-worthy distinctions. In addition, the company often issued figures with intentional variations, such as Signs of Spring featuring, alternately, a girl with one bare foot or wearing two shoes. Whether the differences were deliberate or production inconsistencies mattered little to collectors, who simply knew they had to purchase more than one of the same figure. "Many collectors think nothing of springing for a dozen versions of the same Hummel," noted one article. "A favorite pastime" of Hummel Club members was "for everyone to bring in a designated Hummel, arrange the figures next to one another as if in a police lineup, and then sit down and study the infinitesimal differences."[25] While many Hummel collectors did collect because they simply liked the anodyne figures, many others approached collecting with a more serious connoisseur's eye. That people bought twelve variations of the same figurine belied the assertion that they were merely participating in a casual hobby or treating the figures as memory objects that called up personal associations. Most considered Hummels a serious investment, too.[26]

Making Precious Moments

Precious Moments figurines followed Hummels as Staffordshire's next generation of bastard children. Born in 1979, they became even more popular, selling at lower price points and featuring cloyingly cute, pastel-hued neonates with oversized heads and large, teardrop-shaped eyes (plate 9).[27] Each figurine carried an overtly spiritual message of banal good tidings: God's Speed, Blessed Are the Pure in Heart, Jesus Loves Me, Bless This House, Forgiving Is Forgetting, and so on. "Those things have power," wrote one columnist in 1986. "They have enough sentimental energy to melt the hearts of hundreds of thousands of people across the nation."[28] The sentimental bromides and Bible maxims helped people forget, or deny, that, like Hummels, these wee figures were actually objects of the market.

Also like Hummels, Precious Moments were infused with resonant, spiritually informed autobiographies. They, too, began as heaven-sent sketches, channeled through the medium of Sam Butcher, a lay minister and amateur artist, whose original line of greeting cards featured bug-eyed children delivering messages such as God Is Love, Prayer Changes Things, and Jesus Loves Me. Butcher's official story, presented in the lushly illustrated coffee-table book *The Precious Moments Story: Collectors' Edition* (1986) and punctuated with "miracles" and "fateful" events," explained that his mission was "guided by the Lord."[29] Like Hummels, Precious Moments possessed a quasi-sacral status.

Buyers could both pay obeisance to Butcher and help spread the good word by purchasing Precious Moments, collecting them, and giving them as gifts. They were (until Beanie Babies) the most popular and successful line of intentional collectibles ever produced, largely because their intense "sentimental energy" erased their reality as mass-produced things. This proved a delicate balance, and industry analysts worried that too much success—in the form of licensed products and spin-offs—would, in fact, undermine the company. "The risk," one expert noted, "is people start to see the commerce in it and not the inspiration."[30]

Precious Moments' marketing strategies continued along the Hummel model and were closely followed by makers of other intentional collectibles. A dedicated publication, *Precious Moments Insights*, reached some ten thousand collectors. Members of the Precious Moments Club, boasting over half a million members by the early 1990s, received a welcome kit, chances to purchase exclusive figures, newsletters, and a "registry booklet" with a list of all the merchandise lines.[31] Less focused on connoisseurship and distinction than the Hummel Club, members gathered "mainly to discuss the precious moments they've had with their Precious Moments."[32] There were special events for club members, too, such as an exclusive cruise to the Bahamas hosted by Butcher and singer Pat Boone, plus the opportunity to meet up with the celebrities again at a Precious Moments convention.

Like Hummel figurines that reminded people of (unblemished versions of) their children, Precious Moments also enabled collectors to construct idealized versions of their pasts and to create objects of memory retroactively, even though they had no actual connections to the pasts the figurines embodied. Each new item entered a collection partly because of its imagined backstory and partly because of its extreme cuteness. In this way, people could even anticipate constructing the past with a figurine they did not yet own. One collector explained, "We talk about what a certain piece means to us. My husband's mother passed away some time ago and there is a piece coming out—'No Tears Past the Gate,' a beautiful piece— and to us it symbolizes the fact that now that his mother's in Heaven, she has only joy and happiness." However tenuous, the connections collectors made with their objects were quite real. As another collector explained, "Some of my pieces remind me of my kids. . . . [One] is a little boy cutting off a little girl's pigtail, and that happened once to my daughter. Another one is a nun. It is 'Make a Habit Out of Prayer' and so I bought that and put it in my bathroom to remind me to pray."[33]

This was an intriguing bit of mental gymnastics, especially given that such intensely personal memory objects looked exactly like those of other Precious Moments collectors. But because the figurines encouraged such imaginative musings it was easy for collectors to see them as existing beyond the market and therefore unique. Purchasers looked upon their aggregations not as a series of purchased things but as small communities populated with "children" who were alive with emotional sentiment and spiritual animus. Like Hummels, these, too, captured homogenous and pristine worlds inhabited by the innocent—children (almost exclusively white), plus angels, and the occasional dog.[34] These were not places sullied by the materialism of the market or the vagaries of difference.

Collectibles' imagined distance from the market was the very thing that enabled them, both ironically and intentionally, to be highly profitable for manufacturers, distributors, and retailers. As one newspaper noted at the height of Precious Moments' popularity, manufacturer Enesco had "made a bundle selling these things." The company periodically "retired" figurines in order to "support price appreciation" on the secondary market. In this way, the company ginned up enthusiasm for new purchases within the primary market and created a sense of scarcity within the secondary market. That they were "made cheaply in the Far East" was either hidden from or irrelevant to collectors. For them, the figurines' transcendent messages of godliness gave them a certain unassailability. Because people *so* imbued these things with emotional import and religiosity, to suggest that they were cheaply made would have been to impugn their message, which must be tainted and cheap by association. Unable to disaggregate the material from the sentimental and spiritual, serious collectors would not and could not countenance this kind of critique.[35]

Finally, Enesco took what might otherwise have been the figures' dubious provenance and turned it into yet another promotional opportunity. The company's official *Precious Moments Story* included an extensive section on the production process, featuring descriptions, accompanied by illustrations, of teams of Japanese technicians, artisans, and craftspeople working diligently under the "closely" watchful eye of Butcher himself. This was followed by a chapter on Butcher's missionary work in the Philippines, where he established a Precious Moments doll factory to give jobs to students attending a local Bible school. "I believe that God used Precious Moments subjects to open the door of Christian service whereby we could enter and touch the lives of many," Butcher testified.[36] Here, the book justified collectors' own mission of acquiring by drawing upon well-worn racial hierarchies that placed the "first" world (white, civilized, consumers) against and above the "third" world (nonwhite, non-Christian, laborers).

By the late twentieth century, Hummel and Precious Moments figurines existed within what had become a much larger ecosystem of intentional collectibles supporting both individual collecting interests and the pursuit of mass-produced collectibles more broadly. Distributors hawking collectible wares fashioned themselves as sophisticates whose very names implied exclusivity, rarity, and value: Modern Masters Ltd., Collectible Resource Group, Ernst Limited Editions, American Imports Company, the Worthington Collection, Heirloom Porcelain, and countless others. Many sold their merchandise through mail order, but independent retail shops, from the Honeycomb Gift Shoppe in Wakefield, Massachusetts and Caren's Ltd. Fine Art Gallery & Limited Edition Collectibles in Bath, Ohio, to Lena's Limited Editions Gift Gallery in San Mateo, California, also reported a brisk trade in mass-produced collectibles.

Collectors also attended a growing number of organized gatherings, from national expositions for such collectibles luminaries as Hummel figurines to more modest regional meet-ups, like the South Bend Plate Convention. They also organized clubs, including, among many, many others, The Sebastian Miniature Collectors Society, Southern California Association of Plate Collectors Clubs, Annalee Doll Society, Angel Collectors' Club of America, Hallmark Keepsake Ornament Collectors Club, Elfin Glen Collectors Guild, and Rockwell Society of America. Groups gathered to share information and buy new offerings. Perhaps most important, clubs helped sustain interest and faith in their collectibles' vitality in the secondary market, since the promise of the resale trade drove primary retail sales. A trade analysis in 1990 reported that "the proliferation of collector's clubs—well over 50 presently—is a barometer of the thriving collectibles industry."[37] By 1992 an estimated 1.5 million collectors belonged to more than a hundred different clubs, reliable "profit builders" for retailers who expected them to help attract new members who would become new customers. At one time, the Precious Moments Club alone boasted over half a million members.[38]

Bear Markets

These myriad self-reinforcing efforts—the collectors clubs, the price guides and magazines, the incessant advertising, the slick marketing—helped create a collecting culture centered around intentional collectibles and a collecting bubble that swelled from the 1970s through the 1990s. The intentional collectibles market was thus especially primed for the introduction of Ty Inc.'s Beanie Babies in 1993. Ty either brilliantly capitalized on or cynically exploited the forces that had been driving this market for more than two decades. The charming beanbag creatures were "cheaper than the Cabbage Patch dolls, cuter than trolls . . . and of 'higher caliber' than the pet rock," noted one observer. They were the rare collectible that crossed the typical age and gender divides. Children loved them as toys. Selling for about $5 (or free with McDonald's Happy Meal purchases), they were affordable and easy to amass. Adults loved them because they made good gifts and could "bring a smile to your face."[39]

By 1995 Ty Inc. began to turn these understuffed beanbag plush toys—some called them "roadkill"—into serious collectibles. The company created false scarcity by limiting the production of each new design, without telling collectors that "limited production," like the "limited firing days" of collectible plates, might still mean tens of thousands flown in from China and Korea by the literal planeload.[40] Although Ty Inc. did not disclose its production figures—since uncertainty was an inherent part of the marketing—experts estimated that between a hundred thousand and five million of each figure were made, not "limited" in any meaningful way.[41] Ty also sold the toys through select retailers, avoiding the commodified taint of big-box chains, and controlled the number of each design a store received. The company also regularly launched and discontinued designs, corporate decisions they announced, with great import, at strategic press briefings. Rumors about scarcity only drove up demand, and figures' retirements became occasions for good publicity.[42]

For example, Clubby, the bear launched to commemorate the new official Beanie Babies collectors club, was introduced to the nation on television's *Today* show in 1998. Beanie Baby spokesperson Pat Brady explained to skeptical host Katie Couric that club membership, $10 per year, enabled collectors to have "access" to Clubby, meaning the *opportunity* to purchase a "rare" bear—the bear itself cost extra. What were the other benefits of membership, Couric wanted to know. "Well, it's packed full of fun. . . . You get the official membership club kit. And in there is a gold card, 136 Beanie Baby stickers, a checklist, a giant poster." Still baffled, Couric asked Brady to account for the popularity of these things. "It's—it's so weird, isn't it, in a way?" The aspirational humbuggery of Ty's marketing hype was apparent, but only to those who, like Couric, were not yet true believers:

> Brady: Well, Ty [Warner, the company founder] has made a leap. There's better value. He's provided a five dollar toy that is absolutely adorable.
>
> Couric: I know. But it really goes beyond that, don't you think, Pat, with these people getting on the Internet and selling them on the secondary market, I mentioned, for hundreds and hundreds of dollars. And the rare ones and the more desirable ones.
>
> Brady: Well, they're—they're collectibles. They're—they're—it's a lot of fun for everyone. It's part of the hunt. . . .
>
> [cohost Ann] Curry: I have to say, I love them. There's—there's no question. But I can't imagine spending $400 for one of these things. It's obscene. . . .
>
> Brady: Well, also, you know, Ty wanted it to be fun for kids. And he's very

focused on fun for kids. And the secondary market is nice because it
makes them collectible. Kids can buy one and they know it will increase
in value. But it's . . .

Couric: But you don't encourage that.

Brady: No.

Couric: But, at the same time, don't you think, like, offering this Beanie Baby
official club dealy and then having the limited edition kind of encourages
the secondary market?

Brady: No, actually, it was to make sure that all the kids could have access to
one of those rare bears.

Couric: Yeah, but then later on you can sell the bear for a lot of jack, right?

Brady: No, no, you keep it, you keep it. And for years to come, enjoy it. And if
you need the jack, you go for it when it's college time.

Couric: I got you. It's a good investment for a college education.

Brady: That's right.[43]

The spokeswoman's conflation of entertainment value (a fun hobby) and
monetary value (an investment for college) was purposeful. It kept the
company from breaking the law by making claims that Beanie Babies
would appreciate in value, while at the same time encouraging collectors to
believe it was so. By so closely allying entertainment and monetary value,
the company made it difficult for collectors to distinguish the two (and easy
to justify their hobby). This approach, familiar to other purveyors of intentional collectibles, was quite successful. In 1998 the company's net profits
were some $700 million, more than its two key toy competitors, Hasbro
and Mattel, combined.[44] At one point the sale of Beanie Babies accounted
for 10 percent of all sales on eBay.[45]

In this way, Ty successfully created a massive collecting craze for a fairly
crappy piece of merchandise produced in staggering numbers. In addition to manufacturing scarcity, the company convinced collectors that they
were engaging in connoisseurship. Collectors often encased their darlings,
with their pristine heart-shaped tags still dangling (and enclosed in their
own protective plastic covers), in specially designed clear plastic boxes.
They examined the toys for the minutest differences. As with stamps and
coins, "mistake" examples—that is, defective ones—often fetched the
highest prices. Special issue pieces, such as the Princess Diana bear, were
especially prized: "Mix the Beanie Baby craze with the Princess Diana

memorabilia rush and pandemonium is assured," remarked *USA Today*.[46] Correspondents to the "Ask Dr. Beanie" newspaper column sought advice on everything from the importance of original tags (very) and whether Squealer the pig would "be worth a lot of money when he retires" (no) to how and when the next group of Beanies would be retired and new ones launched (nobody knows).[47]

Even in the face of criticisms from antiques and collectibles experts, and the soft caveats offered by companies themselves, collectors were seduced by the idea that they could compete in serious material contests, applying their expertise about intentional collectibles in savvy ways, just as high-end collectors did. Dedicated collectors saw Ty Warner not as the leader of a multimillion-dollar toy company but as a benevolent force like Berta Hummel and Sam Butcher. "I think that we have a whole group of baby boomers that love Beanies," said one devotee optimistically. She believed Warner when he said, "Others are into this for the quick buck; I'm in it for the longevity." A skeptical antiques expert responded, "You want to prove you want a Beanie Baby for fun? Take a scissors and cut those tags off, hug them, take them to bed with you, play with them. That's fine with me. I don't have any problem with that. When you stick them on a shelf and bow down and worship them, you've got a reality check problem."[48]

As in any other bubble, people's passions for these things, for a time at least, were monetized. There was a burgeoning black market in limp beanbag toys in the late 1990s: over $30,000 of them were stolen in five thefts in Syracuse, New York; over $5,000 taken from stores in a Chicago suburb; and $12,000 stolen from a shop in Kansas City.[49] People devised insurance scams involving Beanie Babies. And a judge overseeing a divorce case forced the couple, who refused to part with their prized possessions, to split up their Beanie Babies collection in open court. "It's ridiculous and embarrassing," confessed the wife, who moments later was "squatting on the courtroom floor alongside her ex-husband to choose first from a pile of dozens of stuffed toys."[50]

Perhaps Ty's most ingenious move was to make the ultimate Beanie Babies retirement announcement, declaring that the company would stop producing the toys altogether at the end of 1999. This alone caused a spike in sales, upwards of 300 percent in some stores, as people tried desperately to fill collection gaps and make their final investments.[51] The timing was right, as sales had been lagging for over a year due to the toys' market saturation, unsustainably high prices on the secondary market, and convincing counterfeits from China—essentially the same products and very likely made in the same factories. Still, many figures remained unsold, jeopardizing both primary retail sales and the secondary market.[52]

These were "nervous times" for Beanie Babies collectors. Becky Phillips, editor of the price guide *Beanie Mania*, remained optimistic about Ty's "cryptic" news, believing that he would only retire existing lines and introduce new ones in their place. She said, as if divining meaning from an oracle, "It all depends on how you interpret the message that he put on the Internet. For me, I see this message as very positive. I think that Ty Warner has something in store for us for the new millennium." Collectibles expert Harry Rinker, however, held a more jaundiced view: "We already know the Beanie Baby markets hit the doldrums; sales are down," adding, "Ty has got a warehouse full of stuff. Who's the winner here? The empty warehouse, money in your pocket. The man can't lose."[53] Indeed, after initiating this run on existing Beanie Babies, Warner also created a new revenue stream when he announced that, due to the pleas of desperate Beanie Baby fans, he would let collectors vote online (at fifty cents a pop) on whether or not to retire all the figures. "You make the decision. You have inspired the Beanie Babies line through your devotion to them," he flattered his fans.[54]

Despite Ty Warner's Barnum-esque marketing and publicity efforts, it was inevitable that the bubble would burst. While issuing a black bear called, not so subtly, The End supposedly marked the line's official retirement, the company kept producing Beanie Babies, thus irrevocably shaking the confidence collectors had placed in Ty the man and Ty Inc. the company (fig. 10.4). In 2003 an article declared that while once hoarded and scarce, Beanie Babies were now "past their prime." Their prices had "plummeted." So many bears, moneys, ducks, crabs, dogs, ladybugs, piglets, and other sad plush were "languishing on store shelves across America." Store owners who at first could not keep Beanie Babies in stock were now glad to get rid of them at any price.[55] Increasingly, newspapers published heart-warming stories of Beanie Baby collections being auctioned off for charity or donated to worthy causes. Some were carried in the pockets of US soldiers, to be given to the Afghani and Iraqi children they encountered during patrols, "thus bringing new smiles to children's faces, and protecting our troops in dangerous places."[56] A more cynical interpretation was that the only way to deal with the surplus of these embar-

Figure 10.4. The special edition Beanie Babies bear "The End" was apparently not. Tim Tiebout Photography. www.timtiebout.com.

rassing objects was to exile them to far-off war zones, guaranteeing they would never return.[57] And people continue to struggle to rid themselves of the sagging pests. Someone trying to sell a lot of seventy-seven of them on eBay in 2016 listed them under the heading, "C'MON, WILL SOMEBODY PLEASE BUY THESE FREAKING BEANIE BABIES!!!!"

Other intentional collectibles suffered the same fate. By the 1990s, the industry as a whole was "sailing in troubled waters," stock in collectibles companies was at an all-time low, and industry experts saw "no anticipation of growth at all."[58] For decades, avid buyers had ridden the intentional collectibles wave, investing in Hummel figurines, Franklin Mint commemoratives, Beanie Babies, and all manner of other collectible crap. In the late 1980s Hummels had been commanding "big bucks"—as much as $20,000 each.[59] Serious collectors were paying premiums to "pickers" who located hard-to-find pieces. In a little over a decade, one woman spent nearly $50,000 on Precious Moments figurines and auxiliary products like dolls, ornaments, plates, buttons, wrapping paper, and greeting cards. Another woman's husband purpose-built an extension on their home—an updated version of the Victorian "best room," shaped like a chapel and illuminated by stained-glass Gothic windows—to house her collection of more than a thousand figurines.[60] Yet another spent so much money on Precious Moments that she bankrupted her family (and allegedly murdered her husband when he found out).[61]

There were many reasons why the bubble finally burst. The growing popularity of eBay and other online sites erased information asymmetries about price and value, not only enabling collectors to see where the market stood at any given moment but also helping them better apprehend the entirety of the collectibles universe beyond their local communities. They could see if supply outweighed demand and could recalibrate the inflated prices that had once prevailed on the secondary market. In addition, due to overproduction, the intentional collectibles market had become saturated (and its value "diluted").[62] "Used too loosely," the already dubious term "collectible" had been drained of any real meaning.[63] Changing demographics affected the market as well. Avid collectors of intentional collectibles tended to be older than thirty, and many were pushing retirement age. Having filled their houses with stuff, they were running out of space to add more. Others, with an eye toward downsizing, were beginning to divest. As parents and grandparents died off, even more collectibles flowed back into the market, especially since younger generations were not interested in this kind of stuff.[64]

People encountered the same bad news with commemorative plates, collectible figurines, and most other born collectibles. The secondary market that drove the primary market was drying up. By the early 2000s companies were restructuring and laying off workers. Enesco cut its workforce by 14 percent in 2001 and, despite producing new Harry Potter collectibles, was posting millions of dollars in net losses.[65] Four years later, the company, too reliant on the "ailing collectibles market," stopped distributing figurines for its parent company, Precious Moments Inc., citing 35 percent in losses and an overall decline in the figurine market from $2.4 billion in 2000 to $1.2 billion in 2003.[66]

Collectors and their descendants came face to face with intentional collectibles' inherent contradictions and obfuscations when it was time to liquidate all those figurines, plates, dolls, and other what-nots, bric-a-brac, and dust-catchers, which people had spent decades of their lives amassing. There, dispassionate economic value prevailed over the sentimental, recreational, and display value that these things once possessed. Personal accounts often bordered on the tragic, illuminating the disparity between the persuasive fantasies created by corporate marketing efforts and the stark realities of the market. Just as their compatriots who collected commemoratives had, these collectors, too, learned the hard way that scarcity could not be manufactured, and, unless you were someone like Ty Warner, the market would not transmute all these mass-produced collectibles into gold.

PART 6

But Wait, There's More

11

JOKE'S ON YOU

People love crap for many reasons, as I hope I have shown. Despite their cheapness and disposability, dime store goods were affordable, accessible, and available in abundant variety. However insincere, gifts were better still, since they were free. While they were often inefficient or created more work, if they worked at all, gadgets at least promised to ease the burdens of labor. Over time, knickknacks, tchotchkes, and bric-a-brac formed a rich material vocabulary that helped people express identity, status, and distinction, even if they were doing it like everyone else. And intentional collectibles, although mass-produced, constructed of inferior materials, and ultimately poor investments, made for pleasing displays and a satisfying hobby.

There are no such apparent explanations for novelty goods. Ironically, making sense of fake dog poop, exploding cigars, Whoopie Cushions, and Joy Buzzers—the crappiest crap of all—requires perhaps the most sophisticated interpretive framework of all. Like the other kinds of crap, novelty goods are complicated things embodying myriad deceptions and contradictions. Their very triviality, evanescence, and disposability may actually show them to be profoundly revealing of the human condition.

The Latest Novelties

Novelty came late to Americans, and with some trepidation. In the eighteenth and early nineteenth centuries, new ideas and experiences were met with both excitement and anxiety. The quest for the new, some argued, "[kept] the mind in a continual gadding." The young, especially, indulged in "ardent longings after new play-things," and as soon as something had "grown familiar," it was "loathed and thrown aside."[1] Novelties, too, could be subversive, exposing people to the extraordinary, remarkable, and provocative: "Hindoo" marriages, snowstorms in South Carolina, mysterious sea serpents sighted off the coast of Massachusetts.[2]

Novel entertainments and observations could open portals to imagined worlds. Early itinerant puppeteers, conjurers, magicians, plate spinners, rope dancers, and magic lantern projectionists profited from Americans' perpetual quest for the new, tramping about the countryside performing magic tricks and demonstrating strange contraptions that they often sold as "rational amusements."[3] Trojan Pillars, Operations in Papyromance, Sympathetic Clocks, and other curiosities engendered wonderment and could "strike the beholder's eye with astonishment" (fig. 11.1).[4]

Americans' quest for the new helped usher in an age of mass consumption in the antebellum era. By equating newness with desirability, advertisers encouraged consumers to rethink their traditional relationship to the material world, encouraging them to cast aside older possessions in favor of the latest things. As one writer observed, "*Novelty*—Has charms that our minds can hardly withstand. The most valuable things, if they have for a long while appeared among us, do not make an impression as they are good, but give us a distaste as they are old."[5] The embrace of novelty marked a fundamental shift in popular attitudes about consumer culture.

By the closing years of the Civil War, Americans were able to purchase not just new experiences and goods in new styles but *entirely new* things that fell outside of established conceptual categories: "novelty" now referred to both the physical thing itself and its state of being as a new thing. Deeply weird, Lightning Sausages, Chinese Finger Traps, Resurrection Plants, and Mystery Boxes found a ready market in the later decades of the nineteenth century for several reasons. The second industrial revolution provided more people with work, increasing disposable income, filling the market with more stuff, and creating a nation of consumers. At the same time, the nature of much of that work—unskilled and in the service of faceless manufacturers—increasingly alienated people from their livelihoods. And so people escaped through entertainment, which they more often sought out in saloons, theaters, and public gardens. Humor, which came to "occupy a distinctive niche in national life," was an important part

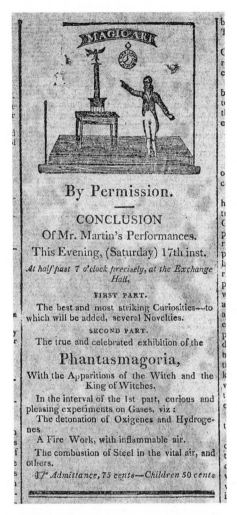

Figure 11.1. Magician Mr. Martin was one of countless traveling performers presenting "novelties" and "curiosities" to early American audiences. Advertisement in the *Republican and Savannah Evening Ledger*, March 17, 1810. Courtesy American Antiquarian Society.

of this, commodified through minstrel shows, humorous presentations, publications like joke books and comic almanacs, and, soon enough, novelty goods.[6] Teams of sales agents working for general merchandisers in the late nineteenth century brought novelties, jokes, and gag goods to the mass market. Quickly becoming a "competitive industry," according to one

Figure 11.2. By the late nineteenth century, mass merchandisers were offering various lines of jokes and gags like Snake Boxes alongside more "practical novelties" like Pocket Stoves and animatronic thermometers. A. Coulter & Co., *Wholesale Price List: Novelties and Notions*, [1883].

contemporary account, novelties were sold alongside other cheap variety goods such as household gadgets and bric-a-brac (fig. 11.2).[7]

Enter, Lightning Sausages and Other Strange Stuff

Novelty goods quickly became popular despite what was often their utter inscrutability. Marketed to adults and children alike, they existed simply because they could: flashlights shaped like snakes' heads and smoking pipes shaped like revolvers; miniature telescopes; giant eyeglasses; and so on. It fell to sales catalogs to explain—to both sales agents and consumers— not only what these things were and what they did but why people needed to have them. Promotional literature, rich in word and image—an "aesthetic of abundance"—helped make sense of these things.[8] The Lightning Sausage (likely an early version of a snake in a can) offered by the Eureka

Novelty Company in 1876, for instance, required an extensive explanation that made no sense without the accompanying illustration (fig. 11.3).

Despite or perhaps because of the surplus descriptions, novelties remained largely incomprehensible to both consumers and distributors. Advertising copy for the Great Japanese Mystery admitted frankly, "No one has yet been able to explain what makes it act as it does, and we are not able here to give you any idea of the strange actions of this mysterious article." The Fargo Novelty Company introduced its Jonah and the Whale novelty by stating, "We show you here a very funny patented article. We cannot describe it very well."[9] And the best the marketers could do for the perennial but similarly obscure "The 'What-Is-It?,'" "a most comical surprise," was to dodge entirely, naming it after the famous P. T. Barnum humbug.[10]

Profoundly useless, novelties, jokes, and gags quickly became known in the trade as "sure sellers." By the first decades of the twentieth century novelty companies and general merchandisers were expanding in number and offering greater varieties of stock, mostly imported, through networks of specialized wholesalers, distributors, and retailers in all parts of

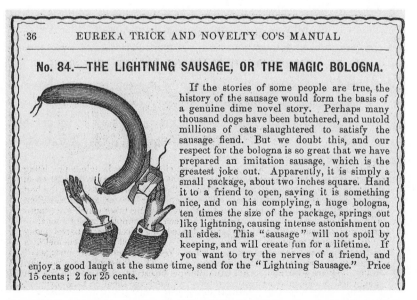

Figure 11.3. Entirely new things, many early novelties like The Lightning Sausage, a.k.a The Magic Bologna, required not only detailed textual explanations but also illustrations that would both show and tell what they would do. Eureka Trick and Novelty Co., *Illustrated Manual of Tricks, Novelties . . .* , 1876.

the country, from A. Coulter & Co. in Chicago (one of the first, est. 1865) to Fargo Novelty in Frenchtown, New Jersey, to Universal Distributors in Stamford, Connecticut.[11] Indianapolis-based Kipp Brothers began in 1893 and by the 1930s had become the leading supplier of cheap carnival prizes. A growing number of retail stores, too, dealt solely in novelties and magic tricks. In 1902 alone Kipp Brothers had invested over $487,000 in merchandise and paid another $66,800 in salesmen's salaries.[12]

But why did people buy these utterly impractical, strange, and inexplicable things? Many consumers were attracted to novelty for the sake of novelty. The "ardent longings after new play-things" decried in the eighteenth century had become an essential selling point by the early twentieth. As with older performances, newer novelties conjured the weird, mysterious, and foreign. By pointing out their exotic origins—Great Egyptian Mysteries, Beautiful Mermaids, Perfumed Shells from Ceylon, and Japanese Tricks—companies offered up "fantastic possibilities" while also obscuring the overseas sweatshops whence this merchandise really came.[13] Alluring provenances included often elaborate backstories. A. Coulter, for instance, claimed that the novel technology for its Stanhope Photo-Microscopic Ring received an honorable mention at the Exposition Universelle in 1867 and was "heretofore, a rare article," and "expensive," but the company was able to import it "direct from Paris."[14] Likewise, novelty wholesaler Bennet & Co. noted that it procured merchandise from around the world: Austria (scrapbooks and surprise boxes); Germany (party supplies, fake mustaches, farting cushions, novelty teeth); Bohemia (novelty picture frames of plush and velvet); and Japan (giant spiders, rubber mice, paper goods).[15] Going to "great trouble and expense" for its customers, the Fargo Novelty Company acquired a "great quantity" of Wonderful Lucky Bug Pens, from "natives" in Brazil's interior, some "thousands of miles away."[16] Consumers could buy this weird foreignness on the cheap.

Meanings of Life

More than simply being cheap and new, some novelties seemed to summon life forces by defying death, thereby concealing profundity in a frivolous disguise. The Resurrection Plant (a.k.a. the Biblical Rose of Jericho), for instance, was a lycopod that could survive in a dormant state for years. When exposed to water, the desiccated plant became lush and green, rising

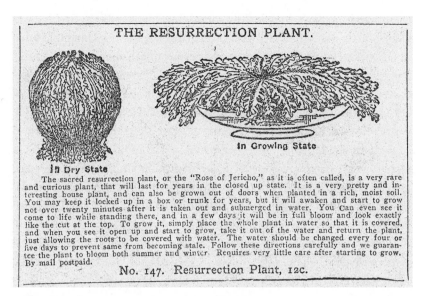

THE RESURRECTION PLANT.

In Growing State

In Dry State

The sacred resurrection plant, or the "Rose of Jericho," as it is often called, is a very rare and curious plant, that will last for years in the closed up state. It is a very pretty and interesting house plant, and can also be grown out of doors when planted in a rich, moist soil. You may keep it locked up in a box or trunk for years, but it will awaken and start to grow not over twenty minutes after it is taken out and submerged in water. You can even see it come to life while standing there, and in a few days it will be in full bloom and look exactly like the cut at the top. To grow it, simply place the whole plant in water so that it is covered, and when you see it open up and start to grow, take it out of the water and return the plant, just allowing the roots to be covered with water. The water should be changed every four or five days to prevent same from becoming stale. Follow these directions carefully and we guarantee the plant to bloom both summer and winter. Requires very little care after starting to grow. By mail postpaid.

No. 147. Resurrection Plant, 12c.

Figure 11.4. Some novelties summoned life forces and seemed to defy death. The Resurrection Plant was one such novelty that could be reanimated. Universal Distributors, *Illustrated Catalogue of Novelties*, ca. 1915.

from the dead as its name promised (fig. 11.4).[17] "One of the greatest wonders of the plant kingdom," it came from an exotic, faraway place—maybe the Holy Land, maybe Mexico, maybe the desert West. Dead but alive and a "rare and beautiful curiosity," it could be planted in soil or stashed away in a box.[18] People became reanimators, bringing things to life "while standing there."[19] (The same impulses drove the later popularity of Magic Rocks—invented in 1940 and first marketed under the name Magic Isle Undersea Garden—and Sea Monkeys, introduced in 1960 as "Instant Life." Magic Rocks sprang up "like magic," and Sea Monkeys, a form of brine shrimp, were "so eager to please, they can even be trained." They, too, can live in states of suspended animation; cryptobiosis makes them a perfect commodity, both alive and dead.[20])

Descendants of the Resurrection Plant and its cousin the Perpetual Rose Bush, Chia Pets also enabled consumers to create spontaneous life, becoming gods of their modest domains. Although they became wildly popular in the Pet Rock era of the 1970s, the Chia Pet, like many other novelties, has a much longer history. In first decade of the twentieth century, ingenious novelty manufacturers realized they could turn the literal act of watching grass grow into a profitable product line. The ur-Chia was likely Murro, das Wunderschwein (Murro, the Wonder Pig), manufactured in Germany (fig. 11.5). The best novelties promised spontaneous drama, and Murro was no exception: "In a short time," the copy noted, "the pig is covered with a verdant fur . . . to the delight of the audience," making viewers a part of the performance and Murro's caretaker its master.[21] Murro's verdant species had enduring commercial appeal. By the 1940s American companies like Morton Pottery were making, in addition to their lines of teapots and mixing bowls for the five-and-dime market, more specialized novelty objects like Paddy O'Hair, on whose ceramic head grass grew.[22] Later in the decade Johnson Smith & Co. was selling its Sunny Jim head for fifty-nine cents. His emerging afro and eyebrows were a "performance" that could be witnessed *in process*, transforming nature into a cultural commodity for human entertainment. "These heads," the company noted, "have been favorites for many years all around the world."[23]

Democratic Surrealism

Novelty goods created alternate realities. Some, like Resurrection Plants and Chia Pets, literally so. Others, like Long Tongues, Mammoth Bow Ties, Giant Thumbs, Burlesque Diamond Rings, and Pop Eye Glasses, used scale to disorient, forcing people to recalibrate how their bodies interacted with the physical world.[24] The pages of novelty catalogs presented wonderfully riotous and disorienting juxtapositions of words and images—each set of associations uncannier than the next—the demon spawn of Salvador Dalí and Montgomery Ward. One double-page spread of the catalog the Zubeck Novelty Company issued in the 1910s, for instance, featured not only a distorting pocket mirror making one look too fat and too thin but imitation fly pins, Mexican jumping beans, a pun box labeled "Black Kids" (containing not fine gloves but "a pair of tiny nude dolls"), a trick cigar, and, more practically, a tube of See Clear to prevent windows from fogging up. It was a bewildering assortment of the odd and exotic and practical.[25] Goo-Goo Teeth, Black Cat Pins, Rooters, Funniscopes, and Boer Snappers made little sense either as individual things or in the aggregate.[26] That was their point. Novelty pioneer S. S. Adams said he was "proud of making anything that is absolutely useless, offensive, or prone to cause shock or embarrassment."[27]

Murro, das Wunderschwein,

| vor der Aussaat, |

| ca. 8 Tage nach der Aussaat, |

| ca. 14 Tage nach der Aussaat, |

Dieses Schwein, ein allerliebstes Schmuck-
stück des Zimmers, ist aus Ton an-
gefertigt; in den Ton sind Rillen ein-
gegraben.

In diese Rillen sät man Agrostisgras und
fühlt darauf das Schwein mit Wasser
— das Schwein ist hohl und hat oben
eine Öffnung.

Das Agrostisgras beginnt nun in diesen
Rillen zu wachsen und in kurzer Zeit
ist das Schwein mit einem grünen Pelz
bedeckt, der zum Ergötzen der Zu-
schauer von Tag zu Tag dichter wird.

Das Besäen des Schweines kann man
das Jahr hindurch an einer und der-
selben Figur wiederholt vornehmen.

1 Schwein
und ein Paket Agrostisgrassamen ge-
nügen, um das Schwein wiederholt zu besäen, ... **95** Pf

1 Paar Schweine **1**⁸⁵
und 2 Pakete Saat M

3 Schweine **2**⁷⁵
und 3 Pakete Saat M

5 Schweine **4**⁶⁵
und 5 Pakete Saat M

☞ Hoher Rabatt für Wiederverkäufer.

M. Peterseims Blumengärtnereien,
Erfurt.

Gratis und franko verlange man unsern
Haupt-Katalog über Gemüsesamen,
Saatkartoffeln, Rosen, Obstbäume.

Figure 11.5. For over a century companies have been able to get people to pay to watch grass grow. The ur-Chia was perhaps Murro, das Wunderschwein, advertised in *Fliegende Blatter*, March 11, 1904.

By turns weird and perplexing, novelty goods were liberating: they encouraged buyers to "gad about" from one odd consumable to the next, rewarding fickleness and surplus with ever more new things and fresh experiences. Although they borrowed strategies used by mainstream mail order catalogs, novelty purveyors made no effort to create order out of chaos by organizing inventories into different "departments" and "categories," instead creating disorder intentionally.[28] Since printed catalogs left lasting impressions, companies went to great lengths to convey an engrossing sense of material confusion. One of the Eureka Novelty Company's greatest operating expenses in 1900 was producing its own catalogs: $30 on a printing press and more than twice that ($82.50) on illustrated cuts.[29] The Fargo Novelty Company pointed out to customers that the illustrations in its 1908 catalog were, "as far as practical, engraved from photographs taken directly from the article itself."[30]

Novelty seller Johnson Smith & Co. used the power of print to even greater effect, cramming material excess into thick catalogs issued directly to customers rather than sales agents, dispensing with the middleman.[31] The founder's son, Paul Smith, explained that his father's approach was influenced by Sears, Roebuck (even the company name was meant to echo the famous mail order house) and that they spent a great deal of time, effort, and money on creating high-quality merchandise illustrations—which, he pointed out, "would have more detail than a photograph."[32] Similarly, premier novelty manufacturer S. S. Adams, who supplied Johnson Smith and its competitors with countless novelty lines, hired special cartoonists to illustrate its goods in action. These images circulated beyond the catalogs, making cameo appearances in the advertising sections of popular magazines and comic books. By exploiting visually intense word-and-image juxtapositions, novelty sellers created terrifically deranged and immersive experiences, turning rationality upside down and making money along the way.

In the process of embracing rather than rejecting all of those weird new rubber pretzels, squirting cigarettes, fake noses, x-ray glasses, and trick packs of chewing gum, American consumers were also embracing the tenets of surrealism, decades before the Museum of Modern Art canonized the movement in its ground-breaking 1936 exhibition Fantastic Art, Dada, Surrealism (fig. 11.6). Critics referred to the show as "humorous fantasy," a "captivating diversion," a "deliberate cult of nonsense and confusion . . .

Figure 11.6. Novelty purveyors like Johnson Smith were selling surrealism to the public over a decade before the aesthetic movement was embraced by the art world. *Popular Mechanics*, December 1923.

an effort not to understand objective reality but to escape it."[33] Surrealism was popular, according to an advertising executive, because "it capitalizes fear, disgust, wonder, and uses the eye-catching, bewildering devices."[34] He could have been referring to a revolutionary art movement or a novelty catalog.

By the interwar period, America had become a country of alienated workers. Factory work had at last overtaken agricultural occupations. Most men were no longer able to be their own bosses; nor could they even entertain that possibility. The pastoral ideal was fast fading in the motorcar's rearview mirror, now just a fantasy indulged by the deluded or naïve (including fans of the Colonial Revival).[35] The surrealists provoked people toward a more revolutionary way of thinking, asking them to reconsider the commodity culture that increasingly dominated and defined their lives. Surrealists' preoccupations, in fact, aligned with the very things that animated novelty goods: the serious purpose of "humorous fantasy"; the "deliberate cult of nonsense and confusion"; the presentation of "strange viewpoints" that brought to the fore "the severity of alienation within social life"; the desire to be liberated through "our powers of imagination"; the achievement of "collective emancipation by means of insubordination, sabotage, and a total revolt against capitalist social relations."[36]

Revolution, in other words. "Clicking solidly" off American production lines in 1940, according to a market report, was subversion itself:

Christmas laugh items handled by H. Fishlove & Company include the Goose That Failed, a magnetic novelty in which a woman knocks off a man's hat; Hotcha Girl, Mystic Glasses, and El Ropo Cigars, made of rope done up in cellophane and silver foil like the best of cigars. Richard Appel, Inc., is featuring snow tablets and stinking plugs, along with a general line of trick and joke items. Eagle Magic Factory has trick gift boxes and other items, in addition to comic Christmas cards. Magnotrix Novelty Corporation reports early orders for comic diplomas, giant thumbs, loony letters, fake doughnuts, solid whisky, and similar items. The firm also offers funny Christmas cards in three colors in folders, regular and baronial sizes.[37]

Rethinking the world of goods by breathing new life into them, much as one might reanimate a Resurrection Plant, people tapped new imaginative dream worlds. Surrealists believed that this process, ideally, would encourage the populace to question existing hierarchies and consider alternative political structures. To this end, they played perceptual tricks on their audiences. In René Magritte's famous paintings *The Key to Dreams* (*La clé des songes*), from the 1930s (fig. 11.7a–b), he placed ordinary objects in unordinary juxtapositions, sometimes calling them by their correct names and sometimes not. This discontinuity of representation and perception, in his words, "projects us into a world of ideas and images, draws us towards a mysterious point on the horizon of the mind, where we encounter strange marvels and come back loaded with them."[38] He could very easily have been talking about novelty goods—their obscure purposes, unsettling associations, and odd names reshaping what people thought they knew to be real: This is not a boutonniere but an exploding flower. This is not a pistol but a water gun. These are not real teeth, nor even real fake teeth like dentures, but simulacra fundamentally like and unlike both.

Novelties and the Commodity Form

Novelty goods did not simply encourage their users to reconsider the material world surrounding them but rather *forced* them to do so, embodying the potential for revolutionary action and thought even more effectively than surrealist artworks. Not only did novelties have hidden "secrets," but they were also not commodities in the familiar sense, with stable and obvious exchange value. What was their purpose? What was their worth?

People enjoyed novelties because they purposely confused established categories of value. There was no way to use a novelty for its intended purpose without also using it up: it would no longer shock, distract, or entertain once it was deployed, because people then knew what to expect. The joke was no longer funny, the surprise no longer surprising. Often, it was not only the experience the novelty produced that was used up but *the novelty itself,* like exploding cigars that burned up and the smoke bombs and itching powders that literally vanished into thin air.[39] Novelties shared surrealism's essential nihilism.

At the same time they defied the process of commodification, however, novelties were quintessential commodities. In ways even more effective than surrealist art, novelty goods simultaneously encapsulated modern alienation and offered its antidote. Trick rubber pretzels, cheeses made of soap, cigars concealing paper fans, cocktail sets shaped like cannonballs,

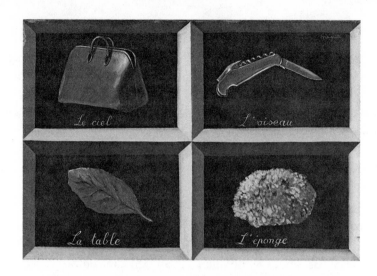

Figure 11.7a–b. Things are not what they seem to be. René Magritte, *The Key to Dreams* (*La clé des songes*), oil on canvas, 1927–30, © 2019 C. Herscovici/Artists Rights Society (ARS), New York; Johnson Smith & Co., *Supplementary Catalogue of Surprising Novelties, Puzzles, Tricks, Joke Goods, Useful Articles, Etc.*, ca. 1930.

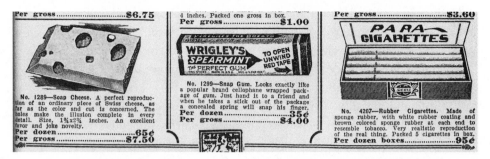

Figure 11.8. Cheese made out of soap and cigarettes made out of rubber were not all that different from the fur-lined teacups of the surrealists. Gellman Bros., *Annual Buyer's Guide Catalog for 1937.*

and coins that squirted water were many things at once—surprises, simulacra, sometimes useful things, and sometimes things that would dissolve into nothing. Wobbly Match Boxes, Wobbly Cigarette Packs, and Wobbly Wedges of Cheese that vibrated were surprising and unsettling. They behaved in weird ways and wrong ways, thwarting understanding about the fundamental nature of the most ordinary and familiar objects. Novelties prevented people from being able to take for granted that which they considered factual and real and tangible—the very realities that the forces of commodity capitalism strived so hard to establish and reinforce. Their meaning was therefore more immediate and accessible than, say, the meaning of surrealist pieces such as Meret Oppenheim's 1936 *Objet (Le déjeuner en fourrure)*, the famous fur-lined cup and saucer. In like fashion, novelties' myriad complications and complexities enabled perpetrators to momentarily create chaos, prompting their victims to question their understanding of reality itself, and were thus more potent and, ironically, more meaningful, than any other consumer goods on the market (fig. 11.8).

Theaters of Aggression

Novelty goods offered users more, though. Jokes and gags were also enlisted as cooperative partners to get even, settle scores, take others down a peg or two. Inherently performative, novelties took center stage as props in "theaters of aggression." The cast included perpetrators, victims, and an audience.[40] Not just fun and games, Chinese finger traps, oversized spiders, and biting snake boxes, animated by thinly veiled cruelty, were part of a longer tradition of American humor that combined the comic and tragic, the funny and malicious. Americans often wielded humor like a knife, to show power through good-natured threats of violence.[41]

This role for novelties, too, can be linked to the rise of capitalism and commodity culture. Americans frequently used humor to express anxieties about the expanding market and to reconcile the aggressive pursuit of profit and morality.[42] Nineteenth-century popular culture routinely depicted financial losers as witless dupes, greenhorns, and gulls whose bad fortune and poor decisions invited jesting derision rather than sympathy. New York's late-century financial elite often engaged in "childish practical jokes" that exposed the economic and cultural backwardness of "these gullible country folk."[43] Others characterized pranking boys—always boys—as "capitalists in training." In the 1883 book *Peck's Bad Boy and His Pa*, George Peck argued that the "best" boys are "full of tricks." He explained, "Those who are the readiest to play innocent jokes . . . are the most apt to turn out to be first-class business men."[44] Pranking showed a high-spiritedness, a willingness to take risks, and, most important, the drive to show oneself as top dog. Perpetrating practical jokes on each other was a way for the best and the brightest to establish hierarchies within bonded groups in the guise of good-natured fun. People also used pranks to demonstrably identify those who were not members of the group by embarrassing them in front of others.[45] As novelty purveyor S. S. Adams explained, "When I am fooling around with a new idea, I try to picture Mr. Average Man sitting around a cocktail lounge or in somebody's house before their weekly game of poker, and I try to ask myself if this new item will go in that sort of group, so if Person A pulls the gag on Person B, Person B will get a kick out of waiting for Person C to walk in and get the surprise of his life."[46]

The pranks themselves were violations; the "conspiracy" that brought about public embarrassment made them doubly so. The point of performances within theaters of aggression was to distinguish the people who were in on the joke—witnesses and collaborators—from the humiliated, who were not.[47] For instance, the Eureka Novelty Company's What-Is-It?, when dropped into someone's lap, would elicit a "look of horror" from the victim and "cause roars of laughter" among witnesses. That was a big payoff for a mere twenty-five cents.[48] Likewise, boys could have fun "scaring your mother, your uncle and aunts and the neighbors" with the Tarantula (a.k.a. Mexican Spider).[49] Just when "the victim" was starting to smile, a button on the Royal Novelty Company's Squirting Camera could be pushed, re-

leasing "a good squirt of water." "Boys, this is the very best joke and causes
no end of fun and laughter," the company promised (fig. 11.9).[50] When
thrust in someone's face to "wiggle and twist realistically," Gellman & Co.'s
Novelty Snake was especially humorous.[51] The "fun commenced" with the
Girl Catcher only after a girl inserted her finger into one end: "No matter
how hard she pulls she cannot get away! The harder she pulls the tighter it
holds. When you are ready to release her she can get her finger out easily,
but not before" (fig. 11.10).[52]

Jokes and gags of this sort created "disruptive" and "provocative" situations that ultimately reinforced the status quo.[53] Upending social conventions provided the opportunity to rearticulate and reinforce them by
eliciting responses that would build consensus and reestablish norms.
Hence, many jokes and gags involved a distinct gender component. Jokes
like Girl Catchers and Squirting Cameras gave boys license to dominate
girls as they would later dominate women. What was more, these jokes
made girls complicit in their own humiliation, since perpetrators needed
ready victims who, because of habits of politeness and deference, would
have to be good sports.

Although there was nothing preventing girls from buying and deploying

Figure 11.9. Novelties separated people into perpetrators and victims. Royal Novelty Co., *Illustrated Catalogue*, ca. 1910.

Figure 11.10. Jokes and gags often reinforced prescribed gender roles. Royal Novelty Co., *Illustrated Catalogue*, ca. 1910.

novelty goods, they were told in so many ways that this was not their world. Humor and play belonged more to boys than to them. Despite the promise of "fun for all," joke and gag articles were in truth a boy's (and man's) game, and pranking reinforced gender hierarchies. Boys (and men) did things while girls (and women) looked on. Further, boys (and men) had the power—physically, culturally—to do things to girls (and women). That pranking was conducted under the guise of good-natured fun robbed its victims of the right to protest, denying them something to protest against. Finally, pranking reinforced the widely held assumption that the fairer sex did not possess an innate sense of humor; women didn't even have a legitimate reason for engaging in humorous activities in the first place.[54]

By dividing their offerings by gender, novelty manufacturers and merchandisers continually reinforced this message.[55] Purveyors offered boys all sorts of mind-bending merchandise, like water pistols, the Boy Printer printing press, laughing cameras, kinematograph viewers, Bugaboo Watches, Fighting Roosters, and books about boxing techniques and making wooden toys. Girls, on the other hand, could spend their pocket money on a relatively paltry selection of cheap jewelry, beaded handbags, miniature kitchens, or fabric scraps and the sewing kits needed to put them together.[56]

Not really toys so much as child-size training devices, these small consumables opened boys to expansive worlds of limitless possibilities far beyond the confines of the home and its stifling domesticity. Merchandise evoking wonderment, curiosity, and acquisitiveness put within boys' reach, often quite literally, fun, exciting, new, and unapologetically frivolous experiences. In contrast, the toys available to girls prepared them for the domestic work they would be performing for the rest of their lives (fig. 11.11). Advertisements in girls' magazines dutifully promoted toys offering only inward-looking experiences, such as poetry books, curling irons, tatting kits, and dollhouses. Meanwhile, boys' literature—from comic books to popular science magazines—promoted toys and games that looked outward, toward adventure, the frontier, and anything else that might seize the imagination—things like good luck puzzles, magic cigar cases, Norwegian mice, squirt badges, magnetic tops, and Anarchist Stink Bombs.[57]

More sophisticated jokes and gags were predicated on the prescribed roles of girls and women as domestic and domesticated caretakers. Doubly cruel, they not only reinforced women's inferiority but also exploited their

Figure 11.11. Boys play cards while girls, tending to their baby carriages, look on. Lewis Hine, "Sidewalk Card Game," ca. 1910. Photography Collection, New York Public Library.

submission for a laugh. Women's caretaking sympathies made them susceptible to gags like the false ear bandage and the fake chipped tooth. Likewise, their charge to maintain a clean household provided the fodder for many fake ink spills on fine linens. The missives on handwritten cards were not love letters but literally dirty notes written in soot ink that blackened the fingers. Because an imitation cigarette pretended to obliterate a woman's careful work and ruin her furnishings, it was, apparently, hilarious: "Place it on top of a nice polished table, and see the lady of the house look daggers at you. Or lay it any place where a lighted butt would be likely to cause damage and note the effect" (fig. 11.12a–c).[58]

The humor of novelties was not just "transgressive" and "subversive."[59] It was also mean-spirited and corrosive, used to demean and embarrass, "at the expense" of someone else.[60] In order to act out "infinite aggressions," perpetrators of jokes needed not just gags and pranks but also victims to serve as the "butt" of the joke. There was a "hidden hostility" in this kind of humor, marked only by the perpetrator "being *theoretically* at least the one person present *who does not laugh*."[61] Novelty inventor S. S. Adams did not laugh when he aimed an exploding snake at a client's face. But he got what

2052.
FALSE TOOTH.

Made of china. Many ways in which to use them will readily suggest themselves to the practical joker. Price, postpaid, 10c each, 3 for 25c.

2072.
BANDAGED EAR.

A perfect representation of a bandaged ear which has been badly cut. Will cause you to receive a lot of sympathy. A very good joke. Price, postpaid, 25c.

2064.
MAGIC INK SPOT AND BOTTLE.

Just imagine placing one of these on a clean table cloth, everyone will rush for salt, towels, anything to prevent the new table cloth from being ruined. We suggest this as one of the thousand ways in which this can be used. One of the best fun makers ever gotten up. Price, postpaid, 25c.

Figure 11.12a–c. Women made for easy victims of jokes and pranks. C. J. Felsman, *Novelties, Jokes, Tricks, Puzzles, Magic from All Over the World and Every Where Else*, ca. 1915.

Figure 11.13. S. S. Adams demonstrates his exploding can of snakes to a prospective client in 1939; he is not laughing. By permission of author William V. Rauscher and 1878 Press, Oxford, CT.

he wanted—the victim in retreat, with his mouth fixed in an obligatory face-saving half-smile. This kind of humor was dead serious (fig. 11.13).

It Hurts to Laugh

The deviousness of early novelty goods turned to violence around the First World War, their nihilism laid bare. Centuries before, Pennsylvania Germans made trick boxes concealing spring-loaded snakes whose heads held a sharp nail and were ready to strike when opened; the playful assassins were called "biting boxes." By the late nineteenth century, companies were selling not just updated versions of biting boxes but novelties that snapped, popped, and exploded. Some were surprisingly violent (and many have since been outlawed). A bejeweled Joke Box containing a percussion cap and detonating powder unexpectedly issued "a loud report" when opened.[62] Exploding cigars, among the most iconic of the violent novelties, were initially laced with traces of chemical explosives.[63] Eventually the cigars exploded via mechanical means, though the shock factor remained. A 1908 patent for an improved, spring-loaded version

Figure 11.14. "You will have the grand HA HA on your friend." The Exploding Cigar, Fargo Novelty Co., ca. 1900.

explained, "Cigar will spread out and have the appearance of an old paint brush shortly after being lighted, without danger of the parts flying in the smoker's face" (fig. 11.14).[64]

Inevitably, violence begat violence. By the 1930s novelty producers offered countless cheap exploding things. Available for purchase from the Bengor Products Co. in 1936 were, among other things, Auto Whizz Bangs, Bango Cigarettes, Bango Shooting Devices, Bingo Hand Shakers, Buzzer Letters, Charm Pistols, Explosive Matches, Exploding Cigarettes, Exploding Cigars, Joy Buzzers, Shooting Match Boxes, Plate Lifters, Shooting Books, Shooting Playing Cards, Shooting Jewel Boxes, Shooting Match Stands, and Snap Gum.[65] The spirit of the biting box lived on in snakes that jumped out of everything from fake tulips and boxes of playing cards to jars of strawberry jam and fountain pens.[66] All manner of seemingly real commodities were behaving in unexpected and often transgressive ways, talking—or rather, snapping—back at their owners. The more commodities, the more opportunities for their doppelgängers to behave badly (figs. 11.15, 11.16).

Cruel and violent novelties assaulted all of the senses, unsettling what people understood about sound, feel, and taste. The well-hidden holes of the Dribble Glass (invented in 1909) enabled its liquid contents to "trickle down the chin and shirtfront of anyone who drinks from it. The victims usually think that the fault is their own; they will wipe off their cheeks and invariably will try again and again."[67] It was, apparently, a big hit at the President's Table at the 1912 New York Paper Dealers' Annual Banquet, where it "caused much merriment, and many shirt fronts were soon ready for the laundry."[68] Speartint Chewing Gum turned one's mouth red, while Pepper Candy made it burn. Itching powder brought about an "incessant" itching, which only became worse with scratching. The Auto Whiz Bang caused drivers to think their cars were exploding, delivering a harrowing if "harmless" experience: "Simply attach to spark plug and when 'victim' steps on starter the 'fireworks' begin. It shoots, whistles, screams and shoots again, followed by a big cloud of smoke."[69] (S. S. Adams admitted that "it may be a trifle violent . . . for sensitive souls with delicate nerves."[70]) The Anarchists' Bomb was a "liquid form of a chemical which produces a most horrible odor." Cachoo Powder instigated a "sneezing craze" across the country when Adams introduced it in 1904. Netting $15,000 in its first year alone, the entrepreneur was able to turn "coal dust into gold dust."[71] All of these things were incredibly popular.

Many novelties were domesticated versions of warfare's agents of death and destruction, whether guns, explosives, or poisonous gases.[72] Such was the intimate relationship between violence and gag humor that many novelty technologies were appropriated for warfare and vice versa. The Bingo exploding device, invented in 1907, was a mousetrap-like mechanism designed to set off percussive caps. Considered "much surer than more elaborate devices," it was adapted by the Ordnance Department for road and personnel mines in World War I.[73]

During World War II, British spies incorporated novelties' tricks and subterfuges into covert operations. Agents sprinkled itching powder in their enemies' underwear, condoms, and cans of foot powder. They transported grenades in fake logs realistically crafted of papier-mâché, and concealed explosives in bottles of chianti and rat carcasses. Lumps of fake cow

Figure 11.15. Explosive novelties became all the rage just before World War II. Gellman Bros., *Annual Buyer's Guide Catalog for 1937.*

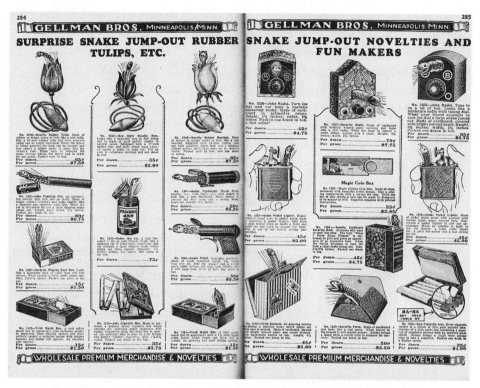

Figure 11.16. The descendants of biting box snakes found new habitats, and surprise snakes jumped from any number of commodities. Gellman Bros., *Annual Buyer's Guide Catalog for 1937.*

manure held hidden messages and bombs.[74] (This is not a log. This is not a rat. This is not poop.)[75] And for decades, the Pentagon has experimented with developing stink bombs.[76] A nonlethal weapon, an effective stink bomb nevertheless would be able to incite "fear, panic, and an overwhelming urge to run away," creating an odor "so repulsive it's truly terrifying." Smell, in fact, is powerfully linked to emotion, which makes the innocuous stink bomb so funny and so fearsome. According to olfactory experts, "unfamiliar smells are more likely to prompt panic than even the nastiest odours [people have] smelled before."[77]

Novelty companies borrowed liberally from wartime technologies, too. The radium used for glow-in-the-dark instrument dials on aircraft and ships was adopted for children's novelty stickers and adults' novelty face creams and, somewhat shockingly, toothpastes. Novelty makers adapted improvements in rubber processing, used for soldiers' waterproof garments, to create more durable Whoopie Cushions, more convincing fake body parts, and more gigantic vermin replicas. The Tear Gas Fountain Pen, which incorporated official police shells, was both novelty and personal defense weapon: "Just point it—release safety trigger and—project a cloud of Blinding TEAR GAS. Instantly Stops, Stuns, and incapacitates the most vicious Man or Beast."[78] Novelties helped domesticate all of this violence. In the field The Bingo was triggering land mines, while at home it was making matchbooks, drink coasters, dinner plates, books, and cigarette cases jump and erupt, "fill[ing] a demand that seemingly goes on forever."[79]

That people needed novelties—the most frivolous frivolities, the crappiest crap—especially in times of duress became quite evident during the Great Depression. By then, leading novelty purveyor Johnson Smith was issuing regular supplements to its 800-page catalog—which people paid for!—to keep up with demand.[80] H. Fishlove's "great laugh-novelty," a miniature toilet bowl affixed to an ashtray, "hit Old Man Depression right between the eyes!"[81] S. S. Adams introduced a product in 1932 that was weird and utterly useless: the Joy Buzzer created a vibrating handshake using a spring-loaded mechanism. A complex article, comprising some thirty springs, gears, pins, vibrating assemblies, punched cases, and covers, it was neither simple nor cheap to produce. Made in America rather than imported, it retailed for fifty cents, equivalent to about two and a half dozen eggs, a pair of silk hose, or five gallons of gas (fig. 11.17).[82]

Although launched during the depths of the Depression, the Joy Buzzer proved to be Adams's most successful novelty, earning him the sobriquets "the Thomas Alva Edison of practical jokesmithing" and "the Henry Ford of his industry." (Ford himself loved the device.)[83] Americans bought an estimated 144,000 Joy Buzzers the first year the item appeared on the market, and more in subsequent years, enabling Adams to not only keep his factory running but maintain his entire workforce at a time when the nation's employment rate approached 25 percent. Over 2.5 million Joy Buzzers were sold domestically in less than a decade, and by 1946 Adams was grossing some $250,000 a year, representing about $2 million in retail sales at four thousand store outlets.[84] Other novelty outfits similarly did well during hard times. The Franco-American Novelty Company was recording $1 million in sales in the mid-1930s, thanks to a brisk trade in Dribble Glasses, Rubber Doughnuts, and three thousand other "articles of nonsense."[85]

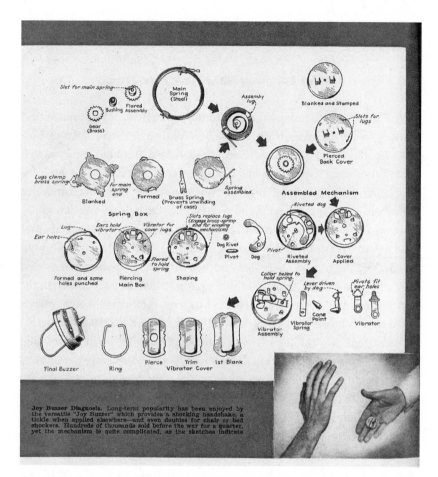

Figure 11.17. The inner workings of the Joy Buzzer. *American Machinist*, August 1946. Reprinted by permission of Informa Media Inc.

In the 1990s the owner of Johnson Smith acknowledged that the success of his company over the years was due mostly to Americans' unwavering enjoyment of disparagement humor: "There's an element of sadism in almost any practical joke. A leveler. My way of bringing you down to my level, or at least pricking your bubble if you're too pompous. To an extent humor is retaliatory."[86] An attorney arguing a product liability case observed that "no kind of cigar is half so funny as the one that explodes very largely and loudly to the person who decided to buy the cigar and give it to somebody else."[87] Neatly eliding humor and aggression, novelty goods were inherently two-faced, complicated, and insincere, the very qualities people loved about them.

Ardent Longings after New Play-Things

Sold alongside stink bombs and Joy Buzzers were lines of "spicy novelties" whose jokes were sexually rather than violently aggressive, such as Bottoms Up glasses, Oola hula girls, and Peek-A-Boo Tumblers. Spicy novelties got their start in America in the middle decades of the nineteenth century, sold by low-profile purveyors of dirty books, erotic prints, condoms, and sex toys.[88] Typical customers were young bachelors in urban areas, who, considering themselves part of the "sporting crowd" thronged cities in search of a decent livelihood and a good time.[89] These tawdry items enabled libertines to fulfill personal gratifications and also show others that they were in the know. A pair of Emblematic Sleeve Buttons offered in the 1870s, for example, incorporated a rebus with pictures of a can, an eye, and a screw surrounded by the letter *U*, which when decoded "can easily be construed into asking a very *pointed* question."[90] One imagines that these were shown off to friends more often than to prospective paramours.

The trade in spicy novelties, imported mostly from France, had been active since before the Civil War, and the war itself only encouraged it. Mail order merchandisers obligingly met the demands of soldiers in the field by offering a full complement of obscene literature and graphic pictures known as "barracks favorites."[91] As young men, soldiers were already predisposed to having a taste for the slightly obscene; away from the prying eyes of relatives and neighbors and with plenty of idle time on their hands, they had new opportunities to indulge in risqué entertainments. Plus, they needed to seek escape from the fighting and the carnage.

Purveyors like Philo's Purchasing Agency were able to offer customers a sprawling selection of goods, which they sent to camps in plain brown wrappers, by express. Soldiers bought not just dirty books and cheap jewelry but X-rated watch papers (disposable linings for the inside of watch cases), "transparent" cards that revealed secret things when held to the light, "mechanicals" of figures engaged in energetic congress, and other similarly titillating things.[92] The boon to novelty sellers became a bane for commanders, however, since spicy novelties threatened to disrupt the discipline and comportment of a good fighting force. Marsena Patrick, provost marshal general of the Army of the Potomac, wearily noted how often he had to confiscate this kind of contraband: "I have seized upon & now hold, large amounts of Bogus Jewelry, Watches, etc. all from the same houses that furnish the vilest of Obscene Books, of which I have made a great haul lately." He burned this "large quantity of Obscene books" a few nights later.[93] He would have to do it again, no doubt, as there was no escaping the primal admixture of sex and death, especially in the guise of a ten-cent dirty gimcrack.

The Comstock Laws of the 1870s put most of the smut peddlers and sellers of X-rated novelties out of business, at least for a time.[94] But by the first decades of the twentieth century, crude and coarse novelties came back. They could be found advertised alongside celebrity profiles in gossip rags, in the pages of pulpy sensational literature like exposés on white slavery, and in the back pages of advice manuals on the art of wooing women. Along with get-rich-quick schemes, readers could send away for "Spicy, piquant entertainment" like Hotsy Totsy Fan Dancers, Bath House Girls, and Naughty Nudies tumblers. "Making a hit with the boys," the ad copy noted, "her innocence disappears and she is revealed in nature's own!"[95] Against the rise of cheap and accessible commodity culture and a nation of hungry consumers, the Comstock laws had no chance.

Like other novelty entrepreneurs, the H. Fishlove Company, founded by Ukrainian émigré Chaim Fishlove (who also marketed Yakity-Yak Teeth), capitalized on selling salacious and sensational novelties to men (who, when younger, might have purchased Girl Traps). The Bottoms Up glass, for instance, was "a knockout" during the mid-1930s. It was a shot glass molded in the shape of a languid woman's naked figure whose curved rear end formed the bottom of the glass; she had to go "Bottoms Up" to be put down.[96] The Oola, "the hottest action novelty," was a rubber bulb shaped like a woman's torso whose breasts inflated when squeezed. The company had to triple its production because nightclubs and conventions were "going wild" over her.[97]

More popular still were gag boxes, which remained sure sellers for decades. Launched in the 1920s by Fishlove, and initially the size of small matchboxes, gag boxes were like three-dimensional mildly raunchy greeting cards (fig. 11.18). The setup to the joke was printed on the box's label; when opened, the box revealed the punch line. For instance, a box illustrated with a silhouette of a man and woman snuggling by moonlight with

Figure 11.18. Gag boxes were three-dimensional versions of raunchy greeting cards, and H. Fishlove made a fortune off them. Tim Tiebout Photography, www.timtiebout.com.

the caption "A Hammock Built for Two" contained a miniature bra. A gag box made exclusively for the 1933 World's Fair was embellished with the label "A Century of Progress." Inside were miniature replicas of a chamber pot ("1833") and a gold toilet ("1933").[98] During World War II, Fishlove's boxes became morale boosters sent by mail to soldiers overseas.[99]

Related novelties, too, helped maintain the gendered milieus that had, by the mid-twentieth century, become fairly well calcified. Enterprises like the Game Room gift shop helped men accessorize and define their domains, whether the golf course, executive suite, or household den. Life Size Pinups could be used as wallpaper for, winkingly, "all men of discernment." Cocktail napkins were imprinted with bawdy cartoons of "sophisticated buffoonery," including the theme of "Sexual Misbehavior in the Human Female" (a "whimsy on Kinsey").[100] While peep show glasses had been around since the 1920s, and became even more popular in the following decades, they continued to enjoy terrific popularity in the postwar era. All the better for men returning home to reassert their authority as successful businessmen and domestic patriarchs by literally putting their hands around women who were, also quite literally, objectified.

There were, needless to say, no humorous equivalents for women consumers. As in the past, the world of humor—especially sexualized humor—was simply not available to them. The Johnson Smith catalog from 1958, for instance, contained only a few products specifically for its female customers: wind chimes, thermal socks, tweezers, and a handful of cheap wedding bands.[101] Just about everything else, from miniature boxing gloves and giant flies to mystery cigarettes and bags of shredded currency, was pitched to boys and men, who spent hours poring over the pages. There were racier items, too, like the Shimmie Minnie, Fancy Dance, Skirt Dancer Bank, and Oui Oui Doll (described as "a very funny item" because water squirted from her nipples).[102] The world of novelties acknowledged women only as sex objects or as victims of pranks. One could even buy a pair of inflatable Life-Like Lady's Legs that "creates a riot when poking out from under the bed, sofa, trunk of auto," turning even women's corpses into hilarious jokes.[103]

Novelty's spectacle of fun and games helped convey deeper messages about power and its close allies—sex, procreation, and regeneration. Commodities of misogyny, spicy novelties were sold alongside violent ones that made victims out of women, but alongside pornography and contraceptives, too. In addition to Snap Gum and Vanishing Watches and Bleeding Fingers and hundreds of other novelties, the Bengor Company, for instance, also sold condoms.[104] Presented on a business card was the "Complete Line" of Jack's in Boston (a.k.a. "The Greatest Blues Chaser"), which included novelties, spicy novelties, and high-grade sanitary rubber goods. The back of the card told a joke suggesting what various girls might say "THE MORNING AFTER."[105] Some novelty companies also became smut publishers, printing cheap risqué booklets like *Bust Humor*, *Breastypes* (by Ches T. Broad), and *Male Orders Taken Here*.[106] Calling them "Stag Fun Packages," Philadelphia's Stag Novelty Co. sold envelopes containing cheaply printed joke cards that worked blue.[107] Novelty distributors routinely placed advertisements in erotic serials like *Spicy Stories*, aimed at men who might also be ordering nude photos from Paris, manhood invigorating treatments, books on sex techniques and eugenics, cards and dice, lonely hearts club memberships, and hair loss treatments.[108]

The boys who devoured jokes and pranks as kids grew up to be men who did the same. The artist who drew the Bazooka Joe comic strip for Topps bubblegum—*the* hot product for boys—also illustrated Tijuana bibles.[109]

S. S. Adams said that his most avid customers were "mature men past thirty." "The biggest repeat business," he remarked, "comes from salesmen and sales executives who are *wild* about joke novelties. Theatrical entertainers, dentists and fraternity men are also incurable addicts."[110] Members of the "lonely crowd," these men exchanged novelties and pranked one another within pseudo-social situations that demanded they show their masculine prowess and skills at mixing pleasure and business. Employing the funny-not-funny language of jokes and gags was a way to demonstrate one's facility with the coded language of power.

Shit Gets Real

Novelties helped people come to grips with their own fallibility. Beyond sex, some, like Whammy Eyes, Red Hot Lips, Goo Goo Teeth, Giant Feet, and Schnozolas, were preoccupied with deviancy. Yet others, making light of bodily functions and the messy corporeality of the human body, were darker still. Horrible Accident Masks and Bleeding Fingers, Ugly Boils and Weepy Eyes, Snoring Machines and Nose Blowers, Poo Poo Cushions, Imitation Vomit, Toilet Ashtrays, and the like were reminders of people's persistent state of decay and inevitable death. It was easier to joke than to talk about these things.

Although in 1930 S. S. Adams rejected the idea of the Whoopie Cushion as "too vulgar" to produce, Americans had actually been buying fart- and shit-themed novelties since at least the late nineteenth century.[111] When sat upon, the bellows hidden in Chair Seat Squawkers and Musical Cushions, for instance, were intended to "make an awful noise and create trouble in the vicinity."[112] Vials of Fart Powder became popular around the same time, as did chinaware knickknacks shaped like toilets and countless things incorporating outhouses. Fake poop—human, canine, avian—had become a reliable novelty staple by the time Adams rejected the Whoopie Cushion.

Plastic vomit came much later—not, presumably, because it suffered from a lack of market demand but because materials technologies had not yet caught up to consumers' incipient needs. Much like exploding cigars and vials of fake ink, efflorescent novelties had to be convincing simulations, which presented logistical challenges. Lifelike plastic vomit, called Whoops, was not brought to market until 1959 because it took that long for manufacturer H. Fishlove to figure out how to create it. Advancements in plastics technologies enabled latex and sponge to be married together to produce just the right kind of texture that blopped into just the right kind of shape. Ingeniously, Fishlove also devised a process by which the liquid rubber, usually white, turned a realistic bile-yellow when it cured. Because workers poured the plastic slurry out of buckets, mimicking the trajectory of real vomit, the shapes of Whoops looked real. Older factory photos show assembly lines of fake vomit laid out to dry, like so many cookies coming out of the oven (fig. 11.19). The point was to make it real enough that it was, as Fishlove's creative team said, "disgusting" and "sick." They thought, in fact, there was "nothing funny about it."[113] But the consuming public disagreed, and from the moment the company introduced Whoops, it was able to sell hundreds of thousands of the putrid things annually. People "just went after it," according to novelty expert Stan Timm. Spin-off products included Plop, "albatross size" fake bird poop, and Glop, fake dog vomit (plate 10). It was funny *because* it was sick.

Yet plastic vomit was really just an old thing in a new form—a novel and material iteration of carnivalesque traditions dating back to the Middle Ages, if not earlier. Postwar American consumers, surrounded as they were with their new-fangled appliances and automobiles, all sleek lines and chrome, may have found something liberating in pedestrian and vulgar

Figure 11.19. Tossing cookies at the H. Fishlove plastic vomit factory. Courtesy of Stan and Mardi Timm.

commodities that evoked more distant and ribald pasts. Rabelais's *Gargantua and Pantagruel*, written in the mid-1500s, vividly depicted the controlled chaos and contained liberation of the carnivalesque—spaces within which rules are suspended and hierarchies of power and status are subverted. Masked and anonymous, participants temporarily shed their identities, freed to behave in peculiar ways.[114] Up is down, down is up; inhibition is exchanged for exhibition, the confining bonds of culture let loose, propriety and civility rendered irrelevant.

Patties of plastic vomit—and vials of scented Fart Powder, Whoopie Cushions, and, for that matter, fake bird crap—were the modern, commodified props that summoned Rabelais's carnivalesque world. His characters ate until they burst, threw feces at one other, and drenched themselves in urine. Their scatalogical humor and "stubborn resistance to reform" appeared again in the English jestbooks of the sixteenth and seventeenth centuries.[115] Never one to shy from robustly crude humor, Benjamin Franklin, too, took great delight in the body's many emanations and excrescences, finding humor in farting, defecating, pissing, and ejaculating.[116] Among other things, he facetiously proposed to the Royal Academy of Brussels in 1781 that it should award a scientific prize to the person who could make the smell of farts not undetectable but more aromatic.[117]

Others also tried to make the best of inconvenient truths. Early Americans obsessively cataloged their excretory processes, since bowel movements and putrescence were indices of wellness and sickness. Elizabeth Drinker's eighteenth-century diaries, for instance, are loaded with references to her household's gastrointestinal issues, and she was perennially experimenting with new curatives for indigestion, constipation, and other complaints. Popular home medical treatises from the time addressed the same issues. When recommended diets failed to work, readers could choose from a number of emetics, laxatives, and suppositories to induce and contain.

The more serious the malady, the more it seemed to lend itself to humorous treatment. Although dysentery was one of the worst diseases afflicting American Civil War soldiers, French tobacco boxes shaped like "human manure" were quite popular among the fighting forces and easily acquired from dealers in dirty books and erotic pictures.[118] By the 1920s people could buy joke matchboxes containing miniature coils of poop fash-

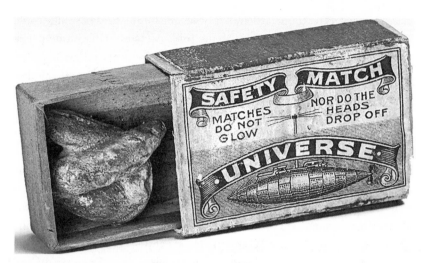

Figure 11.20. Fake poop could be concealed in an innocent box of matches (ca. 1930s). Tim Tiebout Photography, www.timtiebout.com.

ioned of plaster of Paris (fig. 11.20). This was the same era when the market was introduced to the fake dog crap Doggonit, a product one retailer noted was already "too well known to give any details. Very lifelike. Individually boxed." Another explained that if placed on the floor, it would "creat[e] a riot of laughs."[119] By the mid-twentieth century some of H. Fishlove's gag boxes incorporated fake turds, too (see fig. 11.18).

Despite the persistent efforts of nineteenth-century etiquette mavens to civilize and reform the unwashed masses, Americans still exhibited a particular enthusiasm for humor that was dirty, raw, and crude. That scatology was so thoroughly banned from the parlor meant it was an ever-present threat requiring vigilant policing. Foulness that was contained through rules and constraints meant foulness that could not be completely conquered. Anxieties about the improprieties of their more animal selves continued to vex the Victorians. The extensive complement of rules pertaining to eating, drinking, sitting, standing, walking, and just about everything else could only temporarily hold people's baser selves at bay: the body's weaknesses might reveal themselves at any time, through mouths, nostrils, anuses.

Seeking to attain a state of "refinement," nineteenth-century middle-class Americans restrained their carnivalesque impulses as best they could in an attempt to deny the more shameful aspects of being human. The 1870 *Bazar Book of Decorum,* one of countless titles in the etiquette genre, discoursed on ear pulling ("ungracious and dangerous"), hangnails, clumsy drinking, "wriggling" noses, and more. The typically unrefined fellow, the book observed, "generally daubs himself with soup and grease," with a napkin stuck through his buttonhole. Guzzling his drink, he "coughs in his glass and besprinkles the company." His "strange gestures" included not just nose picking but examining the result, "so as to make the company sick."[120] Novelty sellers and general merchandisers responded to such anxieties by catering to people's inner devils and angels both, appealing to the conflicting yet simultaneous impulses of control and release: one person's shame was another's entertainment. Their catalogs offered, often cheek by jowl, mustache protectors and false mustaches, "fine" linen towels and the "trumpet in handkerchief" novelty (a noisy nose blower), gold-plated Spencerian pens and disappearing ink, and even dueling cheap book titles like *The Etiquette of Courtship* and *A Cart-Load of Fun.*[121]

Victorians found bodily transformations—marked most clearly by eating, evacuation, and sex—incredibly problematic not only because they reminded them of their base animality but also, and more to the point, because they provided clear evidence that bodies cannot be controlled. We can only ever master our physical selves temporarily, since bodies are in a perpetual state of change and corruption that leads, inevitably, to death itself. As Franklin certainly would have appreciated, the beauty of novelties like stink bombs, Fart Powder, Whoopee Cushions, and fake turds—*crap in its most literal form*—is that these objects made visible, often in gleeful and unapologetic ways, the very things people tried hardest to control, to obscure, and to deny.

The Ridiculousness of Being Human

Because of their seeming frivolity, ephemerality, and uselessness, novelties have, over time, helped provide cover for explorations of these most profound issues, articulated in the language of commodities within a larger conversation about modern consumer culture. Victorian conventions persisted, and people, understandably, remained embarrassed about how—

whether belching, farting, or vomiting—the body betrayed. This was (and is) an essential part of being human.

Perhaps more than any other commodities, novelties have enabled people to come face to face with the otherwise taboo, feared, and banished. Fake poop and plastic vomit are (not) funny. As folklorist Gershon Legman observed, scatological jokes can evoke "anger, terror, shock, offense, or laughter—that is to say, *humor*." Jokes having to do with excrement or vomit are "always," according to Legman, an "assault" on the receiver of the joke—not because the jokes are shocking in the way that explosive pranks are but because they force people to confront their bodily effluvia: "The excretory acts themselves are of daily familiarity, and only their appearance in public is taboo," he writes.[122] Body-themed novelties have simply given the taboos cover. The Rooter (or Cold in the Head), for example, "imitates the blowing of the nose exactly, except that the noise is magnified at least a dozen times, sound[ing] like the bass horn in a German band."[123] Bleeding Fingers, featuring gauze bandages "with big drops of blood," not only gross people out but turn sympathy into "howls of laughter when the joke is out."[124] The Doggonit Floor Novelty can "create a riot of laughs."[125] Johnson Smith promoted Whoops fake vomit to its customers in this way:

> Amazingly realistic PUKE! Looks like someone was SICK, SICK, SICK! almost turns your stomach to use as a joke, it's so realistic. Made of plastic. The "gloppiest" look. Place by baby, dog, dinner table or pretend you've been sick. Most revolting, dirtiest trick we've seen yet. (Created a riot when we tested it!)[126]

The best gross-out humor has the power to not only offend the sensibilities but elicit a physical response. Real enough, products like Whoops promised to "almost turn your stomach." In a good way. The humor lay in the frisson of nausea and surprise: gag reflexes were the physical response of turning the shame and vulnerability of private upset into public spectacle.

For a time, Americans couldn't seem to get enough of the latest novelties. By the 1940s, people could buy their x-ray specs, striptease tumblers, gag boxes, and rubber chickens at any one of four thousand shops across the country.[127] Novelty purveyor Spencer Gifts, which began as a mail order enterprise, opened its flagship chain of retail stores, some 450 of them, in

1963. Fishlove was selling millions of gag boxes a year in the 1960s. By the early 2000s Sea Monkeys had become a multimillion-dollar enterprise, recording sales of $3.4 million in 2006 alone.[128] Ever popular, some five hundred thousand Chia Pets were sold in 2007.[129] (A recent satirical headline read, "Chia Pets Often Euthanized after Novelty Wears Off."[130])

Today, Americans live mostly in a post-novelties world, perhaps because, as one of the few material things that connect us to vestiges of our premodern, corporeal selves, there is no room for them, and for deep thoughts about mortality, in the hyperreal world of air-brushed, eternal youth. People have not turned away from novelty itself; in fact, they seek it out more habitually than ever before. The virtual world at once satisfies and creates perpetual opportunities for the "ardent longings after new playthings." It is an easy and palatable escape from the human condition, a way to avoid thinking existential thoughts. More practically, personal humiliations now occur virtually, online, and quite publicly rather than within fairly closed and cohesive groups; the pranks made possible by novelty goods seem utterly quaint and "perfectly harmless" by comparison.

If, as I have argued, we can identify crappy goods by their inherent dishonesty, insincerity, and cynicism, then what should we make of novelties? Because they seem to be frivolous things, skimming only the surface, they actually help us plumb the depths. By ridiculing our bodily functions, they bring out in the open what we most urgently try to conceal. By making a mockery of sex and virility, they lay bare the status contests that sometimes have life-and-death consequences. By making light of violence, they reveal how humor and aggression are integral parts of each other. Being the most silly, throwaway, ephemeral, and stupid things, novelty goods are ultimately the most profound. When asked who created fake dog poop, Joseph (Bud) Adams, president of the S. S. Adams Company, responded, "I don't know who. How far do dogs go back?"[131] When I asked a friend when he thought we transitioned from a society that didn't need plastic vomit to one that did, he replied, "Wendy, we have *always* needed plastic vomit." He had a point.

A WORLD MADE OF CRAP

Crap is not just a part of America's past but is alive and well in the present and will remain with us for a long time in the future. We live in a perpetual state of material satiation. And yet there is always more to want and need. Take, for instance, The Flair Hair Visor, which is a golf visor with a toupee attached to it. The SkyMall catalog touted it as "an amusing gift—especially for those who have everything." (Except, presumably, hair.) Shouldn't the perfect gift for those "who have everything" be nothing? (fig. 12.1).

Those who have everything can always have something else, especially something crappy. The Flair Hair Visor perfectly embodies our relationship with consumer culture and late capitalism. It has no purpose and no reason for being. It is neither high concept nor well made. It is probably not even a welcome gift.

We might laugh, as I often do, at the bald absurdities and proud stupidities of crappy stuff: the countless Baconizers and Miracle Pants Fasteners and Winged Messenger figurines. It takes creativity to generate new kinds of crap in this age of total surplus. True visionaries can imagine utterly new things, whether the mustache cups that hit the market in the late nineteenth century or something more recent, like the Menorahment, the Christmas tree topper shaped like the Star of David. The ingenious are able to figure out how to make these things—integrating crappiness into their very designs and maximizing economies of scale to produce as cheaply as possible. Not only making but convincingly marketing crap can be a challenging proposition.

But however much we might take perverse pleasure in the absurdity of the latest bullshit gadget or the surreal excesses of the newest lines of mass-produced collectibles, or are simply "happy with crappy" because it is

311

It's a laugh a minute with this spiked hair hat

Figure 12.1. What you get for the "person who has everything." The Flair Hair Visor, SkyMall catalog, 2011.

cheap, crap has always come at a cost.[1] And so we need to ask ourselves as individuals and as a society: Is it all worth it? What would it be like to live in a crapless world? Is it a world we can even imagine? Consider:

Manufacturing processes involving plastics—the preferred material of twenty-first-century crap—pollute the environment by releasing toxic chemicals into the air and waterways. Meant to be disposable, merchandise made with cheap plastics breaks down more readily but never completely disappears. Not long ago, Americans were recyclers and repairers and reusers. Now we are shoppers and wasters. Older crap was at least made out of materials that could be repurposed—metal, wood, glass, paper, bone, and rubber.[2] Newer crap is at once more ephemeral and more long-lived, made as it is from synthetics that survive in perpetuity but cannot be refashioned into anything useful. The worthless carcasses of so many crappy goods clot landfills and contribute to the metastasizing trash gyres cycling in our oceans. The Great Pacific Garbage Patch, for example, is made up of 1.8 trillion pieces of plastic and weighs about 88,000 tons. It is the largest of five such gyres, which, ever growing, consist not just of plastic packaging but also of crap—Boobie Beer Covers, Banana Dogs, and yes, even Flair Hair Visors swirling around in the sea, forever.[3] They will not degrade, as compost does, to contribute to the cycle of life; instead, choking ecospheres, they hasten death. This is one legacy of crap.

Labor exploitation is another cost that crap exacts. Cheap stuff isn't just made of dubious low-cost materials; producers and sellers pay workers as little as possible in order to realize slim profit margins. Most of the time this work is performed overseas, in dismal conditions—the exploitation a long and now familiar story. By the second half of the twentieth century the centers of crap production had shifted from Germany and Japan to Taiwan and, by the 1980s, China, where workers were earning, on average, $45 to $165 a month.[4] By the early twenty-first century crap had remade China in every possible way. Seemingly overnight, farmland once used to grow crops to feed families had become "instant cities" organized and populated solely to feed the insatiable markets for export goods, much of it crap destined for American consumers.[5] These "factory zones" are rural areas that have been converted into centers of commerce supporting factories, workers' housing, and shopping malls for residents and international traders alike. In one town alone, "there's a scarf district, a plastic bag market, an avenue where every shop sells elastic. If you're burned out on buttons," noted one observer, "take a stroll down Binwang Zipper Professional Street."[6]

To make all this stuff, workers must be "willing to eat bitterness."[7] Willing or not, by 2010 Chinese workers were pivotal to turning the country into the world's largest exporter of goods—more than $1.5 trillion worth, as much as 19 percent of the global export market, up from a mere 2 percent in 1998 and nearly zero in the late 1970s.[8] All because it can produce fast and cheap. In fact, China's export prices have typically been about 60 percent lower than those of other developed economies.[9]

Retailers have continued to "gratify the morbid love of cheapness" shared by most consumers. Early variety stores morphed into independent fixed-price stores, presaging the larger chains, which were better capitalized, centrally organized, and able to consolidate their power to command larger market shares. Some of the independents sued the chains for being monopolies, and some localities tried to levy taxes to help smaller stores remain competitive, but most resistance was futile.[10] To cite but one telling statistic, a congressional study of thirty small towns between 1926 and 1931 found that the number of chain stores increased by almost 90 percent, while independents fell by nearly 8 percent.[11]

The legacies are not hard to see, since we are reminded of them every day. Some crap purveyors have, for sure, suffered from changing tastes, markets, and sensibilities. Claims of racism, sexism, and immorality have tamped down the popularity of certain novelty goods. Consumers have boycotted Spencer's, for instance, because it carried anti-Irish t-shirts and anti-Arab Halloween masks and casually sold "adult-oriented" goods alongside those for kids. Some collectibles, too, have become less popular: Goebel, for one, stopped making Hummel figurines in 2008, "due to a steep decline in sales."[12] And the venerable SkyMall catalog, which since 1990 had been entertaining air travelers with pages of crappy merchandise including life-size Yeti replicas and miracle head massagers, ceased publication in 2015.[13]

But others in the crap trade have thrived. Like earlier chain retailers, Walmart, with its big boxes of crap, has crowded out smaller mom-and-pop general merchandise stores (and hardware stores, and beauty salons, and . . .) in both small towns and big cities. The same for dollar store chains. Catering specifically to low-end customers when they were first established in the late 1950s, they have been embraced by all American consumers. By 2004 two in three households were regularly shopping at these retailers, which proudly sold cheap.[14] In the first decades of the twenty-first century, all major dollar store chains saw steady increases in sales. In FY2006, for example, Dollar Tree recorded annual revenues of $3.9 billion, up 17 percent from the year before; by 2015 this figure had risen to $8.6 billion.[15] Family Dollar's revenues grew from $6.8 billion in FY2007 to $10.4 billion just seven years later.[16] That's a lot of crap.

A significant percentage of the stock on discount retailers' shelves is the crappiest crap possible—much of it from China.[17] Like the flammable celluloid flowers of decades past and the sharp plastic toys that also fell apart, a lot of newer merchandise is not just crappy but harmful. In recent years various lines of cheap merchandise have been recalled: "assorted metal jewelry" and plastic Halloween pails with unacceptably high levels of lead; ceramic heaters posing fire hazards; die-cast metal toys with "laceration hazards"; remote-controlled tanks that were potentially "burn hazards"; dart gun sets (linked to two deaths) that were "aspiration hazards"; lollipops with "the possible presence of metal fibers or flakes"; pet food contaminated with the plastic melamine; retractable dog leashes that "posed serious risk of injury"; toy guns that were choking hazards.[18] As in the past, we can continue to exercise our freedom "to be ruined in our own way," an ethos that has become, apparently, as American as prepackaged apple pie in a dollar store freezer case.

Crap has thus remade not just economies and societies overseas but the United States' as well. Once sturdy and vibrant manufacturing towns built on companies employing higher-paid workers producing higher-quality products have now closed up shop; many have become dystopic homes to heroin dens and meth labs.[19] Workers still employed in domestic man-ufacturing have seen their wages stagnate, if not decline, due to foreign competition, the waning influence of unions, and other consequences of outsourcing.[20] People who used to be their own bosses now work for oth-ers for little over minimum wage and often rely on public assistance to fill in the gaps. The irony, of course, is that they can only afford to shop in the same "Everyday Low Price" retail marts that have had a hand in their exploitation.[21] Employees have taken legal action against Dollar Tree, Fam-ily Dollar, and other retailers of crap for wage violations and prohibiting workers from taking meal and bathroom breaks.[22] Many of these things, to be sure, are related to broader, macroeconomic shifts in global commerce. But imagine how systems of exploitation—from sourcing to making to dis-tributing to selling—might be disrupted if we could think and live beyond the Crap Industrial Complex.

Indeed, the logic of crap can, in many ways, be understood as a cycle of degradation, from how things are made to the stuff itself. In order to churn out things fast and cheap, workers have to be exploited, which is why many crap producers are based in developing nations or in officially designated factory zones that can skirt the provisions of international treaties govern-ing labor conditions. The lives of poorer and powerless workers—whether women and children in Nuremberg who fashioned cheap dolls, Japanese families who wove wicker baskets for dime stores, the young boys whirling and dipping Staffordshire pottery, or, more recently, the Filipino teenag-ers crafting Precious Moments figurines—have been put in the service of making stuff for more privileged consumers. That their labor results in crappy rather than nice things makes their exploitation more degrading still. In turn, low-paid workers in the West are then put in the service of selling that crap—stocking shelves and tending cash registers for big-box retailers. Often, too, crappy goods themselves carry overt messages of deg-radation, whether novelties' racist caricatures like Sambo wigs and over-sized Jewish rubber noses or sexually demeaning merchandise like boob ashtrays and striptease tumblers. Our consumption of crap only incen-tivizes the many layers in this system of degradation, obliterating the less powerful by objectifying them or erasing their labor altogether.

There are also the unquantifiable and more abstract costs exacted by the ubiquity of crap. One of the key ways we communicate today is through the language of goods, an often complex system of signs that are uncon-

sciously "read"—the semiotics of objects. The fashions in clothing and furnishings are the signifiers with which we are the most conversant. But just
about every material object carries some kind of abstract meaning that,
internalized, we come to know and understand.

This, too, is not a new practice. In the nineteenth century, Americans,
who increasingly found themselves surrounded by strangers—and new,
ever more affordable market goods—came to rely upon an evolving language of things to help them better understand others' identities and to
shape their own. Through what they wore, possessed, and displayed, people
communicated aspects of themselves such as class, ethnicity, occupation,
and even religion. Belongings, too, increasingly signaled membership in
specific cohorts, serving as a way to forge identity among, and establish status within, particular peer groups: I am like you, I am not like them; I am
your superior, I am your inferior. Over time—and especially with the rise
of mass merchandisers—objects became even more common, and necessary, signs of status and identity.[23] The consumer revolution and explosion
of available goods made this language all the more sophisticated and nuanced. Engaging in the world of things and speaking its material language
became necessary and imperative. People judged you and your things—
and treated you accordingly—whether or not you did the same.

Today, this language of goods is the language of advanced capitalism:
there is no way to speak outside of it or beyond it or even critically of it
without invoking capitalism itself. It has become that totalizing. Today's
consumers actually embrace, quite enthusiastically, things like branded
merchandise, becoming walking advertisements for multinational corpo-
rations, making them part of us. To many, it matters—*a lot*—what kind of
animal is embroidered on a shirt, the stitching pattern on the pockets of a
pair of jeans, the logo imprinted on a purse.

What happens, then, when the language of goods is not just the language of commodity capitalism but a dialect that is particularly cynical,
dishonest, and crappy? We assume our clothes, products of fast fashion,
will not last very long. We will throw stuff away, all torn seams and broken zippers, buying new without thinking. Meant to be worn just once,
some clothes cannot even be laundered. The same goes for our furniture.
IKEA items are notoriously crappy—made, often, not of wood but of easily
compromised particleboard that, once assembled, cannot be taken apart
and put back together without losing some of its integrity. Having actually
killed people, they are also "bad goods."[24]

Even goods we might consider "good goods" are often crap. Today's "artware"—like blown glass vases found in giftware "galleries," for instance— is yet another fiction of late capitalism. People tend to look for unique pieces that bear the markings of an artist's hand, attesting to personal vi- sion and individual creativity. But these objects are often mass-produced overseas, generated by machines able to add intentionally irregular tex- tures in order to mimic the unique and valued traces of human labor. Sig- natures, too, are turned into empty signifiers, merely brands stamped by machine or hand-painted by line workers who do nothing else all day.[25] And artists' unique conceptions become nothing more than commodi- fied "looks"; their minor variations, products of pieceworkers, make them "unique" and "artistic": the same only different.[26]

Handcraftsmanship is big business. No matter how supposedly unique, distinctive, distressed, or handmade, giftware is—and always has been—a product of industrialization and globalization. Today, there are gift indus- try trade shows all over the world, from Hong Kong to Leipzig, Budapest to Milan, representing the Gift and Home Trade Association, the Giftware Association, the Gift Basket Association, the International Gift and Home Furnishings Association, the Balloon Council, the Indian Arts and Crafts Association, the North Staffordshire Table and Giftware Sector, and many, many others.[27] Also supporting the markets in "crafts retailing" and whole- saling are a number of trade publications, including *Gifts & Decorative Accessories*, *Giftware News*, *Gift Beat*, *Souvenirs*, and *Gifts and Novelties*. Gift shop sales in the United States alone were estimated at some $16.7 bil- lion in 2017: $51 for every man, woman, and child. According to one trade observer, the item of handcrafted giftware "expresses the moral, ethical, value system of the craftsperson." But that is actually not the case, since the giftware artist's success relies on the value systems of the market. What the giftware industry is selling ("artware" commodities) and what giftware consumers think they are buying (fine art) are two different things.[28]

Most of the gifts we give each other are chosen from a sea of mass- produced and shoddy things. Not only that, but, perversely, we are more incentivized to make good purchases than to do good works, since we will be rewarded with crap. The practice of distributing advertising special- ties has become even more promiscuous, and obligatory, than in the past; hence the abundance of swag—"stuff we all get." According to the Promo- tional Products Association International, the manufacture, importation, and distribution of free stuff was a $20.8 *trillion* industry in 2015, all built upon logoed ballpoint pens, beer coozies, visors, mugs, key chains, memo pads, stickers, tote bags, desk calendars, tape measures, lunch boxes, bookmarks, and even first aid kits.[29] The Advertising Specialty Institute, linking suppliers with distributors of logoed stuff, has some 23,500 members who traffic in over 950,000 different kinds of crap.[30]

Accepting swag does not make us worthy recipients of gifts but rather promotional agents who act as mobile advertisers carrying the world of commerce on our persons. In 2017 half of the thousand people who took an online survey reported having a promotional item with them at any given time. Advertisers do not even care that 80 percent of recipients reject their gifts by giving them away, because that simply disseminates their messages more widely, "further[ing] the brand's reach," at no additional cost. And we have still been validated by receiving a gift, even if it is cheap, mass-produced, insincere, and something we don't want. Studies have shown that getting free stuff makes us feel "happy," "thankful," "appreciated," and "special."[31] Do we deserve nothing better?

Our goodwill, apparently, is easily bought, because we continue to be "pleased," in the words of Henry Bunting, "out of all proportion to the intrinsic value of the article."[32] The Pavlovian response to free stuff, which can establish lifelong commercial loyalties and relationships, is, ideally, concretized when consumers are young.[33] The number of "child-directed premium expenditures" in the first decades of the twenty-first century was in the billions, to say nothing of "advergaming," online branded entertainment experiences in which kids can earn virtual points to redeem for real-life stuff.[34] At events like the Baseball Think Tank held in Las Vegas each year, executives from Major League Baseball teams gather to learn about "product trends for in-stadium giveaways and ticket renewal programs." Giving away "free" bobblehead figures and other "special" merchandise not only attracts more people to games but enables teams to implement "dynamic pricing," meaning charging more for tickets, and to get excited fans to come to games early, which means, also, selling more concessions.[35]

More than taking advantage of our lizard brains' desire for free stuff, the idea of getting an emotional reward in such an abjectly commercial way has changed the practice of gifting itself. It is falsely nostalgic to believe that the cultures of sentiment and profit were ever completely separate. But the thriving trade in gift-commodities, business gifts, advertising specialties, swag—call these things what you will—suggests something significant has changed. It has become harder to distinguish gifts from commodities and, hence, to determine what their exchange actually means. For companies, business gifts help "place the objective of profitability into a more socially acceptable role," disguising the economic imperative as a present or reward.[36] According to one scholar, one of the "consequences" of using commodities as gifts (rather than, say, making things ourselves) is that "messages sent in the form of gifts are in danger of being lost amidst the constant background 'noise' of things acquired through market transactions." Crappy free stuff is both the gift and the noise.[37]

There are ethical implications, too. What influence might business gifts, even crappy ones, have on important decision-making processes? Knowing that personal exposure positively influences rates of prescriptions, drug companies spend a significant percentage of their advertising budgets — tens of millions of dollars — on crap like pens, visors, coffee mugs, and umbrellas for their reps to give out to doctors.[38] They've even coined a new term for it, "relationship marketing."[39]

"Relationship marketing" is one of countless euphemisms conceived of and necessitated by late capitalism: the production and consumption of crap has reshaped not only our material world but also the way we talk about it. Crappy stuff needs the scaffolding of promotional bullshit to transform it from cheap and useless into something worthwhile, elegantly and efficiently eliding the celebratory and the cynical. Take, for instance, gadgets, which find purchase in the marketplace due to their promotional hype more than their actual function. Veg-O-Matics, Seal-A-Meals, GLH (Good Looking Hair, a.k.a. Hair in a Can), and other devices promise fast, easy, magical, and often entertaining transformations, whether facilitating slices of wafer-thin vegetables or making baldness disappear with a press of a button. The Flowbee haircutting "system" combines hair clippers with a vacuum cleaner, mitigating the mess of errant trimmings, but "victims" of the device "might look like they lost a fight with a ceiling fan." That Be-Dazzlers would "turn tacky clothes into rhinestone extravaganzas" could be considered either a feature or a bug, depending. Ginsu knives claimed to be "able to cut things kitchen cutlery is rarely called upon to cut." Channeling the Electric Spinal Belt of a century earlier, which claimed to alleviate liver complaints, dyspepsia, sciatica, rheumatism, kidney disease, and female complaints by sending "galvanic currents of electricity directly along the spinal cord" is Dr. Ho's Muscle Massage System. It, too, provides electrical muscle stimulation promising to cure "such ailments as migraines, neck pain, foot pain, tendonitis, sciatica, bursitis, menstrual cramps, carpal tunnel syndrome, stress-related insomnia, and many, many more." Users have to be warned not to use it on open wounds.[40]

Faced with the inevitable failure of most gadgets, consumers have not sworn off them but have continued to buy into new forms of hype. Disappointing performances merely encourage us to buy something else, to pin our hopes on the next newest-fangled thing, whether the New Rembrandt Potato Peeler of the past or the Copper Chef XL nonstick pan ("Can make you a great chef") of today. Deviously effective, the puffery has convinced us that gadgets and other forms of crap haven't failed us, but that we have failed them.

But crap *has* failed us, over a long period of time and in myriad ways. The "cheapening mania" has taken its toll: we can't even get decent crap anymore. Plastic vomit, containing multitudes, provides an apt metaphor for our degraded condition. For decades a reliable staple of the novelty market, it is now too crappy to be any good. Online reviewers have scathing words for the latest versions, which are now made in China: "Very disappointing. It does not look at ALL like throw up." "Fake looking fake vomit." "Terrible product. . . . Do not waste your money. Junk." "Would only fool a blind person." "Not as natural as one would hope."[41]

What *would* we hope of crap? Should we actually expect it to provide us with anything other than inferior versions of inferior things? After all, we have embraced this degraded material world, sometimes knowingly, sometimes not. The things we need to live our lives—to do our work, to express ourselves, to understand who we are, and to forge relationships with others—are fundamentally cheap and alienating. Paradoxically, we are impoverished by this surfeit of stuff. Our crappy world is populated with people who are not just connoisseurs of plastic vomit but lovers of Breast Mugs and TruckNutz and Furniture Feet and Huggles and Hamilton Collection figurines. It is as rich in variety and novelty as it is poor in sincerity and gravitas.

Does our impoverished world of goods mean that our ideas and our sentiments are also impoverished? Can we feel sincerity and find meaning and express true love in a world built of nothing but what-nots, thingums, and Cheap Jack? Have we ourselves become crappy?

ACKNOWLEDGMENTS

I have lost count of the number of times, to my great delight, people have told me, "I saw this piece of crap and thought of you." These have been longtime friends and new acquaintances, librarians and curators, colleagues and distant correspondents—all of whom have embraced the spirit of this project. The fact that crap unites us all should be evident in the expansive number of collaborators I have the opportunity to thank here.

This project began as a rather harebrained idea cooked up one evening after a long day at OAH (or was it AHA) between my friend and former editor, Robert Devens, and me, as we careered down the Dan Ryan expressway, his kids singing alt rock standards behind us in their car seats. A fittingly surreal moment. I'm not sure if he meant for me to take crap this seriously, but I did.

Over time, as the book grew, it gained support from and came to rely on an increasingly far-flung network of people and institutions. Those include, first, the many library, archives, and museum professionals at research institutions who are essential to what we do. In the early stage of the project, Marc Brodsky, Reference Archivist at Virginia Tech, provided important information about Civil War-era prize packages; and throughout, Connie King, Chief of Reference and my former colleague at the Library Company, has sent me terrific examples of crappy things from the nineteenth century that otherwise would have escaped my notice. In addition to fielding often bizarre questions about subject headings for novelty goods and vintage synonyms for giftware, Julie Still, my friend and Reference Librarian at the Rutgers-Camden Robeson Library, helped me access analog sources in the digital age, reacquainting me with old but still useful friends, the *Readers' Guide to Periodical Literature* and *Social Sciences* indexes. Glenn Sandberg, Supervisor of Interlibrary Loan at Rutgers, managed to get microfilm copies of the Kresge Papers from Ann Arbor to Camden. Archivist Julie Rossi made my trip to Rochester's Strong Museum to consult the J. Edwin Rich Papers efficient and productive. At the Indiana Historical Society, Theresa Koenigsknecht helped me access the business records for Kipp Brothers, a long-standing purveyor of novelty goods and carnival prizes. Judith Farrar, Associate Librarian, Archives and Special Collections, UMass Dartmouth, generously digitized the E. P. Charleton Papers, which shed light on the management of early twentieth-century dime stores.

Residencies at three institutions, funded by research fellowships, were particularly fruitful thanks to assistance provided by their staff. At the Hagley Library, Max Moeller, Curator of Published Collections, and Linda Gross, Reference Librarian, helped me locate materials related to the early novelties trade, the technologies of gadgetry, and the twentieth-century commerce of Japanese goods. My work at the American Antiquarian Society, particularly under the guidance of Gigi Barnhill, Nan Wolverton, and Thomas Knoles, helped flesh out the early history of crap in America. And the amazing people at Duke University's Hartman Center for Sales, Advertising, and Marketing History contributed in myriad ways, making my research trip more successful—and much more enjoyable—than I could have imagined. Among others, Trudi Abel, Research Services Archivist, Elizabeth Dunn, Research Services Librarian, and Joshua Larkin Rowley, Reference Archivist, embraced my project and happily pulled all sorts of stuff for me, from boxes containing documents related to the Franklin Mint's Princess Diana dolls to vintage videos of gadget infomercials.

Thanks to my friends in the rare book and ephemera world, I did not have to travel far to do a lot of my research; I can also confess to the dubious distinction of possessing, no doubt, one of the most extensive crap-related research collections. The first things I acquired from Carol and Dean Kamin were trade catalogs for party stores and magic shops that showed novelty goods in action; many of these images appear here. Since our first meeting, they have become not just enthusiastic supporters of the book but de facto research assistants, finding obscure yet invaluable ephemera, from postcards and billheads to catalogs and pamphlets. Another essential ally has been Peter Masi, bookseller extraordinaire, who has located all manner of crapological material that has informed every chapter: weird, strange, and really oddball stuff, including books on toilet humor, vintage gadget catalogs, and manuals for traveling sales agents. Curators of the commercial world supplying much of the archival record that underpins this book, they have made an invaluable contribution. I thank them not only for their stuff but for their good company.

I am also grateful for the support I have received from others in the commercial world: in particular, Mr. Ash, proprietor of his eponymous Magic Shop in Chicago, and Douglas Robinson, owner of Monarch Novelties in Washington, DC. Mr. Ash unearthed for me vintage jokes and gags from his deep inventory, and Mr. Robinson schooled me in the arcana of early carnival prizes and gaffed games. They are among the last of their breed, and I wish them many years of good health and continued success.

Fodder to think with, our world of crap is fundamentally material and can't be completely described or understood through words alone. Therefore, the images here are as essential as the text; *Crap* would not have been possible without the efforts and expertise of the dedicated professionals who helped lead me to essential images and assisted in providing publishable-quality versions. In addition to the people above, they include Chief Curator W. Douglas McCombs and Museum Administrator Jessica Lux, Albany Institute; Marie Lamoureaux, Rights and Reproductions Manager, American Antiquarian Society; Robert Zinck, Photographer, Widener Library, Harvard; Caroline Hayden, Rights and Reproductions, Historical Society of Pennsylvania; Concetta Barbera and Ann McShane, the Rights and Reproductions team at the Library Company of Philadelphia; A. Nicholas Powers, Curator of Collections at the Museum of the Shenandoah Valley; Andrea Ko, Registrar, and William Peniston, Librarian/Archivist, Newark Museum; David Kuzma, Library Associate, Special Collections, Dean Meister, Annex Branch Manager, and Jen Reiber, Resource Sharing Coordinator, Rutgers University Libraries; and Amanda J. Hanson, Museum Supervisor at the Skokie Heritage Museum.

The more exotic images found their way to these pages more circuitously, tracked to obscure sources with the assistance of several people who generously made introductions, chased cold trails, and offered parts of their own private collections in order to help me out. Lisa Hix of *Collectors Weekly* put me in touch with kindred spirits Mardi and Stan Timm, the most knowledgeable experts on novelty goods. We have them to thank for the image of the plastic vomit assembly line, which comes from their extensive personal collection. I have appreciated their support and generosity. Mark Newgarden helped me locate William V. Rauscher, from whose seminal book on S. S. Adams comes the picture of the iconic novelty inventor deploying his exploding sausage, yet another essential image. I am deeply grateful for Mr. Rauscher's enthusiasm as well. Amanda Bock, Assistant Curator of Photographs at the Philadelphia Museum of Art, was able to pin down biographical details about photographer Arthur Gerlach, whose works appeared in early issues of *Fortune* magazine and have since, unfortunately, faded into relative obscurity. I'm thankful for her efforts on my behalf and am happy to give Gerlach's work the exposure it deserves. And Tim Tiebout, master photographer and man of good humor whose skills are usually put toward documenting fine arts masterpieces, gamely worked to achieve the perfect shots of Precious Moments figurines and fake dog barf. A generous subvention grant from Rutgers University has made the inclusion of all these terrific images possible.

I am fortunate to inhabit several scholarly communities that have over the years encouraged me and my work and provided invaluable feedback on chapter drafts. These include the McNeil Center for Early American Studies, with encouragement from Dan Richter in particular; the Society for the Historians of the Early American Republic; and the Business History Conference, with a special shout-out to Roger Horowitz, who also invited me to present some of this work at the Hagley Library's Research Seminar. At our own Lees Seminar at Rutgers-Camden, Arwen Mohun thoughtfully critiqued an early version of the project. And I have had the privilege of presenting my work at two conferences organized by graduate students, Virginia Tech's annual Brian Bertoti Conference and the University of Delaware's Center for Material Culture Emerging Scholars Symposium.

My colleagues in the History Department at Rutgers-Camden have not only provided incisive critiques about various aspects of this book but have allowed me to talk crap during many a happy hour. This has been especially true of fellow material culture scholars in our orbit, including Nicole Belolan, Kim Martin, David Nescior, and Matt White. I am especially indebted to Sharon Smith, our Administrative Assistant at Rutgers: no problem has been too big or too small for her to make disappear, as if by magic.

Special thanks also go to the anonymous reviewers of the manuscript, whose feedback helped me clarify my argument and organize my thoughts.

I am also grateful to the team at Chicago, including my editor, Tim Mennel, his stalwart assistant, Susannah Marie Engstrom, and eagle-eyed copy editor India Cooper.

There are many others who deserve thanks for being in my corner over the years. I am especially indebted to Susan Fermer, my favorite high school teacher, who exposed me to the satisfactions of research and writing. My close friends Sharon Hildebrand, Jennifer Woods Rosner, and Sarah Weatherwax have been particularly supportive and encouraging. Others, too, I am grateful to have in my life, including Jerry Bedenk, Bruce Compton, Paul Erickson, Josh Greenberg, Kasey Grier, the late William Helfand (always good for an animated conversation about plastic vomit), Mary Ann Hines, Ted Hobgood, Dave Jacobson, Brian Luskey, Michelle Craig McDonald, Roderick McDonald, Stephen Mihm, Marina Moskowitz, Sharon Murphy, Sharon Portelance, Charles Rosenberg, Terry Snyder, Linda Stanley, Donald Stilwell, JoAnn Stilwell, and Richa Tiwary. Also, thanks to Dan and Stu, my go-to guys at Liberties Parcel.

Finally, to my family, starting with my beloved beagle Cecil, who reminded me there was more to the world than a dull computer screen. You are missed every day. My thanks, too, to my brother, Blake Woloson, and his partner, Melissa Kerr; my mother, Joan Woloson; and my father, Kent Woloson, and his indomitable wife, Linda Woloson. Last and most, to David Miller, who read too many drafts and gamely made late-night runs to Apple Cabin for turts, coodlers, and dirt rockets. I love you, Pup.

Putting names to paper cannot begin to repay or even fully capture the contributions everyone here has made to this work, and the gratitude I feel for your help and encouragement. I raise a dribble glass in thanks to you all.

NOTES

Introduction

1   "One for the Angels," *The Twilight Zone*, season 1, episode 2, written by Rod Serling and directed by Robert Parrish; first aired on CBS, October 9, 1959.

2   On early consumer revolutions in America, see Richard Bushman, *The Refinement of America: Persons, Houses, Cities* (New York: Vintage, 1993); Carey Carson, Ronald Hoffman, and Peter J. Albert, eds., *Of Consuming Interests: The Style of Life in the Eighteenth Century* (Charlottesville: University of Virginia Press, 1994); T. H. Breen, *The Marketplace of Revolution: How Consumer Politics Shaped American Independence* (New York: Oxford University Press, 2004).

3   By the late 1860s and early 1870s many inventors were patenting designs for "combination tools" that looked quite similar to the multi-tools and Swiss Army Knives of more recent vintage. One early representative device was patented by Aurestus S. Perham, of Washington, DC, on February 1, 1870. His "Improvement in Compound Tool," US Patent #99,470, united a hammer, axe, chisel, can opener, screwdriver, saw, and tape measure, in order to "combine several mechanical tools, such as are more commonly used by persons in every branch of business, in shop and in household, in one neat compact body."

4   "How New-Yorkers Live: Visits to the Homes of the Poor in the First and Fourth Wards," *New York Times*, June 20, 1859. The poor may have become even more sentimentally attached to such goods, since they could afford to replace them much less often. Magazines sometimes published complicated instructions for repairing cherished cheap statuettes that had become "battered" over time. See, e.g., "How to Bronze Plaster-of-Paris Figures," *Hearth and Home* 3.19 (May 13, 1871), 370.

5   "Art. IV.—The Study and Practice of Art in America," *Christian Examiner* 71.1 (July 1861), 67.

6   "The Image Peddler," *Youth's Penny Gazette* 8.7 (March 28, 1855), 25.

7   This conceptualization borrows from Igor Kopytoff's "The Cultural Biography of Things: Commoditization as Process," in *The Social Life of Things: Commodities in Cultural Perspective*, ed. Arjun Appadurai (Cambridge: Cambridge University Press, 1986).

8   Gillo Dorfles, *Kitsch: The World of Bad Taste* (New York: Bell, 1969), 15–16.

9   Karl Marx, "The Fetishism of Commodities and the Secret Thereof," in *Capital*, vol. 1 (London: J. M. Dent & Sons, 1930).

10   Heidi Ann Von Recklinghausen, *The Official M. I. Hummel Price Guide: Figurines and Plates*, 2nd ed. (Iola, WI: Krause Publications, 2013), 5.

11   Scholars have expanded on William James's observation that "a man's Self is the sum total of all that he CAN call his," including not just his family members but all of his stuff. Whether a yacht or a child, "all these things give him the same emotions. If they wax and prosper, he feels triumphant; if they dwindle and die away, he feels cast down,—not necessarily in the same degree for each thing, but in much the same way for all." See Russell K. Belk, "Possessions and the Extended Self," *Journal of Consumer Research* 15.2 (September 1988), 139–68, quote at 139.

12   Some of the earliest appearances of "crap" in the English language, dating to the early fifteenth century, referred to detritus and refuse from food preparation, such as chaff from wheat processing and tallow from rendering meat. By the mid-nineteenth century people were using the word "crap" to refer to bodily waste, and soon thereafter as a synonym for "nonsense" and "lies"—the excrement issuing from the mouths of the insincere and dishonest.

13   As Harry Frankfurt points out, these things are not just the products of "laxity . . . or carelessness or inattention to detail." In seeming contradiction, they are not necessarily shoddy but can be "carefully wrought," their lies often quite "expansive" and artful and baroque. *On Bullshit* (Princeton, NJ: Princeton University Press, 2005), 22, 24, 53.

14   Quoted in Miles Orvell, *The Real Thing: Imitation and Authenticity in American Culture* (Chapel Hill: University of North Carolina Press, 1989), 19.

15   Stuart Chase and Frederick J. Schlink, *Your Money's Worth: A Study in the Waste of the Consumer's Dollar* (1927; New York: MacMillan, 1936), 12–13.

Chapter 1

1   There is an extensive literature on the history of peddlers. The most recent and insightful works are David Jaffee, "Peddlers of Progress and the Transformation of the Rural North," *Journal of American History* 78.2 (September 1991), 511–35; and Joseph T. Rainer, "The 'Sharper' Image: Yankee Peddlers, Southern Consumers, and the Market Revolution," *Business and Economic History* 26.1 (Fall 1997), 27–44. A writer described Yankee peddlers in 1837 as "insinuating wheedling characters, dealing for the most part in goods of an inferior description . . . [who] find but little difficulty in selling or bartering their wares amongst the country-people . . ." "Yankee Pedlers, and Peddling in America," *Penny Magazine of the Society for the Diffusion of Useful Knowledge* 6 (July 15, 1837), 270.

2   John B. McConnel, *Western Characters; or, Types of Border Life in the Western States* (New York: Redfield, 1853), 275.

3   Nathaniel Hawthorne, *American Note-Books*, vol. 1 (London: Smith, Elder, 1868), 202, 203.

4   See, for instance, the myriad "Nick-Nackatories" offered by early peddlers, enumerated in "A Satirical Harangue, in the Person of a Hawker," *Gentleman's Magazine*, February 1734, 84–5.

5   Jaffee, "Peddlers of Progress," 512.

6   Selah Norton, "Variety Store, in Ashfield," *Hampshire Gazette* (Northampton, MA), January 1, 1794.

7   "New Variety Store!," *Vermont Republican* (Windsor), December 25, 1815.

8   Edward Vernon, "Cheap, Cheap, Cheap, for Cash," *Patriot* (Utica, NY), February 9, 1819.

9   See, for instance, "John Chandler & Brothers," *Greenfield* (MA) *Gazette*, January 11, 1800; and "Cheap! Cheap! L. Williams & Co.," *Hampshire Gazette* (Northampton, MA), February 1, 1820.

10   [Edward Clay?], *Or Fair samples of MILKY DUMPLINGS offered for CORNBREAD* ([Philadelphia?, ca. 1822–36]), gouache and ink on cardstock. Library Company of Philadelphia.

11   A Plain Practical Man, *Remarks upon the Auction System, as Practised in New-York . . .* , (New York: n.p., 1829), 22, 23.

12   For example, "a lace [ribbon] of half an inch wide will contain 50 or 60 yards, while a rich one of two or three inches wide, and of which half a yard is artfully left dangling below, that it may be observed, has only five or six yards." A Plain Practical Man, *Remarks upon the Auction System*, 37.

13   A Plain Practical Man, *Remarks upon the Auction System*, 35. Advertisements shown in fig. 1.2 from *Impartial Observer* (Providence, RI), January 16, 1802 (Ebeling vol. 70, Houghton Library, Harvard University); *New Hampshire Observer* (Concord), May 27, 1822; *Evening Post* (New York), October 18, 1824; *Augusta* (GA) *Chronicle*, March 11, 1826; *Torch Light* (Hagerstown, MD), May 15, 1828; *Washington* (PA) *Review and Examiner*, August 1, 1829; *Boston Traveller*, December 5, 1834. Courtesy American Antiquarian Society.

14   "Cheap! Cheap!! Cheap!!!," *Connecticut Herald* (New Haven), April 16, 1816, original emphasis.

15   Robert M. Schindler, "The Excitement of Getting a Bargain: Some Hypotheses Concerning the Origins and Effects of Smart-Shopper Feelings," *Advances in Consumer Research* 16 (1989), 449. I thank him for sharing his research findings with me. For more on smart shopper feelings, see Ellen Gibson, "Stores Succeed by Turning Shopping into a 'Hunt,'" NBC News, July 5, 2011, http://www.nbcnews.com/id/43644360/ns/business-retail/t/stores-succeed-turning-shopping-hunt/#.Wvt5_cgh3sk; and Gonca P. Soysal and Lakshman Krishnamurthi, "Demand Dynamics in the Seasonal Goods Industry: An Empirical Analysis," *Marketing Science* 31.2 (March–April 2012), 293–316.

16   Prof. W. J. Walter, "The Scotch Pedlar," *Godey's* 23 (December 1841), 241.

17   Jackson Lears, *Fables of Abundance: A Cultural History of Advertising in America* (New York: Basic Books, 1994), 71.

18   "Cheap Jack," *Anglo-American*, December 12, 1846, 172.

19   "Bargain Hunters," *Spirit of the Times*, September 13, 1845, 342.

20   "Cheap Jack," 171–72.

21   "John Brown's Variety Store," quoted in *Miscellanies Selected from the Public Journals*, vol. 1 (Boston: Joseph T. Buckingham, 1822), 135–6.

22   "Great Variety Store," *Rural Repository* 20.10 (December 30, 1843), 79; "'The Temple of Fancy,'" *Farmer and Mechanic*, n.s., 2.14 (April 4, 1848), 169.

23   "Van Schaack's Mammoth Variety Store," *Gavel* 1.8 (April 1845), 223. See also S. Wilson, comp., *1844 Albany City Guide* (Albany: C. Wendell, 1844), 101. The advertisement addressed a typical variety store's many customers, from tourists who should "spend an hour or two at this establishment" in order to "gratify their curiosity" to country merchants wishing to find articles "they cannot obtain elsewhere." See also "New Variety Store," *Institute Omnibus and School-Day Gleaner* 1.16 (March 16, 1849), 63.

24   "John Brown's Variety Store," 137.

25   *Holden & Cutter, Importers and Wholesale Dealers in French, German and English Fancy Goods and Toys and Articles of American Manufacture* (Boston: R. Stiles, ca. 1840).

26   *Dominicus Hanson's Catalogue of Apothecary, Book and Variety Store* (Dover, NH: Dover Gazette Power-Press, 1854), 4.

27   *Georgia Journal*, April 5, 1832, quoted in Rainer, "The 'Sharper' Image," 36. As Jackson Lears has pointed out, "the stimulation of fantasy was central to the expansion of the consumer market." *Fables of Abundance*, 51.

28   McConnel, *Western Characters*, 269.

29   Wilson, *1844 Albany City Guide*, 99–100. For more on R. H. Pease, see Thomas Nelson, "Pease's Great Variety Store and America's First Christmas Card," *Ephemera Journal* 17.2 (January 2015), 1–8.

30   "GOODS AT THE BOSTON CENT STORE ALMOST GIVEN AWAY," *Sun* (Baltimore, MD), October 18, 1845.

31   "ALWAYS SOMETHING SELLING CHEAP . . . ," *Sun* (Baltimore, MD), September 20, 1845.

32   "Auction Notice.—Thos. Bell, Auctioneer . . . ," *New York Herald*, March 24, 1858.

33   "Make Way for 100 Cases and Bales," *Connecticut Courant* (Hartford), September 19, 1857.

34   W. W. Palmer & Co., "GREAT WANT OF MONEY!!—AND DRY GOODS CHEAP!!!," *Salem* (MA) *Register*, October 15, 1857.

35   *Directory of Pittsburgh and Allegheny Cities* (Pittsburgh: George H. Thurston, 1862), 157.

36   "What 25 Cents Will Do!" *Patriot* (Harrisburg, PA), April 16, 1866.

37   "Now Then Store!," *Boston Daily Journal*, December 23, 1868.

38   "Go to the Great One Two Three Dollar Store," *Tete-a-Tete* 3.2 (November 16, 1869), 4.

39   Consumer psychologists call this the "scarcity principle," whereby people tend to value more highly items that seem to be in limited supply. Desirability increases more with competition, which can be encouraged if a few truly good bargains are included with the rest of the merchandise. The effect, witnessed during every Black Friday riot, has been likened to the feeding frenzy created by fishermen's chum. Robert B. Cialdini, *Influence: The Psychology of Persuasion* (New York: Collins Business, 2007 [1984]), 237–71. Merchandise that "has been rummaged through" also "create[s] a sense of value." Patricia Sabatini, "'Buyout,' Beware," *Philadelphia Inquirer*, June 11, 2017.

40   F. W. Woolworth, "From One Store to Six Hundred," *Business* 29.2 (August 1912), 62, my emphasis.

41   "The corner of St. Andrew and Magazine streets . . . ," *Daily Picayune* (New Orleans), November 4, 1866; "Unparalleled Success of Braselman's Twenty-Five Cent Counter," *Daily Picayune* (New Orleans), December 6, 1866.

42   Watson's China Store, Wilmington, NC, ca. 1880s, trade card. Advertising Ephemera Collection, John W. Hartman Center for Sales, Advertising and Marketing History, David M. Rubenstein Library, Duke University (hereafter, Hartman Center, Duke University).

43   "5 Cent Counter Supplies!" *Portland* (ME) *Daily Press*, March 24, 1879.

44   Woolworth, "From One Store," 103.

45   John P. Nichols, *Skyline Queen and the Merchant Prince: The Woolworth Story* (New York: Trident Press, 1973), 28, 33.

46   "Leisure Moments," *Cultivator and Country Gentleman* 41.1199 (January 20, 1876), 47.

47 Nichols, *Skyline Queen*, 30, 34, 54–5.

48 "Go to the Eureka 50 Cent Store," *Quincy* (IL) *Daily Whig*, December 13, 1872; "Do You Want to Get Your Money's Worth?" *Cincinnati Daily Enquirer*, December 21, 1869; "Valentines at the 99 Cent Store," *Omaha Herald*, February 4, 1879; "New York Fifty Cent Store," *Chicago Tribune*, October 9, 1870; Munson's 99 Cent Store, *See What 99 Cents Will Buy at Munson's New 99 Cent Store* ([Boston: n.p., ca. 1870]).

49 "Literary and Trade News," *Publishers' Weekly*, 7.164 (March 6, 1875), 265.

50 "90 CENT GOODS!," *Lowell* (MA) *Daily Citizen and News*, July 18, 1873.

51 "Surprise Store," *Arkansas Gazette* (Little Rock), May 29, 1881.

52 "Review 4," *Literary World* 16.2 (January 24, 1885), 26.

53 "Barrels of Booty," *Puck*, March 6, 1915, 23.

54 "Bargain-Hunting," *Household Monthly* 2.2 (May 1859), 120.

55 "Miscellaneous," *Geo. P. Rowell's American Newspaper Reporter* 9.13 (March 29, 1875), 258.

56 "Business and Commercial Items," *Christian Standard* 6.7 (February 18, 1871), 56.

57 "The Dollar Sale Men," *Advertisers Gazette* 3.4 (February 1869), 4; "The Dollar Sales," *Advance* 2.72 (February 25, 1869), 4. Dollar stores were on the rise at the same time as dollar sales schemes, and the public often confused the two. The dollar sales scheme, based on chance, was a form of lottery or gift exchange. See "The Dollar Store," *Ladies' Pearl* 8.2 (May 1874), 81; and Wendy A. Woloson, "Wishful Thinking: Retail Premiums in Mid-19th-Century America," *Enterprise & Society* 13.4 (December 2012), 790–831.

58 "Underselling at Dollar Stores," *Publishers' Weekly* 7.8 (February 20, 1875), 217, original emphasis.

59 "The Boston Lottery," *Publishers' Weekly* 8.14 (October 9, 1875), 562; "If the publishers . . . ," *Publishers' Weekly* 8.21 (December 4, 1875), 883. Critics also referred to them as "panic" concerns. See, e.g., "After the Battle," *Publishers' Weekly* 9.2 (January 8, 1876), 31; and "Fourth Session—Thursday Morning," *Publishers' Weekly* 8.4 (July 24, 1875), 222.

60 "After the Battle," 31; "The Underselling Shops," *Publishers' Weekly* 8.22 (December 11, 1875), 907.

61 "The following we copy . . . ," *Godey's Lady's Book* 82.489 (March 1871), 295. For more on dollar store wedding gifts, see "The Silver Wedding of Mose Skinner," *New England Homestead* 3.41 (February 18, 1871), 326; and *Godey's Lady's Book* (January 1873), 96: "Wedding cards are now issued with the notice, 'No plated ware,' printed in one corner. Would it not be as well to add, 'No presents from the Dollar stores put up in Caldwell's or Bailey's boxes'?"

62 "When Your Aunt Comes to Visit," *New York Clipper* 24.10 (June 3, 1876), 76.

63 "Facetiae. Anticipation versus Realization," *Harper's Bazaar* 19.2 (January 9, 1886), 32. See also "Fixing Up Around Home," *Hearth and Home* (July 18, 1874), 71.

64 "If one should catch . . . ," *Chronicle: A Weekly Journal* 12.13 (September 25, 1873), 202.

65 "Cheap Literature," *Magenta* 4.6 (December 4, 1874), 64.

66 Quoted in Nichols, *Skyline Queen*, 38.

67 Quoted in T. F. Bradshaw, "Superior Methods Created the Early Chain Store," *Bulletin of the Business Historical Society* 17 (1943), 41.

68 "Dollar Stores and Dollar Sales," *Haney's Journal of Useful Information* 2.16 (April 1869), 59. For more on dollar sales, see "The '$1 Sale' Swindles," *Phreno-*

logical Journal and Life Illustrated 49.8 (August 1869), 309; "Humbugs! Humbugs!" *Journal of Agriculture* 6.7 (August 14, 1869), 106; and "Corey O'Lanus' Epistle," *New York Fireside Companion* 4.103 (October 19, 1869), 4.

69 *An Examination of the Reasons, Why the Present System of Auctions Ought to Be Abolished* . . . (Boston: Beals, Homer, 1828), 15, original emphasis.

70 "Economy Is Wealth," *Worcester* (MA) *Daily Spy*, March 15, 1878.

71 Emma Cleveland Ward, "Mrs. Brown Visits the Capital," *Century Magazine* 53 (1896), 159.

72 "Bargains," *Puck* (February 15, 1911), 5; "How to Display 5–10–25¢ Goods in the General Store," *The Butler Brothers Way for the General Merchant* 1 (January 1913), 2.

73 Henry William Hancmann, "The Fatal Lure of the Whim-Wham," *Life* 76.1966 (July 8, 1920), 81.

74 Oscar Lovell Triggs, "Arts and Crafts," *Brush and Pencil* 1.3 (December 1897), 48.

Chapter 2

1 For more on the carnivalesque and its importance in the modern retail environment, see Jackson Lears, *Fables of Abundance: A Cultural History of Advertising in America* (New York: Basic Books, 1994).

2 Ada Louise Huxtable, "The Death of the Five-and-Ten," *New York Times*, November 8, 1979.

3 T. F. Bradshaw, "Superior Methods Created the Early Chain Store," *Bulletin of the Business Historical Society* 17 (1943), 38.

4 "Increase Your Profits. They Can!," *The Butler Way for the General Merchant* 2.1 (January 1913), 14.

5 *"Jim Lane" The Price Wrecking Fool in Charge* (Philadelphia: Long Publishing, ca. 1920–5]), my emphasis.

6 "Nickels, Dimes and Quarters," *The Butler Way for the General Merchant* 2.1 (January 1913), 14–15.

7 Sears, Roebuck, *The Bargain Counter: Low Price Sale of Odds and Ends* (Philadelphia: [Sears, Roebuck, 1921]), 2.

8 Researchers have found that consumers process textual information more logically and take more time doing it. In contrast, they apply "gestalt processing" to assortments of items presented visually (and, I would argue by extension, materially), meaning that both individual items and entire groups are evaluated, and emotionally rather than rationally. See Claudia Townsend and Barbara E. Kahn, "The 'Visual Preference Heuristic': The Influence of Visual versus Verbal Depiction on Assortment Processing, Perceived Variety, and Choice Overload," *Journal of Consumer Research* 40.5 (February 2014), 993–1015.

9 Butler Brothers, *Manual of Variety Storekeeping* (New York: Butler Brothers, 1925), 61.

10 Fleura Bardhi, "Thrill of the Hunt: Thrift Shopping for Pleasure," *Advances in Consumer Research* 30 (2003), 175. She cites specifically the "disorganized, anonymous ambience" in thrift stores and explains, "Thrift shoppers tend to search every corner of the store each time they shop . . . [and] engage in what they call 'hunting for the jewel'" (175). See also Fleura Bardhi and Eric J. Arnould, "Thrift Shopping: Combining Utilitarian Thrift and Hedonic Treat Benefits," *Journal of Consumer Behaviour* 4.4 (2005), 223–33.

11 E. P. Charlton to Simon Kapstein, Fall River, MA, May 26, 1910. University of

Massachusetts Dartmouth, Claire T. Carney Library Archives and Special Collections, MC 31, Earle P. Charlton Family Papers, 1889–1995.

12   Butler Brothers, *Manual of Variety Storekeeping*), 62.

13   Malcolm P. McNair, *Expenses and Profits of Limited Price Variety Chains in 1936* (Boston: Harvard University Graduate School of Business Administration Bureau of Business Research, 1937), 9, 26. This was true even though apparel and accessories, selling for more than thirty cents, comprised on average over 54 percent of sales.

14   G. E. Hubbard and Denzil Baring, *Eastern Industrialization and Its Effect on the West* (London: Oxford University Press, 1935); Elizabeth Boody Schumpeter, ed., *The Industrialization of Japan and Manchukuo, 1930–1940* (New York: Macmillan, 1940); Edwin P. Reubens, "Small-Scale Industry in Japan," *Quarterly Journal of Economics* 61.4 (August 1947), 577–605; Shin'Ichi Yonekawa and Hideki Yoshihara, *Business History of General Trading Companies*, International Conference on Business History 13, Proceedings of the Fugi Conference (Tokyo: University of Tokyo Press, 1987); Masayuki Tanimoto, "From Peasant Economy to Urban Agglomeration: The Transformation of 'Labour-Intensive Industrialization' in Modern Japan," in *Labour-Intensive Industrialization in Global History*), ed. Gareth Austin and Kaora Sugihara (London and New York: Routledge, 2013), 144–75.

15   Quoted in John P. Nichols, *Skyline Queen and the Merchant Prince: The Woolworth Story* (New York: Trident Press, 1973), 44.

16   Nichols, *Skyline Queen*, 49.

17   This pattern held true for other variety chains as well. According to Godfrey Lebhar, W. T. Grant went from one store in 1907 doing $99,500 in sales to 691 stores in 1957 doing over $406 million in sales. In 1933 H. L. Green's 182 stores did a total of $28.9 million, and by 1957 the company had 224 stores doing $110.6 million in sales. McCrory's 20 stores, doing half a million dollars in sales in 1901, had grown to 215 stores with annual sales of $111.8 million in 1957. J. J. Newberry's first store in 1912 reported just over $32,000 in sales; by 1957 its 476 stores were doing total annual sales of $213 million. *Chain Stores in America, 1859–1959* (New York: Chain Store Publishing Corporation, 1959).

18   Catherine Hackett, "Why We Women Won't Buy," *Forum and Century* 58.6 (December 1932), 345.

19   Hackett, "Why We Women Won't Buy," 345, 346.

20   Since at least 1927, Woolworth's sales had remained flat because of increased competition and rising rents. Decreases in revenues were due not to sales volume but to increasing commodity prices and the inelasticity of rents. See, e.g., "Variety Chains Face Growing Problems," *Bradstreet's Weekly* 60.2844 (December 31, 1932), 1732. According to one contemporary analysis, "The comparatively better showing of variety chains [over department stores] during the years when business was at a low ebb was partly attributed to the attraction of their low fixed price limits during a period when purchasing power was on the decline." McNair, *Expenses and Profits*, 2.

21   Woolworth's raised its price ceiling to twenty cents in 1932 because of inflation. "Woolworth Goes to 20 Cents," *Printers' Ink* 158 (February 25, 1932), 12; Ronald P. Hartwell, "Adjusting Business Policies to Changing Conditions," *Magazine of Wall Street*, March 19, 1932, 670–1, 694.

22   Andrew M. Howe, "10 Cents a Sleeve—5 Cents a Button," *Printers' Ink* 158 (March 17, 1932), 102.

23   "Face-Lifting the Dime Stores," *Business Week*, March 26, 1938, 39.
24   On rising dime store prices, see McNair, *Expenses and Profits*; Neil H. Borden, *The Economic Effects of Advertising* (Chicago: Richard D. Irwin, 1942), esp. 589–605; Lawrence R. Kahn, "Changing Status of the Variety Chains," *CFA Institute* 11.2 (May 1955), 31–3; "Woolworth Goes to 20 Cents." At first Woolworth's introduced twenty-cent goods as a trial in a hundred of its stores, and 80 percent of the merchandise was still at the lower price points. See Hartwell, "Adjusting Business Policies to Changing Conditions," 670–1, 694.
25   Francis Bourne, "How to Sell the '5 & 10's,'" *Advertising and Selling* 23 (June 21, 1934), 26. Another merchandising expert noted that any ten-cent good that required over a minute of a salesperson's time had "proved its inability to sell itself and must go out for good." Howard McLellan, "Turning Slow Movers into Best Sellers," *American Business* 7 (July 1937), 26.
26   McLellan, "Turning Slow Movers," 27.
27   In 1933 the average per capita expenditure in departments stores was $20, compared with just $5 in variety stores. Wilford L. White, "The Consumer's Dollar," *Credit and Financial Management* 39 (March 1937), 6. The stores' very merchandise suppliers admitted having an anti-chain bias until they began stocking "better quality, new items." Howard McLellan, "New Sales Opportunities in the Variety-Store Boom," *American Business* 7 (May 1937), 19.
28   McLellan, "Turning Slow Movers," 49. This particular retailing gambit proved so confounding to consumers that in 1936 the Federal Trade Commission determined it to be an unfair competitive practice. See, e.g., *Annual Report of the Federal Trade Commission for the Fiscal Year Ended June 30, 1936* (Washington: Government Printing Office, 1936), 38. See also Borden, *The Economic Effects of Advertising*, which discusses the many ways that consumers have difficulty making decisions, esp. chap. 23, "Advertising as a Guide to Consumption."
29   James M. Rock and Brian W. Peckham, "Recession, Depression, and War: The Wisconsin Aluminum Cookware Industry, 1920–1941," *Wisconsin Magazine of History* 73.3 (Spring 1990), 217.
30   Quoted in Rock and Peckham, "Recession, Depression, and War," 217–8.
31   *Annual Report of the Federal Trade Commission for the Fiscal Year Ended June 30, 1925* (Washington: Government Printing Office, 1925). The Bernard Hewitt & Co. mail order business was charged with using the words "Silk," "Satin," "Pongee," "Cotton Pongee," "Tussah Silk," "Art Silk," "New Silk," "Silkoline," "Silk Faille Poplin," "French Art Rayon Silk," "Mercerized Pongee," "Silk Bengaline," and "Neutrisilk" to describe articles not made of silk. The company also used "wool" and "wool mix" to describe fabrics that were not wool, "Fine Grade Tan Alligator Leather" for products not made of alligator skin, and "Silverine" and "Nickel Silverine" for watches made of punk metal. See *Annual Report of the Federal Trade Commission for the Fiscal Year Ended June 30, 1929* (Washington: Government Printing Office, 1929).
32   In *The Economic Effects of Advertising*, Neil Borden acknowledged the "difficulties met by the consumer in making wise choices when faced by a multiplicity of goods and services." Variety store managers could choose from a core list of items approved by top executives—some four thousand to five thousand different things, amounting to some twenty thousand considering all available variations. See "Woolworth's $250,000,000 Trick," *Fortune* 8.5 (November 1933), 67.
33   "Toys Made in Japan Go to World Markets," *Trans-Pacific* 21 (December 28, 1933), 18.
34   Dorothy J. Orchard, "An Analysis of Japan's Cheap Labor," *Political Science Quarterly* 44.2 (June 1929), 215–58.

35   Rock and Peckham, "Recession, Depression, and War," 213. Increasing duties led to a decrease in imports, to $72,100 by 1927. Sales of imported aluminum cookware increased from $1,855 in 1919 to $672,239 in 1921.

36   "In 1924–28 raw silk formed 83%, by value, of American imports from Japan; by 1937 its share had dropped to 49%." Nathan M. Becker, "The Anti-Japanese Boycott in the United States," *Far Eastern Survey* 8.5 (March 1, 1939), 50.

37   Becker, "The Anti-Japanese Boycott," 49.

38   Tariffs of, e.g., 60 percent ad valorem and one cent for each celluloid toy that moved; 50 percent ad valorem and one cent for each stationary celluloid toy. Japanese manufacturers benefited in part from Jewish retailers' boycott of German-made toys during Hitler's rise to power.

39   "The Made-in-Japan Christmas in the United States," *China Weekly Review* 63.6 (January 7, 1933), 280.

40   General Electric, "Let LAMPS Be Gay . . . Not Gay Deceivers!" *New York Times*, December 4, 1932.

41   Becker, "The Anti-Japanese Boycott," 49.

42   "The Made-In-Japan Christmas," 280.

43   McLellan, "New Sales Opportunities," 20.

44   *Annual Report of the Federal Trade Commission for the Fiscal Year Ended June 30, 1936* (Washington: Government Printing Office, 1936), 51.

45   Jean Lyon, "Shopping Guide for Boycotters," *Nation* 145.17 (October 23, 1937), 427–8.

46   Becker, "The Anti-Japanese Boycott," 51, 52.

47   Lyon, "Shopping Guide," 428.

48   Becker, "The Anti-Japanese Boycott," 50–1.

49   "Axis Goods 'Out,'" *Business Week*, December 20, 1941, 37; "Boycott Applied," *Business Week*, December 20, 1941, 46, 47.

50   "Japan's Exports Fail to Hit Par," *Business Week*, December 6, 1947, 121. See also Meghan Warner Mettler, "Gimcracks, Dollar Blouses, and Transistors: American Reactions to Imported Japanese Products, 1945–1964," *Pacific Historical Review* 79.2 (May 2010), esp. 205–9.

51   Mettler, "Gimcracks," 210.

52   Mettler, "Gimcracks," 209–10; Sidney Shalett, "Why We're Trading with the Enemy," *Saturday Evening Post*, July 12, 1947, 25.

53   George Rosen, "Japanese Industry since the War," *Quarterly Journal of Economics* 67.3 (August 1953), 454. See also G. C. Allen, "Japanese Industry: Its Organization and Development to 1937," in *The Industrialization of Japan and Manchukuo, 1930–1940*), ed. Elizabeth Boody Schumpeter (New York: Macmillan, 1940), esp. 543–66.

54   "When You Buy Japanese, Double-Check the Goods," *Business Week*, December 17, 1949, 106.

55   Akira Nagashima, "A Comparison of Japanese and U.S. Attitudes towards Foreign Products," *Journal of Marketing* 34.1 (January 1970), 73.

56   Curtis C. Reierson, "Attitude Changes toward Foreign Products," *Journal of Marketing Research* 4.4 (November 1967), 386–7.

57   Curtis C. Reierson, "Are Foreign Products Seen as National Stereotypes?" *Journal of Retailing* 42.3 (Fall 1966), 33–40.

58   United States Tariff Commission, *United States Imports from Japan and Their Relation to the Defense Program and to the Economy of the Country* (Washington, DC: United States Tariff Commission, 1941), 16.

59   United States Tariff Commission, *United States Imports from Japan*, 19, 194–5, 200, 201, 205, 206, 208, 230.

60   Most of the goods represented in *Merchandise That Japan Offers*, from 1955, "excepting the top class," were "those of medium and small scale industry," of operations employing a hundred people or fewer that were "anxious to introduce their products to overseas markets." Federation of Foreign Trade Promotion Institutes of Japan, *Merchandise That Japan Offers, 1955* (Printed in Japan: Nippon Seihan, [1954?]), i.

61   Rosen, "Japanese Industry since the War," 452.

62   *Guide to Japanese Products for U.S. and Canadian Importers* (Tokyo: Japan External Trade Recovery Organization, [1958]), 9, 12.

63   *Guide to Japanese Products for U.S. and Canadian Importers*, 16.

64   John Sasso and Michael A. Brown Jr., *Plastics in Practice* (New York: McGraw-Hill, 1945), 134, 135, 137. See also Wilford L. White, "The Situation in Chain-Store Distribution," *Southern Economic Journal* 3.4 (April 1937), 411–26.

65   *Guide to Japanese Products for U.S. and Canadian Importers*, 40, 42.

66   "Woolworth's $250,000,000 Trick," 65–6.

67   *Guide to Japanese Products for U.S. and Canadian Importers*, 40.

68   Japanese Rubber Manufacturers' Association, *Japan's '57 Rubber Goods* (Tokyo: Japanese Rubber Manufacturers' Association, [1957?]), 34.

69   Kahn, "Changing Status of the Variety Chains," 31; Spencer Gifts catalog (Atlantic City, NJ: [The Company], 1964).

70   David Auw, "Making 'MIT' Mean Quality," *Free China Journal*, September 7, 1989.

71   Auw, "Making 'MIT' Mean Quality"; Andrew Quinn, "'Made in Taiwan' Brands Want to Buy a Little Respect," Reuters News, September 13, 1989. See also Peter C. DuBois, "Taiwan: A View from Snake Alley," *Barron's*, December 19, 1988.

72   "Hong Kong, Thanks to China, Tops Toy Market," *Australian Financial Review*, January 18, 1989.

73   Craig Wolff, "Consumers' World: Copies of Popular Toys Are Often Hazardous," *New York Times*, November 21, 1987.

74   Wolff, "Consumers' World."

75   Wolff, "Consumers' World." For contemporary articles on designer knockoffs in the fashion industry, see, e.g., Elaine Williams, "World Watch Industry: Switzerland Survives—at a Price," *Financial Times*, October 2, 1982; William Kazer, "Asia Takes Real Bite at Lucrative Faking Business," Reuters, January 12, 1988; Katina Alexander, "You Can't Always Tell the Real Thing from the 'Hideous' Gucci Counterfeits," *Orange County Register*, February 17, 1989; and Alberto Arebalos, "Smuggling Is a Way of Life," Reuters, February 20, 1989.

76   Steven Husted and Shuichiro Nishioka, "China's Fare Share? The Growth of Chinese Exports in World Trade," *Review of World Economies* 149.3 (2013), 565–6. They note that in 2010, 16.7 percent of all merchandise imports to the US came from China (56–9). Jianqing Ruan and Xiaobo Zhang, "Low-Quality Crisis and Quality Improvement: The Case of Industrial Clusters in Zhejian Province," in *Industrial Districts in History and the Developing World*, ed. Tomoko Hashino and Keijiro Otsuka (Singapore: Springer Science+Business Media, 2016), 170. Peter K. Schott, "The Relative Sophistication of Chinese Exports," *Economic Policy* 23.1 (January 2008), 21. Schott concludes that "the substantial price disparities observed between Chinese and OECD exports within product markets suggests Chinese exports may be of lower quality," 35.

77   Husted and Nishioka, "China's Fare Share?," 569–70.

78   For more on the history of Walmart, see the 2005 documentary film *Wal-Mart: The High Cost of Low Price*, directed by Robert Greenwald and produced by Brave New Films.

79   MarketLine, *Company Profile Dollar Tree, Inc.*, January 8, 2016, 4, https://www.marketline.com, reference code E7F38462-EF1E-4309-904C-496EA358DBA1, 4, 5.

80   MarketLine, *Company Profile Dollar Tree, Inc.*, 6.

81   Datamonitor, *Family Dollar Stores Company Profile*, July 21, 2008, 6 (my emphasis); *International Directory of Company Histories*, vol. 62, "Family Dollar Stores, Inc." (St. James, MO: St. James Press, 2004), 133–6.

82   MarketLine, *Company Profile Dollar Tree, Inc.*, 8.

83   MarketLine, *Company Profile Family Dollar Stores, Inc.*, August 29, 2012, 7 https://www.marketline.com, reference code 9EC011A2-C7DA-4A45-B933-E74BC6E64EBB.

84   Datamonitor, *Dollar General Corporation Company Profile*, May 22, 2007, 6–7.

85   Datamonitor, *Dollar General Corporation Company Profile*, January 31, 2011, 7; MarketLine, *Company Profile Dollar General Corporation*, September 27, 2013, 7, and February 5, 2016, 7, https://www.marketline.com, reference code 4922A2BE-CD57-4475-A5B6-AD9D3F0322DA.

Chapter 3

1   Eugene S. Ferguson called their creators "radical mechanics," because they invented things that "wrought radical changes." "The American-ness of American Technology," *Technology and Culture* 20.1 (January 1979), 6.

2   "Old Times," *North American Review* 13 (May 1817), 4–11.

3   A Man Born out of Season, "A Complaint against Convenience," *New-York Mirror* 17.13 (September 21, 1839), 100.

4   "A Useful Lesson: Old Humphrey and the Farmer," *Mirror and Keystone*, August 31, 1853, 274.

5   "Yankee Ingenuity," *Yankee Farmer and New England Cultivator* 5.5 (February 2, 1839), 36.

6   "Yankee Ingenuity," *New England Farmer* 2.42 (May 15, 1824), 334.

7   "The Fecundity of Yankee Ingenuity," *New-York Organ and Temperance Safeguard* 8.47 (May 19, 1849), 346.

8   "Yankee Ingenuity," *Maine Farmer and Journal of the Useful Arts* 1.44 (November 16, 1833), 351.

9   "Yankee Ingenuity," *Kentucky New Era* (June 10, 1858), 1; "Yankee Ingenuity," *Illustrated New York News*, June 21, 1851, 22. For more on this attitude, see David Jaffee, "Peddlers of Progress and the Transformation of the Rural North," *Journal of American History* 78.2 (September 1991), 511–35; and Joseph T. Rainer, "The 'Sharper' Image: Yankee Peddlers, Southern Consumers, and the Market Revolution," *Business and Economic History* 26.1 (Fall 1997), 27–44.

10   "Machine for Wringing Clothes," *American Agriculturist* 19.8 (August 1860), 247.

11   Algernon Sidney Johnston, *Memoirs of a Nullifier* (Columbia, SC: Printed and Published at the Telescope Office, 1832), 8.

12   Quoted in Rainer, "'Sharper' Image," 39–40.

13   "Jonathan's Patent Labor-Saving, Self-Adjusting Hog Regenerator," *Yankee Notions* 2.5 (May 1853), 155. For an example of how the concept of Yankee ingenuity was politicized during the Civil War, see "The Yankee Nut-Crackers," *Scientific American* 12.7 (April 22, 1865), 265.

14   For contemporary accounts of hog slaughter on farms, see H. D. Richardson, *The Hog, His Origin and Varieties* (New York: C. M. Saxton, 1856), chap. 10 "Slaughtering and Curing," 55–64.

15 William Youatt, *The Pig: A Treatise on the Breeds, Management, Feeding, and Medical Treatment of Swine* (New York: C. M. Saxton, 1852), 153.

16 According to Dominic A. Pacyga, it took a skilled butcher between eight and ten hours to slaughter and dress a steer on the farm and just thirty-five minutes in Chicago slaughterhouses. Anne Bramley, "How Chicago's Slaughterhouse Spectacles Paved the Way for Big Meat," National Public Radio, December 3, 2015, https://www.npr.org/sections/thesalt/2015/12/03/458314767/how-chicago-s -slaughterhouse-spectacles-paved-the-way-for-big-meat. See also Dominic A. Pacyga, *Slaughterhouse: Chicago's Union Stock Yard and the World It Made* (Chicago: University of Chicago Press, 2015).

17 Old Lady, "Economy of Labor-Saving Utensils in a Kitchen or on a Farm," *American Agriculturist* 6.1 (May 1847), 158.

18 These contradictions are outlined in two seminal works: Susan Strasser, *Never Done: A History of American Housework* (New York: Pantheon, 1982); and Ruth Schwartz Cowan, *More Work for Mother: The Ironies of Household Technology from the Open Hearth to the Microwave* (New York: Basic Books, 1983).

19 "Kitchen Song," *Massachusetts Cataract* 6.2 (March 30, 1848), 5.

20 Mrs. M. L. Rayne, "John Merrill's Theory," *Prairie Farmer* 18.12 (September 22, 1866), 12.

21 There is a large body of literature dedicated to the history of domestic technologies in America. In addition to Strasser and Cowan, see, for instance: Elizabeth Faulkner Baker, *Technology and Woman's Work* (New York: Columbia University Press, 1964); Siegfried Giedion, *Mechanization Takes Command: A Contribution to Anonymous History* (1948; New York: W. W. Norton, 1975); Sarah F. Berk, ed., *Women and Household Labor* (Beverly Hills, CA: Sage Publications, 1983); and Glenna Matthews, *"Just a Housewife": The Rise and Fall of Domesticity in America* (New York: Oxford University Press, 1987).

22 George Basalla, *The Evolution of Technology* (New York: Cambridge University Press, 1988), 69. On the ratio of "useful" to "trivial" patents, see B. Zorina Khan, "Property Rights and Patent Litigation in Early Nineteenth-Century America," *Journal of Economic History* 55.1 (March 1995), 60. For more on patenting in the nineteenth century, see B. Zorina Khan, "'Not for Ornament': Patenting Activity by Nineteenth-Century Women Inventors," *Journal of Interdisciplinary History* 31.2 (Autumn 2000), 159–95; Jacob Schmookler, *Invention and Economic Growth* (Cambridge, MA: Harvard University Press, 1966); Kenneth L. Sokoloff, "Inventive Activity in Early Industrial America: Evidence from Patent Records, 1790–1846," *Journal of Economic History* 48 (1988), 813–50; Sokoloff and Khan, "The Democratization of Invention during Early Industrialization: Evidence from the United States, 1790–1846," *Journal of Economic History* 50 (1990), 363–78.

23 "Useless Patents," *Mechanic*, November 1834, 344. For more on the burgeoning of popular invention and patents during the nineteenth century, see Naomi R. Lamoreaux, Kennel L. Sokoloff, and Dhanoos Sutthiphisal, "Patent Alchemy: The Market for Technology in US History," *Business History Review* 87 (Spring 2013), 3–38. They note that, for example, the number of registered patent attorneys per million people rose from 10.7 in 1883 to 74.5 in 1910 (16).

24 "Labor-Saving Implements," *Working Farmer*, July 1857, 112, 113; "Facts for the Curious. Curiosities of the Patent Office," *Portland Transcript*, April 20, 1867, 20–1.

25 E. W. Slade, "A Glance at Patent Churns: Gault's Rotary Churn," *Ohio Cultivator* 5.2 (January 15, 1849), 17.

26 "Should Farmers Buy Patent Rights?," *Prairie Farmer* 18.7 (August 18, 1866), 102.

27 "The Inventor—Some of the Obstacles to Success," *Scientific American* 17.20 (November 16, 1867), 313. Patent agents, too, profited from innovations in gadgetry. A few years later, the magazine issued "A Caution" about patent agents, particularly the "vast horde of self-styled 'solicitors'" who "hang about" in Washington, DC, representing themselves as men who "understand the 'ropes'" but "have only windy pretensions." Their circulars, like the new-fangled gadgets they puffed, were of "wondrous length and thundering sound . . . promising the most brilliant results." See "Patents for Inventions," *New York Mercantile Register for 1848–49* (New York: John P. Prall, 1848), 29; "Inventors and Patent Agents," *Scientific American* 5.23 (February 23, 1850), 182; "Patent Agents—A Caution," *Scientific American* 8.40 (June 18, 1853), 317; and "Scoundrelism in Patent Agents," *Scientific American* 10.1 (September 16, 1854), 5.

28 "The 'Gullibility' of Farmers," *Maine Farmer* 41.10 (February 8, 1873), 1.

29 Home Manufacturing Company, *Description, Testimonials and Directions of the Celebrated Home Washer!* (New York: E. S. Dodge, ca. 1869). Hartman Center, Duke University.

30 Vandergrift Manufacturing Company, *Catalogue of the Vandergrift Manufacturing Co.* (Buffalo: Gies, ca. 1880). Hartman Center, Duke University.

31 Stratton & Terstegge Co., *Stratton & Terstegge Co., Manufacturers of Pieced and Stamped Tinware, Japanned Ware, Galvanized Ware, Sheet Iron Ware . . .* ([Louisville, KY?]: n.p., [1906]). Hartman Center, Duke University.

32 Glenda Riley, "In or Out of the Historical Kitchen? Interpretations of Minnesota Rural Women," *Minnesota History* 52.2 (Summer 1990), 65.

33 Riley, "In or Out of the Historical Kitchen?," 66.

34 "Labor-Saving Implements," rpt. in *Working Farmer*, July 1857, 113. For examples of newspaper "advertorials," see Hudson River Wire Works, *New Patent White Wire Clothes Line. $10,000 Reward for Superior Article* ([New York]: n.p., ca. 1868). Hartman Center, Duke University. Whether these testimonials were complete fabrications or just edited in uniform fashion is unclear. See Edward Slavishak, "'The Ten Year Club': Artificial Limbs and Testimonials at the Turn of the Twentieth Century," in *Testimonial Advertising in the American Marketplace: Emulation, Identity, Community*, ed. Marlis Schweitzer and Marina Moskowitz (New York: Palgrave Macmillan, 2009), 96.

35 Home Manufacturing Company, *Description, Testimonials and Directions*.

36 Leach Roaster and Baker Co., *The Improved Roaster and Baker* ([Birmingham, AL?]: n.p., ca. 1891). Hartman Center, Duke University.

37 "Potato Steamers—A Yankee Invention," *Life Illustrated*, April 15, 1857, 195.

38 J. E. Shepard & Co., *Catalogue of J. E. Shepard & Co. Manufacturers of Household Novelties and Specialties for Canvassing Agents* ([Cincinnati?]: n.p., 1884), 4; J.C. Tilton, *150,000 Already Sold* (Pittsburgh: n.p., [1873]). Hartman Center, Duke University.

39 W. H. Baird & Co., *Manufacturers of Household Necessities!* ([Pittsburgh]: n.p., ca. 1890), no pagination; original emphases in bold and italics.

40 Riley, "In or Out of the Historical Kitchen?," 64.

41 The idea of "extravagant futility" has its roots in kitsch theory. See Vittorio Gregotti, "Kitsch and Architecture," in *Kitsch: The World of Bad Taste*, comp. Gillo Dorfles (New York: Bell Publishing, 1968), 263.

42 Charles Babbage, *On the Economy of Machinery and Manufactures* (London: J. Murray, 1846), 121, 122. Paraphrased from Michael Zakim, "Importing the Crystal Palace," in *Capitalism: New Histories*, ed. Sven Beckert and Christine

Desan (New York: Columbia University Press, 2018), 343, who writes of "the means for realizing those means."

43　Quoted in Rainer, "The 'Sharper' Image," 39, citing the *Connecticut Courant* of November 1, 1812.

44　M. D. Leggett, *Subject-Matter Index of Patents for Inventions Issued by the United States Patent Office from 1790 to 1873, Inclusive*, 3 vols. (Washington, DC: Government Printing Office, 1874).

45　P. T. Barnum, *The Life of P. T. Barnum, Written by Himself* (New York: Redfield, 1855), 25.

46　*M. Young's Monthly Publication of New Inventions* (New York: M. Young, 1875), 1.

47　Thomas Manufacturing Company, *Thousands of Agents Are Making Big Money Selling The Full Nickel-Plated Washington* ([Dayton, OH?]: n.p., ca. 1900?), my emphasis. Hartman Center, Duke University.

48　Such "complex" merchandise possessed several different "product virtues," which could be, according to one advertising theorist, "antagonistic if not directly contradictory; if a product is 'best' in one respect, it commonly follows that the product cannot be best in some other respects." Neil Borden, *The Economic Effects of Advertising* (Chicago: Irwin, 1942), 649.

49　Todd Timmons, *Science and Technology in 19th-Century America* (Westport, CT: Greenwood Press, 2005).

50　Walter A. Friedman, *Birth of a Salesman: The Transformation of Selling in America* (Cambridge, MA: Harvard University Press, 2004), 35.

51　On flexible manufacturing, see Philip Scranton, *Endless Novelty: Specialty Production and American Industrialization, 1865–1925* (Princeton, NJ: Princeton University Press, 2000).

52　*Official Catalogue."Novelties" Exhibition. Philadelphia, Commencing September 15, 1885, under the Direction of the Franklin Institute . . .* (Philadelphia: Burk & McFetridge, 1885).

53　See Samuel Hopkins Adams, "The Great American Fraud. Quacks and Quackery I.—The Sure Cure School," *Collier's* 37 (July 4, 1906), 12. Charlatans' success was due, in his words, to the "pitiful gropers after relief from suffering," who would try any nostrum or device that guaranteed to "raise you from your coffin and restore you to your astonished and admiring friends."

54　L. Shaw, *How to Be Beautiful! Ladies' Manual* (New York: Terwilliger & Peck, ca. 1886), 21–2, my emphasis. Hartman Center, Duke University.

55　American Dentaphone Company, *The Testimony of One Hundred Living Witnesses* ([Cincinnati?]: n.p., ca. 1882), 1. Hartman Center, Duke University. For more on these devices, see Albert Mudry and Anders Tjellström, "Historical Background of Bone Conduction Hearing Devices and Bone Conduction Hearing Aids," in *Implantable Bone Conduction Hearing Aids*, ed. Martin Kompis and Marco-Domenico Caversaccio (Basel: Karger, 2011), 1–9.

56　Adams, "Great American Fraud," 12.

57　Advertisement in Butter Improvement Company, *Hints to Butter-Makers and Book of Reference* (Buffalo: Butter Improvement Co., 1879), 24. Hartman Center, Duke University.

58　US Patent 357,647, Peter H. Vander Weyde, February 15, 1887; American Galvanic Company, "Nervous Debility Cured without Medicine," *Educational Weekly* 8.169 (September 16, 1880), 122.

59　P & M Agency, *A Triumph of Modern Ingenuity and Science! The "Conquerer" Electric Ring Leads All Others!* ([Palmyra, PA?]: n.p., ca. 1890s). Hartman Center, Duke University.

60 American Electrocure Co., *Twentieth-Century Electrocure* (Vineland, NJ?: n.p., ca. 1898), 5. Hartman Center, Duke University.

61 Andrew Chrystal, *Catalogue of Professor Chrystal's Electric Belts and Appliances* ([Marshall, MI?]: n.p., [1897]), 4.

Chapter 4

1 On Americans' desire to see for themselves how things worked, see Neil Harris, "The Operational Aesthetic," in *Humbug: The Art of P. T. Barnum* (Chicago: University of Chicago Press, 1981), 59–90.

2 United Manufacturing Company, *Information for Your Benefit*, ca. 1910. Hartman Center, Duke University.

3 Letter from Harold [?] G. Wolf to Henry W. Jones, Leipsic, OH, July 14, 1910. H. W. Jones Collection, David M. Rubenstein Library, Duke University (hereafter, Jones Collection, Rubenstein Library).

4 Letter from Harold [?] G. Wolff to Henry W. Jones, Leipsic, OH, July 30, 1910. Jones Collection, Rubenstein Library.

5 United Manufacturing Company, *Instructions to Salesmen: A Confidential, Man-to-Man Talk with our Representatives by the General Manager* (Toledo, OH: Franklin, ca. 1910), 6, Hartman Center, Duke University.

6 Niagara Merchandising Company, form letter to prospective sales agents, Lockport, NY, ca. 1910. Jones Collection, Rubenstein Library.

7 [Niagara Merchandising Company], *The Cinch Tire Repair Kit* ([Buffalo?]: n.p., ca. 1910), [8]. Hartman Center, Duke University.

8 Letter from William A. Heacock to H. W. Jones, Lockport, NY, September 5, 1912. Jones Collection, Rubenstein Library.

9 This was similar to United's fire extinguisher, which only worked with the company's exclusive "chemicals." Letter from Heacock to Jones, September 7, 1912. Jones Collection, Rubenstein Library.

10 For more on this, see Sigfried Giedion, *Mechanization Takes Command: A Contribution to Anonymous History* (1948; New York: W. W. Norton, 1975), esp. 516–56.

11 Christine Frederick, *Household Engineering: Scientific Management in the Home* (1915; Chicago: American School of Home Economics, 1921), 100, 101. Frederick remarked that a salesman's in-house performance was an opportunity to see and experience a product, since it was "better than trusting entirely to circulars" (104).

12 Frederick, *Household Engineering*, 105.

13 Frederick, *Household Engineering*, 105, 107, original emphasis. Frederick, in fact, argued that more efficient hand labor, like washing dishes, was superior to using mechanical appliances.

14 "Good Housekeeping Institute: A Little Story of Its Growth and Service," *Good Housekeeping* 55.2 (August 1912), 278, 280, 281.

15 "The Dictatorship of Inanimate Things," *Independent* 116.3953 (March 6, 1926), 259.

16 G[urney]W[illiams], "Queerespondence: A Study in Our National Absurdities," *Life* 102.2604 (July 1935), 40, my emphasis.

17 Stanley Walker, "Come Back, Appleknocker!," *Forum and Century* 86.6 (December 1931), 378.

18 Frances Drewry McMullen, "New Jobs for Women," *North American Review* 234.2 (August 1932), 134.

19   Peter Marzio, *Rube Goldberg: His Life and Work* (New York: Harper & Row, 1973), 145, 177, 179.

20   George Kent, "The Answer to Their Prayers," *Life* 97.2522 (March 6, 1931), 6.

21   Gurney Williams, "Gadgets Wanted," *Life* 101.2589 (April 1934), 14.

22   T. W. S. "The Woman's Slant: This Month's Madnesses," *Life* 100.2580 (July 1933), 44.

23   See also Williams, "Queerespondence," 32. On gadget purchases by gender, see Richardson Wright, "The Fascination of Gadgets," *House & Garden* 56 (August 1929), 60, who claimed that men "are the ones who first feel the fascination of gadgets."

24   Neil H. Borden, *The Economic Effects of Advertising* (Chicago: Richard D. Irwin, 1942), 614. He notes that truly significant innovations often take decades, if not centuries, to at last become "worth while" (615–29).

25   Borden, *Economic Effects of Advertising*, 644, 645.

26   Catherine Hackett, "Why We Women Won't Buy," *Forum and Century* 58.6 (December 1932), 347.

27   Borden, *Economic Effects of Advertising*, 642.

28   Hackett, "Why We Women Won't Buy," 347–8.

29   Bancroft's, *Bancroft's Out of this World Selections* ([Chicago]: n.p., ca. 1950s), 43, 50.

30   United States Postal Service, "America's Mailing Industry: Hammacher Schlemmer," https://postalmuseum.si.edu/americasmailingindustry/Hammacher -Schlemmer.html. Pop-up toasters were introduced in 1930, the electric toothbrush in 1955, and answering machines in 1968. In 1980 Hammacher Schlemmer was bought by J. Roderick MacArthur, who founded the collectible crap company Bradford Exchange in 1973.

31   Examples taken from Hammacher Schlemmer catalogs from autumn 1967 and autumn 1973.

32   Elaine Tyler May, *Homeward Bound: American Families in the Cold War Era* (New York: Basic Books, 1988), 160.

33   Jon Nathanson, "The Economics of Infomercials," Priceonomics, November 14, 2013, https://priceonomics.com/the-economics-of-infomercials/.

34   Charles Lindsley, *Radio and Television Communication* (New York: McGraw Hill, 1952), 349.

35   Lindsley, *Radio and Television Communication*, 347–50.

36   In 1984, as part of Reagan's deregulation efforts, the FCC lifted restrictions on commercials for all channels, enabling long infomercials to air once again on non-cable channels.

37   Nathanson, "The Economics of Infomercials."

38   Topval Corporation, "New Rembrandt Automatic Potato Peeler," advertising circular. ([Lindenhurst, NY?: n.p., ca. 1958).

39   "Gadget-of-the-Month Club Supplies Latest Gimcracks," *Reading Eagle*, December 26, 1948; Robert M. Hyatt, "Ever the Gadgeteer," *Challenge* 2.4 (January 1954), 13–15.

40   Quoted in Timothy Samuelson, *But, Wait! There's More! The Irresistible Appeal and Spiel of Ronco and Popeil* (New York: Rizzoli, 2002), 17.

41   Robert Palmer Corporation, *Selling for Keeps*, leaflet 18, "Dramatize!" ([Santa Barbara, CA: Robert Palmer Corporation, 1950]).

42   Samuelson, *But, Wait! There's More!*, 21.

43   Herman M. Southworth, "Implications of Changing Patterns of Consumption

Preferences and Motivations," *Journal of Farm Economics* 39.5 (December 1957), 1303, 1304, 1305.

44   Samuelson, *But, Wait! There's More!*, 22–3.

45   Samuelson, *But, Wait! There's More!*, 25–7.

46   Timothy R. Hawthorne, *The Complete Guide to Infomercial Marketing* (Lincolnwood, IL: NTC Business Books, 1997), 73.

47   Hawthorne, *Complete Guide to Infomercial Marketing*, 15; Nathanson, "The Economics of Infomercials."

48   Nathanson, "The Economics of Infomercials."

49   Peter Bieler with Suzanne Costas, *"This business has legs": How I Used Infomercial Marketing to Create the $100,000,000 Thighmaster® Exerciser Craze* (New York: John Wiley & Sons, 1996), 68–9. Infomercial marketing became the target of consumer watchdog groups who accused advertisers of making false product claims; in the early 1990s, Congress held hearings because members were "troubled" by advertisements exploiting "consumers' willingness to believe what they hear and see on TV." House of Representatives Committee on Small Business, *Consumer Protection and Infomercial Advertising . . . May 18, 1990*, Serial No. 101-60 (Washington, DC: US Government Printing Office, 1990), 2. See also Joshua Levine, "Entertainment or Deception?," *Forbes*, August 2, 1993, 102. Even broadcasters—who relied on the revenue from infomercials—aired them reluctantly; one confided they were "con jobs." House of Representatives Committee on Small Business, *Consumer Protection and Infomercial Advertising*, 120. See also Federal Trade Commission, "Program Length Commercials," *For Consumers* (Washington, DC: Office of Consumer/Business Education), July 1989.

50   Earning the company a fine of $550,000 from the FTC. Levine, "Entertainment or Deception?"

51   Levine, "Entertainment or Deception?"

52   Hawthorne, *The Complete Guide to Infomercial Marketing*, 62, original emphasis.

53   It was $5 to manufacture and would cost another $2 for packing and warehousing, $3 to staff the phone lines for orders, $2 for miscellaneous expenses, and $3 for profit and royalties, leaving $15 for media if each unit sold for $29.95. Bieler, *"This business has legs,"* 77–8; Nathanson, "The Economics of Infomercials."

54   Bieler, *"This business has legs,"* 77.

55   Bieler, *"This business has legs,"* 76.

56   Bieler, *"This business has legs,"* 120.

57   J. Walter Thompson, "Infomercials," November 1993, presentation notes, sheets 30, 25, 29. J. Walter Thompson Archives, Hartman Center, Duke University.

Chapter 5

1   See, for instance, "Terms," *Christian Advocate*, January 20, 1827.

2   George Meredith, *Effective Merchandising with Premiums* (New York: McGraw-Hill, 1962), 4. For more on premiums, see Matthew Shannon, *100 Years of Premium Promotions, 1851–1951* (New York: Premium Advertising Association, 1951). Perhaps not surprisingly, Babbitt was a friend of P. T. Barnum.

3   He was also known, apparently, as someone who didn't like paying his way. "Evading a Railroad Fare," *Daily Atlas* (Boston), December 3, 1853.

4   For more on soap in America, see Richard L. Bushman and Claudia L. Bush-

man, "The Early History of Cleanliness in America," *Journal of American History* 74.4 (March 1988), 1213–38. Ross is listed in city directories as a "fancy" soap maker—i.e., a purveyor of fine rather than laundry soap. (See, for example, *The Worcester Almanac, Directory, and Business Advertiser* for 1849 and 1850.)

5   Advertising broadside, *Major Ross, the World-Renowned Soap Man!* (Lowell, MA: Vox Populi Print, [1856]).

6   "$15 Per Day Easy $15," *Harper's Weekly*, July 18, 1863, 463. See also S. C. Rickards advertisements "Extraordinary," *Harper's Weekly*, November 22, 1862, 752; and "A Speculation," *Harper's Weekly*, January 17, 1863, 47.

7   "Opposition Prize Package Company," *Harper's Weekly*, November 29, 1862, 768; "Valentine Packages," *Harper's Weekly*, January 24, 1862, 64; "Agents, Something That Beats the World" and "A Free Gift," *Harper's Weekly*, April 5, 1862, 224; "1,000,000 Persons, Clip: Club!! Clinch!!!" *Harper's Weekly*, September 7, 1867, 575.

8   "Agents, Something That Beats the World."

9   See Albert S. Bolles, *Industrial History of the United States: From the Earliest Settlements to the Present Time* (Norwich, CT: Henry Bill Publishing, 1889) on the New England costume jewelry industry: "Ornaments in a thousand patterns were thus produced. . . . Filled jewelry found a wide market from the very first. The universal Yankee peddler sold immense quantities of it, and the manufacture of it increased year by year" (328).

10   "Head-Quarters for Cheap Jewelry," *Harper's Weekly*, May 17, 1862, 320. For more on prize packages, see Wendy A. Woloson, "Wishful Thinking: Retail Premiums in Mid-19th-Century America," *Enterprise & Society* 13.4 (December 2012), esp. 801–8.

11   These gift-book enterprises should not be confused with the genre of gift books, which are described in Frederick W. Faxon, *Literary Annuals and Gift Books: A Bibliography, 1823–1903* (rpt. Pinner, UK: Private Libraries Association, 1973); E. Brucke Kriham and John W. Fink, comps., *Indices to American Literary Annuals and Gift Books, 1825–1865* (New Haven, CT: Research Publications, 1975); and Ralph Thompson, *American Literary Annuals and Gift Books, 1825–1865* (rpt. Hamden, CT: Archon Books, 1967). In this context, "gift" refers to the prize given free with the purchase of a book; in other words, book enterprises giving out gifts.

12   "Forty-Ninth Philadelphia Trade Sale," *American Publishers Circular*, September 26, 1857, 614. My thanks to Michael Winship for this reference. See also the lithograph by Edward Sachse, *Interior View of Evans' Original Gift Book Establishment* (Baltimore: E. Sachse & Son, 1859). Historical Society of Pennsylvania.

13   G. G. Evans, *G. G. Evans & Co.'s Great New England Gift Book Sale!* ([Boston?: n.p., betw. 1858–60]), 2, 5.

14   *State v. Clarke & a.*, Superior Court of Judicature of New Hampshire, 33 N.H. 329; 1856 N.H. Lexis 83, July 1856.

15   Albert Colby, *History of the Gift Book Business: Its Nature and Origin* (Boston: Albert Colby, 1859), 18.

16   G. G. Evans, *New Feature in Trade, and Something Worthy of Your Attention!* ([Philadelphia: G. G. Evans?, 1860]), original emphasis.

17   George G. Evans, printed receipt sent to G. R. Wells, completed in manuscript ([Philadelphia]: n.p., [November 9, 1859]). Library Company of Philadelphia.

18   "Dead-Headed," *Harper's New Monthly Magazine* 42 (May 1871), 921.

19   Roy J. Bullock, "The Early History of the Great Atlantic & Pacific Tea Company," *Harvard Business Review* 11.3 (April 1933), 289–98.

20   Howard Stanger, "The Larkin Clubs of Ten: Consumer Buying Clubs and Mail-Order Commerce, 1890–1940," *Enterprise & Society* 9.1 (March 2008), 134–5.

21   Henry S. Bunting, *The Premium System of Forcing Sales*, 2nd ed. (Chicago: Novelty News Press, 1913), 123, original emphasis.

22   Great London Tea Co., *Price and Premium List with Cash Prices for Premiums* ([Boston: The Company, 1891]), 1.

23   M. W. Savage, *Savage's Free Premiums* (n.p.: M. W. Savage, 1914), front cover.

24   On the history of scrapbooks, see Susan Tucker, Katherine Ott, and Patricia P. Buckler, eds., *The Scrapbook in American Life* (Philadelphia: Temple University Press, 2006); Jessica Helfand, *Scrapbooks: An American History* (New Haven, CT: Yale University Press, 2008); and Ellen Gruber Garvey, *Writing with Scissors: American Scrapbooks from the Civil War to the Harlem Renaissance* (Oxford: Oxford University Press, 2013).

25   Charles K. Johnson, "Trading Stamps and the Retailer" (MS thesis, Kansas State University, 1965), 2.

26   On the early years of retail premium promotions, see Susan Strasser, *Satisfaction Guaranteed: The Making of the American Mass Market* (New York: Pantheon, 1989), 163–78.

27   Albrecht R. Sommer, "Premium Advertising," *Harvard Business Review* 10.2 (January 1932), 206.

28   I. M. Rubinow, "Premiums in Retail Trade," *Journal of Political Economy* 13 (September 1905), 574.

29   Lucy M. Salmon, "The Economics of Spending," *Outlook* 91 (April 17, 1909), 886.

30   Rubinow, "Premiums in Retail Trade," 574–5, 576.

31   Bunting describes this in *The Premium System*, 120–3.

32   "A Flank Movement on the Profit-Sharing Coupon System," *Current Opinion* 58.6 (June 1915), 439.

33   Bunting, *The Premium System*, 40.

34   The trading-stamp and coupon companies—like the Pennfield Merchandise Company, the United Profit-Sharing Corporation, and Sperry & Hutchinson—were the ones that most benefited from coupon systems. The trading-stamp dealer's profit was made on the difference between the price of the equivalent stamps sold to the retailer and the cost of the premium given to the customer, minus advertising and printing expenses. The stamp agent stood to net even more than that, however, when consumers lost or never redeemed their stamps; unredeemed stamps meant fewer premiums sent out. See Rubinow, "Premiums in Retail Trade," esp. 583–5. One early twentieth-century account estimated that printers and sellers of trading stamps and other premium coupons saw profits of more than $100 million each year, while retailers often suffered a net loss, despite sometimes seeing an initial uptick in sales. "A Flank Movement," 439; David Shelton Kennedy, "Are Trading Stamps a Fraud?" *Forum*, August 1917, 247–52.

35   Penfield Trading Company, *Cash Trade* ([Buffalo?]: n.p., ca. 1904), 1.

36   Rubinow, "Premiums in Retail Trade," 581.

37   "Help a Brother," *Railroad Telegrapher* 28.1 (January–June 1911), 62; "An Appeal," *Railroad Telegrapher* 28.1 (January–June 1911), 867; "The following request was published . . . ," *Railroad Telegrapher* 31.1 (January–June 1914), 197; and *Railroad Telegrapher* 30.2 (July–December 1913), 1475.

38   "A Flank Movement," 440.

39   Kennedy, "Are Trading Stamps a Fraud?," 248, original emphasis.

40   By 1937 there were some three hundred carnivals touring the country. See Michael Baers, "Carnivals," in *St. James Encyclopedia of Popular Culture* 5 (2013),

158–60. See also Sam Brown, "How Carnival Games Cheat Customers," *Modern Mechanix*, June 1930.

41   Walter B. Gibson, *The Bunco Book* (Holyoke, MA: Sidney H. Radner, 1946); see also Harry Crews, "Carny," in Nathaniel Knaebel, *Step Right Up: Stories of Carnivals, Sideshows, and the Circus* (New York: Carroll & Graf, 2004); William L. Alderson, "Carnie Talk from the West Coast," *American Speech*, 28.2 (May 1953), 112–19; and Earl Chapin, "How Carnival Racketeers Fleece the Public," *Modern Mechanix*, August 1934. Carnies gaffed games themselves, and companies produced them for the trade.

42   Crews, "Carny," 60. On plush toys, see Margaret Walsh, "Plush Endeavors: An Analysis of the Modern American Soft-Toy Industry," *Business History Review* 66.4 (Winter 1992), 637–70.

43   Baker's Game Shop, *1949–50 Baker's Game Shop Games That Are Games* ([Detroit?]: n.p., [1949]).

44   Gibson, *Bunco Book*, 26, 32.

45   "Samuel Pockar. Veteran Slum Dealer Looks Back 50 Years," *Billboard* 68.14 (April 7, 1956), 96; Irwin Kirby, "Hand-in-Hand: Slum Jewelry, Midways Longtime Partners," *Billboard* 69.15, pt. 2 (April 7, 1958), 38. Slum also became a staple in gumball machines and mechanical "merchandiser" games. The odds of winning anything were so long that courts eventually defined both carnival midway games and merchandisers as forms of gambling. For a brief history of digger games, see the website James Roller's Vintage Amusements, www.jamesroller.com. For a selection of typical merchandise, see Kipp Brothers, *Carnival Catalog, no. 166* ([Indianapolis, IN: The Company, ca. 1940]).

46   Lee Manufacturing Company, *Lee's Wonderful Catalogue of Easy Selling Goods and Premiums* (Chicago: [The Company], 1924), [3].

47   Rubinow, "Premiums in Retail Trade," 584.

48   Wendy A. Woloson, *Refined Tastes: Sugar, Consumers, and Confectionery in Nineteenth-Century America* (Baltimore: Johns Hopkins University Press, 2002), 50. Various lottery/candy operators used this scheme many decades later. See, e.g., *Annual Report of the Federal Trade Commission for the Fiscal Year Ended June 30, 1939* (Washington, DC: Government Printing Office, 1939), 58–72.

49   John M. Miller & Son, *Price List . . . for the Season of 1876* ([Philadelphia?]: n.p., [1876]), [4]. See also offerings in N. Shure Co., *Sure Winner Catalog No. 121, 1933* (Chicago: [The Company, 1933]).

50   Woloson, *Refined Tastes*, 47. A representative ad for W. C. Smith appears in the *Confectioners' Journal*, May 1890, 46.

51   Dowst Bros. Co., "New Novelties Every Week," *International Confectioner*, January 1904, 44; Woloson, *Refined Tastes*, 47–48.

52   Richard Alliger Osmun, "Toys as Inducement Goods—A Rare Opportunity for Publishers, Manufacturers, Jobbers, and Retailers," *Novelty News* 9 (October 1909), 52.

53   Henry Bunting, *Specialty Advertising: A New Way to Build Business* (Chicago: Novelty News Press, 1914), 109.

54   Richard Alliger Osmun, "Simple, Inexpensive Toys Are Suitable and Satisfactory 'Personal Appeal' Mediums," *Novelty News* 9 (December 1909), 26.

55   Osmun, "Toys as Inducement Goods," 54.

56   For surveys of premiums offered by national manufacturers and local retailers, see E. Evalyn Grumbine, *Reaching Juvenile Markets: How to Advertise, Sell, and Merchandise through Boys and Girls* (New York: McGraw-Hill, 1938), 93–

106. She cited 151 national manufacturers who used 226 juvenile premiums offered through radio, newspapers, magazines, and point-of-sale promotions. "Each classification," she noted, "has its stories of amazing results achieved" (92). See also, later, Eugene Gilbert, *Advertising and Marketing to Young People* (Pleasantville, NY: Printers' Ink Books, 1957), 260–69. Companies ranged from shampoo makers to sporting goods retailers, and prizes included everything from play money, t-shirts, toy rifles, and wallets to dolls, comic books, pencil sharpeners, and yo-yos.

57   *Printers' Ink*, February 9, 1922, 121, quoted in Daniel Thomas Cook, "The Other 'Child Study': Figuring Children as Consumers in Market Research, 1910s–1990s," *Sociological Quarterly* 41.3 (Summer 2000), 493.

58   Newton Manufacturing Co., *Gift Advertising for 1923*.

59   Cook, "The Other 'Child Study,'" 488.

60   Neurology and biology professor Robert Sapolsky has found that anticipation releases dopamine in our brains. He stresses that it is not *achieving* the goal or the thing, but the *anticipation* of it, that evokes feelings of happiness and elation. If uncertainty is introduced, then even more dopamine is released. "Same Neurochemistry, One Difference," Dopamine Project, July 24, 2011, http://dopamineproject.org/2011/07/same-neurochemistry-one-difference-dr-robert-sapolsky-on-dopamine/.

61   Grumbine, *Reaching Juvenile Markets*, 108, 109, 115.

62   Meredith, *Effective Merchandising with Premiums*, 96.

63   See Cook, "The Other 'Child Study,'" 495.

64   This is described more fully in Kyle Asquith, "Knowing the Child Consumer through Box Tops: Data Collection, Measurement, and Advertising to Children, 1920–1954," *Critical Studies in Media Communication* 32.2 (June 2015), 112–27.

65   Population statistics based on 1940 US Census, https://1940census.archives.gov/about/.

66   Belena Chapp, *A Surprise Inside! The Work and Wizardry of John Walworth* (Newark: University of Delaware University Gallery, 1990), 6–7.

67   Meredith, *Effective Merchandising with Premiums*, 88.

68   Gilbert, *Advertising and Marketing to Young People*, 244–6.

69   Meredith, *Effective Merchandising with Premiums*, 89. A study from 1966 determined much the same. Chapp, *A Surprise*, 7.

70   Meredith, *Effective Merchandising with Premiums*, 93.

71   This case study is described in Gilbert, *Advertising and Marketing to Young People*, 249–51.

72   Gilbert, *Advertising and Marketing to Young People*, 244.

73   Meredith, *Effective Merchandising with Premiums*, 93.

74   Gilbert, *Advertising and Marketing to Young People*, 252.

75   Grumbine, *Reaching Juvenile Markets*, 112.

76   Children's "materialistic attitudes" are mentioned in Louise A. Heslop and Adrian B. Ryans, "A Second Look at Children and the Advertising of Premiums," *Journal of Consumer Research* 6 (March 1980), 414. People on the other side of the issue, like members of the American Marketing Association, saw premiums as a "true value" that brought "psychic satisfactions" to consumers. Jerry Harwood, "Public Policy and Issues Platform: Seek AMAers Comments on Proposed FTC Ban of Premium Offer Ads to Children," *Marketing News* 8.9 (November 1, 1974), 7.

77   Roland Marchand notes that as the Depression deepened, advertisers turned more often to appeals about economy and practicality. *Advertising the American*

Dream: Making Way for Modernity, 1920–1940 (Berkeley: University of California Press, 1985), esp. 288–9.

78   *Novelty News* 52.4 (April 1931), original emphasis.

79   "Something for Nothing," *Business Week*, March 1, 1933, 9.

80   "Boom in Premiums," *Business Week*, May 14, 1938, 27.

81   "Something for Nothing," 10.

82   "Premiums Prosper," *Business Week*, May 6, 1939, 37

83   "Unconquerable Premiums," *Business Week*, January 5, 1935, 16. The article estimated that premium production was by then double what it had been in 1929.

84   "Premium Folk Happy," *Business Week*, May 8, 1937, 24; "Boom in Premiums," 27.

85   Eugene R. Beem, "Who Profits from Trading Stamps?," *Harvard Business Review* 35 (1957), 123, 128; for a more thorough explanation of the entropy of premiums in the interwar era, along with statistics recording it, see 124.

86   Beem, "Who Profits," 129. For a representative merchandise catalog see Philadelphia Yellow Trading Stamp Company Inc., *Yellow Trading Stamps: The Seal of Approval for 53 Years* (Philadelphia: Horowitz-Kreb, 1957).

Chapter 6

1   "Address by Lewellyn E. Pratt of New York City, Representing 'Specialty and Novelty Advertising,'" *Eighth Annual Convention of the Associated Advertising Clubs of America . . . 1912* (n.p.: Associated Advertising Clubs of America, 1912), 311.

2   George Meredith, *Effective Merchandising with Premiums* (New York: McGraw-Hill, 1962), 266, original emphasis.

3   In his foundational book *Specialty Advertising: The New Way to Build Business* (Chicago: Novelty News Press, 1910), Henry Bunting suggested yet more equally viable terms, like "souvenir advertising," "inducement advertising," "premium advertising," "concrete advertising," "selling publicity," "the silent salesmen media," and "the last word in the advertising appeal" (14).

4   Bunting, *Specialty Advertising*, 14.

5   For work on early retail premiums, see Wendy A. Woloson, "Wishful Thinking: Retail Premiums in Mid-19th-Century America," *Enterprise & Society*, 13.4 (December 2012), 790–831; and Howard Stanger, "The Larkin Clubs of Ten: Consumer Buying Clubs and Mail-Order Commerce, 1890–1940," *Enterprise & Society* 9.1 (2008), 125–64.

6   Ellen Litwicki, "From the 'ornamental and evanescent' to 'good, useful things': Redesigning the Gift in Progressive America," *Journal of the Gilded Age and Progressive Era* 10.4 (October 2011), 472. Two classic works regarding gifting are Nicholas Thomas, *Entangled Objects: Exchange, Material Culture, and Colonialism in the Pacific* (Cambridge, MA: Harvard University Press, 1991); and Lewis Hyde, *The Gift: Imagination and the Erotic Life of Property* (New York: Random House, 1983).

7   Ralph Waldo Emerson, "Gifts," in *Essays: Second Series* (Boston: James Munroe, 1844).

8   On the material culture of capitalism, see, for example, Caitlin Rosenthal, "From Memory to Mastery: Accounting for Control in America," *Enterprise & Society* 14.4 (December 2013), 732–48, and "Storybook-keepers: Numbers and Narratives in Nineteenth-Century America," *Common-place* 12, no. 3 (April 2012); and Christopher Allison, "The Materiality of American Trust: The R. G. Credit Report

Volumes," paper presented at the American Studies Association's 2013 annual conference.

9   On this see, e.g., Susan Strasser, *Satisfaction Guaranteed: The Making of the American Mass Market* (New York: Pantheon, 1989).

10   Litwicki, "From the 'ornamental,'" 497.

11   See also Axel Petrus Johnson, editor of the *Library of Advertising*, who noted that special occasions like anniversaries and birthdays—even for businesses—could be created and monetized. *Library of Advertising*, vol. 4, *Show Window Display and Specialty Advertising* (Chicago: Cree Publishing, 1911), 237.

12   *Nearly Three Hundred Ways to Dress Show Windows* (Baltimore: Show Window Publishing, 1889), advertising section, [32].

13   Bunting, *Specialty Advertising*, 21, 25, 26, original emphasis.

14   For more examples, see Bunting, *Specialty Advertising*, 5–14.

15   One marketing professional reminisced, "There was a day when . . . a specialty salesman could open his grip and become instantly the center of a gaping and admiring crowd." Lewellyn E. Pratt, "Advertising Specialties and Practical Sales Plans," *Advertising and Selling* 26.2 (July 1916), 43. For more on specialty manufactures in general, see Philip Scranton, *Endless Novelty: Specialty Production and American Industrialization, 1865–1925* (Princeton, NJ: Princeton University Press, 1997). The number of businesses involved in advertising specialties decreased during the Great Depression but rebounded after World War II.

16   Meredith, *Effective Merchandising with Premiums*, 268, original emphasis.

17   Of course, as Marcel Mauss has pointed out, all gifts create debt and obligation, which is their point. He characterizes gifts as more than just sincere gestures of kindness and benevolence. As "prestations" they are "in theory voluntary, disinterested and spontaneous, but [which] are in fact obligatory and interested. The form usually taken is that of the gift generously offered; but the accompanying behaviour is formal pretence and social deception, while the transaction itself is based on obligation and economic self-interest." *The Gift: Forms and Functions of Exchange in Archaic Societies*, trans. Ian Cunnison (London: Cohen & West, 1966), 1.

18   M. E. Ream, "Selling Specialty Advertising," in Johnson, *Library of Advertising* 4:202–3, 205.

19   Ream, "Selling Specialty Advertising," 206–7, original emphasis.

20   As James Carrier observers, perhaps obviously, "Americans give different sorts of gifts to different sorts of people on different sorts of occasions, and there are different expectations and ideals against which these are judged. . . . Underlying these differences, however, are a common set of cultural concerns and tensions revolving around the ways that objects are involved in relationships between people." "Gifts in a World of Commodities: The Ideology of the Perfect Gift in American Society," *Social Analysis* 29 (December 1990), 20.

21   National Recovery Administration, *Code of Fair Competition for the Advertising Specialty Manufacturing Industry* (Washington, DC: Government Printing Office, 1933), 1.

22   Bunting, *Specialty Advertising*, 27.

23   Emerson, "Gifts," 174.

24   See Strasser, *Satisfaction Guaranteed*, esp. chap. 2, "The Name on the Label," 29–57; and James Carrier, *Gifts and Commodities: Exchange and Western Capitalism since 1700* (London: Routledge, 1995), esp. 98–105.

25   Carrier argues that the changes in the market at the turn of the twentieth century, characterized by more attenuated and anonymous relationships, "reduce[d]

the likelihood that objects will be possessions rather than commodities" (*Gifts and Commodities*, 100), since people were unlikely to do anything creatively to work with them in order to transform them into more personal possessions. But he doesn't consider the advertising specialty, which was both gift and commodity and by then already an important and ubiquitous category of thing.

26   Bunting, *Specialty Advertising*, 145.

27   "Just a Hint on What to Buy," *Advertising Specialties* 1.1 (October 1929), 22.

28   See, e.g., Barry Schwartz, "The Social Psychology of the Gift," *American Journal of Sociology* 73.1 (July 1967), esp. 4: "The leader takes care not to fall into debt to his followers but to insure, on the contrary, that the benefits he renders unto others are never fully repaid."

29   Ream, "Selling Specialty Advertising," 205.

30   J. A. Hall, "Business Improves in Advertising Specialties," *Associated Advertising* 12.7 (July 1921), 44.

31   Controlling promiscuous distribution in another way, a columnist in the trade journal *Novelty News* noted that some dealers would give Crisco-branded pancake turners only to "occasional" store visitors rather than regulars, as a way to get them to come back. "A Procter & Gamble Sales Turnover on Crisco," *Novelty News* 52.4 (April 1931), 30.

32   Newton Manufacturing Co., *Gift Advertising for 1923* ([Newton, IA?]: n.p., [1923]).

33   Sanders Manufacturing Company, *Price List and Catalogue No. 30 Illustrating a Few of Our Advertising Specialties* ([Nashville?]: [The Company], 1931), 1. Georg Simmel has theorized that the person initiating the gift exchange is in a position of power because s/he is the first to act. The gift, "because it was first, has a voluntary character which no return gift can have. For, to return the benefit we are obliged ethically; we operate under a coercion which, though neither social nor legal but moral, is still a coercion." Quoted in Schwartz, "The Social Psychology of the Gift," 9.

34   John P. Nichols, *Skyline Queen and the Merchant Prince: The Woolworth Story* (New York: Trident Press, 1973), 39.

35   As Schwartz observes, "One expresses unfriendliness through gift giving by breaking the rule of approximate reciprocity (returning a gift in near, but not exact, value of that received). Returning 'tit for tat' transforms the relation into an economic one and expresses a refusal to play the role of grateful recipient. This offense represents a desire to end the relationship or at least define it on an impersonal, non-sentimental level." "The Social Psychology of the Gift," 6.

36   Newton Manufacturing Co., *Gift Advertising for 1923*.

37   As "gifts," they also inserted themselves into an activity that had been, primarily, women's work. Advice literature discussing gift occasions and proper gifts typically addressed women, who were largely responsible for the what, the when, and the who of gifting, since such emotional labor constituted part of their larger role in supporting and maintaining the interpersonal relationships of family and community within the domestic sphere. For a contemporary take on women's emotional labor, which includes a lengthy discussion of the burden of gifts, see "Where's My Cut? On Unpaid Emotional Labor," MetaFilter, July 15, 2015, https://www.metafilter.com/151267/Wheres-My-Cut-On-Unpaid-Emotional-Labor.

38   Lillian James, "We Appeal to Women," *Advertising Specialties* 1.1 (October 1929), 20–1. For more on the penetration of advertising into the homes of nineteenth- and twentieth-century Americans, see Ellen Gruber Garvey, *The Adman in the Parlor: Magazines and the Gendering of Consumer Culture, 1880s–1910s* (Ox-

ford: Oxford University Press, 1996); and Vicki Howard, *Brides, Inc.: American Weddings and the Business of Tradition* (Philadelphia: University of Pennsylvania Press, 2008).

39   Bunting, *Specialty Advertising*, 29.

40   The Chicago-based Greenduck Company, for instance, placed an ad in *Popular Science Monthly* soliciting inventors for "small inventions, wood or metal, suitable for advertising specialties," adding, "must be of such character that they can be sold at 1¢ each or less." "WANTED," *Popular Science Monthly* 92.6 (June 1918), 992.

41   Franklin P. Adams, "Button, Button, Who's Got the Button?," *Nation's Business* 27.7 (July 1939), 52.

42   In 1914 the *Daily Consular and Trade Reports* noted that among the current "foreign trade opportunities" was an agent in Germany who wanted to purchase blanks of cheap advertising specialties, "in quantities of 100 to 1,000 or more," that could be easily imprinted. "Foreign Trade Opportunities," *Daily Consular and Trade Reports* 61 (March 14, 1914), 992.

43   George L. Herpel and Richard A. Collins, *Specialty Advertising in Marketing* (Homewood, IL: Dow Jones–Irwin, 1972), 13.

44   Bunting, *Premium System*, 48.

45   J. B. Carroll Co., *Experts in Developing and Designing New Ideas for the Advertiser* ([Chicago]: n.p., [1941]), n.p.

46   James Rorty, *Our Master's Voice: Advertising* (New York: John Day, 1934), 34.

47   J. B. Carroll Co., *Experts in Developing and Designing New Ideas for the Advertiser*, n.p.

48   Herpel and Collins, *Specialty Advertising*, 83. A typical business's offerings can be found in Standard Advertising and Printing Co., *Catalog No. 40M: Sales Stimulators* (Fort Scott, KS: [The Company], 1940).

49   L. A. Chambliss, "Novelties—The New Advertising Medium," *Bankers' Magazine* 121.1 (July 1930), 124.

50   James Rorty documented the decline in the amount of print advertising, dollar values of advertising revenues for key magazines and newspapers, and wages and numbers of advertising professionals during this time. Radio advertising, an exception, showed an upward trend. Other kinds of advertising, including categories encompassing advertising premiums, remained about the same. "Advertising and the Depression," *Nation* 137.3572 (December 20, 1933), 703.

51   Rorty, *Our Master's Voice*, 220.

52   Albert Jay Nock, "Bogus Era of Good Feeling," *American Mercury* 40.158 (February 1937), 222.

53   *Brown & Bigelow v. Remembrance Advertising Products, Inc., and Elmer B. Usher*, New York Supreme Court, Appellate Division—First Department, Appellant's Brief, Notice of Appeal filed March 6, 1951, 43–6, 50, 55, 57, 174, 175, 312.

54   Census data from https://www2.census.gov/library/publications/decennial/1950/population-volume-2/23760756v2p38.pdf.

55   Meredith, *Effective Merchandising with Premiums*, 272–3; Herpel and Collins, *Specialty Advertising*, 37–42.

56   Louis F. Dow Co., *Presenting the Louis F. Dow Co. Goodwill Advertising St. Paul, Minn., the Outstanding Line for 1956* ([St. Paul: Louis F. Dow Co. Litho U.S.A., 1955]), 3.

57   *Brown & Bigelow v. Remembrance Advertising Products, Inc.*, 61.

58   Of the 248 executives surveyed in 1940, 180 (72.5 percent) gave business presents to both employees and customers at Christmas time. Companies averaged

$6.03 per gift. Henry Bunting, "Givers and their Gifts," *Business Week*, April 27, 1940, 42–3. The breakdown was as follows: cash, 65 firms; edibles, 42; desk or office articles, 37; smoking equipment, 33; leather goods/novelties, 21; wearing apparel/jewelry, 18; liquor, 9; books/magazines, 8.

59 Dur-O-Lite, *Dur-O-Lite Business Gifts Catalog No. 54* ([Chicago]: n.p., [1954]).

60 Meredith, *Effective Merchandising*, 278, original emphasis.

61 Quoted in Herpel and Collins, *Specialty Advertising*, 18.

62 "AMA Surveys Give and Take of Business Xmas Gifts," *Business Week*, December 24, 1955, 94, my emphasis. See also Herpel and Collins, *Specialty Advertising*, 18; and "Personal Business," *Business Week*, November 28, 1959, 161.

63 "Management Briefs," *Business Week*, November 26, 1955, 132.

64 Scholars refer to this as "reciprocity theory"—"giving can lead to a recipient's perceived sense of obligation to return the favor." See Richard F. Beltramini, "Exploring the Effectiveness of Business Gifts: Replication and Extension," *Journal of Advertising* 29.2 (Summer 2000), 75.

65 Herpel and Collins, *Specialty Advertising*, 19.

66 "Personal Business," *Business Week*, November 28, 1959, 161.

67 "Mystery Shopper," *Giftware Business* 59.10 (October 2001), 16.

68 "Personal Business," *Business Week*, November 28, 1959, 161, 162.

69 "Personal Business," *Business Week*, November 28, 1959, 162. In his study on business gifts, Beltramini speculated that some customers receiving business gifts of medium value did not have more positive feelings about the giver because they knew others had received nicer gifts. Disappointed, they took their patronage elsewhere. "Exploring the Effectiveness of Business Gifts," 77.

70 "Personal Business," *Business Week*, December 8, 1962, 120.

71 Scholars of gift culture note that people *do* find ways to personalize commodities by altering them in some way, like covering them in giftwrap. Sometimes, just removing something from its packaging and integrating it into a living space is enough. But these actions do not disguise or negate the fact that the items are still just commodities anonymously produced and exchanged via market transactions.

72 Promotional Products Association International (PPAI), *Mapping Out the Modern Consumer*, 2016, 8, https://advocate.ppai.org/Documents/PPAI%202017%20Consumer%20Study%20Report.pdf. For more on this, see James Carrier, "Reconciling Commodities and Personal Relations in Industrial Society," *Theory and Society* 19.5 (October 1990), 579–98.

73 Emerson, "Gifts," 175.

Chapter 7

1 Cynthia A. Brandimarte, "'To Make the Whole World Homelike': Gender, Space and America's Tea Room Movement," *Winterthur Portfolio* 30.1 (Spring 1995), 2–3.

2 Kate Kirk, "A Modern and Old-Fashioned Tea Room," *Decorator and Furnisher* 1.1 (October 1882), 14.

3 Alice Bradley, *The Tea Room Booklet* (n.p.: Women's Home Companion, 1922), 3.

4 Bradley, *Tea Room Booklet*, 19–20.

5 Bradley, *Tea Room Booklet*, 20–1.

6 Dean MacCannell, *The Tourist: A New Theory of the Leisure Class* (Berkeley: University of California Press, 1999), characterizes this as the "spurious side of the social structure of modernity," wherein "information, memories, images, and

other representations . . . become detached from genuine cultural elements" and "are circulated and accumulated in everyday life" (147, 150).

7   On the early history of automobile touring, see John A. Jakle, *The Tourist: Travel in Twentieth-Century North America* (Lincoln: University of Nebraska Press, 1985), esp. chap. 5; and Warren J. Belasco, *Americans on the Road: From Auto Camp to Motel, 1910–1945* (Cambridge, MA: MIT Press, 1979).

8   Effie Price Gladding, *Across the Continent by the Lincoln Highway* (New York: Brentano's, 1915), 228, 35.

9   Beatrice Larned Massey, *It Might Have Been Worse: A Motor Trip from Coast to Coast* (San Francisco: Harr Wagner, 1920), 14.

10   Grace Knudson, *Gift and Art Shop Merchandising* (Boston: Little, Brown, 1926), 4.

11   Barry Schwartz writes, "Gifts are one of the ways in which the pictures that others have of us in their minds are transmitted." He adds, "The gift imposes an identity upon the giver as well as the receiver." "The Social Psychology of the Gift," *American Journal of Sociology* 73.1 (July 1967), 1, 2.

12   James Carrier, "Gifts in a World of Commodities: The Ideology of the Perfect Gift in American Society," *Social Analysis* 29 (December 1990), 25.

13   Knudson, *Gift and Art Shop Merchandising*, frontispiece.

14   Knudson, *Gift and Art Shop Merchandising*, 48; Grace Knudson, *Through the Gift Shop Door* (New York: Woman's Home Companion, 1923), 15.

15   On the psychology of anticipation and possession, see, e.g., Emily Haisley and George Loewenstein, "It's Not What You Get but When You Get It," *Journal of Marketing Research* 48.1 (February 2011), 103–15; George Loewenstein, "Anticipation and the Valuation of Delayed Consumption," *Economic Journal* 97 (September 1987), 666–84; and Michel Tuan Pham, "Anticipation and Consumer Decision Making," *Advances in Consumer Research* 22 (1995), 275–6.

16   Knudson, *Through the Gift Shop Door*, 15.

17   Knudson, *Through the Gift Shop Door*, 10.

18   Knudson, *Through the Gift Shop Door*, 16

19   Pohlson Galleries, *Gifts* ([Pawtucket, RI?]: n.p., ca. 1925), 1.

20   White's Quaint Shop, *Gifts* ([Westfield, MA?]: n.p., ca. 1920s), i.

21   Arthur J. Peel, *How to Run a Gift Shop* (Boston: Hale, Cushman & Flint, 1941), 32.

22   Pitkin and Brooks, *1906 Winners. Exclusive Price List. Assorted Packages and Open Stock* (n.p.: n.p., 1906), 23, 24, 26.

23   White's Quaint Shop, *Gifts*; Prince's Gift Shop, *The Gift Shop with Moderate Prices* ([Lowell, MA?]: n.p., ca. 1915), [1].

24   Peel, *How to Run a Gift Shop*, 21–22.

25   Peel, *How to Run a Gift Shop*, 21–2.

26   Peel, *How to Run a Gift Shop*, 32.

27   Peel, *How to Run a Gift Shop*, 59, my emphasis.

28   Robert W. Kellogg, *This Is the Town of Kellogg, the Home of the Gift Unusual* (Springfield, MA: F. A. Bassette, [1932]), i.

29   Abigail Carroll, "Of Kettles and Cranes: Colonial Revival Kitchens and the Performance of National Identity," *Winterthur Portfolio* 43.4 (Winter 2009), 336, 337.

30   On colonial kitchens, see Rodris Roth, "The New England, or 'Olde Tyme,' Kitchen Exhibit at Nineteenth-Century Fairs," in *The Colonial Revival in America*, ed. Alan Axelrod (Winterthur, DE: Henry Francis du Pont Winterthur Museum; New York: W. W. Norton, 1985), 59–83.

31   As Marilyn Casto points out, the objects *did* serve a purpose, but not a practical

one, as "social messages and stimulus to the imagination." See "The Concept of Hand Production in Colonial Interiors," in *Re-creating the American Past. Essays on the Colonial Revival*, ed. Richard Guy Wilson, Shaun Eyring, and Kenny Marotta (Charlottesville: University of Virginia Press, 2006), 322.

32   Carroll, "Of Kettles and Cranes," 343.

33   See, e.g., Thomas Andrew Denenberg's discussion of women in Wallace Nutting's photographs in "Wallace Nutting and the Invention of Old America," in Wilson, Eyring, and Marotta, *Re-creating the American Past*, 29–39.

34   "Christmas Gifts of Individuality, Charm, and Lasting Utility," *New Republic*, December 7, 1921, ix.

35   Albany Foundry Company, *Undecorated Grey Iron Castings* ([Albany: n.p.], 1928), 2.

36   Croston, Inc., *Croston, Inc.* ([Boston]: n.p., ca. 1920); see also advertisement for Art Colony Industries, *Survey*, June 11, 1921, 385.

37   Bridget A. May, "Progressivism and the Colonial Revival: The Modern Colonial House, 1900–1920," *Winterthur Portfolio* 26.2/3 (Summer–Autumn 1991), 110.

38   Quoted in William B. Rhoads, "The Long and Unsuccessful Effort to Kill Off the Colonial Revival," in Wilson, Eyring, and Marotta, *Re-creating the American Past*, 18. Michael Kammen observes that the concept of heritage is valuable "to entrepreneurs offering safe havens in a world that is commercial as well as secular, self-indulgent, and intensely concerned with social status." *In the Past Lane: Historical Perspectives on American Culture* (New York: Oxford University Press, 1997), 217. On racism in the built environment, see Robert R. Weyeneth, "The Architecture of Racial Segregation: The Challenges of Preserving the Problematical Past," *Public Historian* 27.4 (Fall 2005), 11–44; the phrase "spatial strategies of white supremacy" appears on 3.

39   Prince's Gift Shop, *The Gift Shop with Moderate Prices*, 3.

40   Fireside Studios, *Objets des Arts by Fireside* ([Adrian, MI?]: n.p., [1931]), 27, 28, 46, 22.

41   If there is any doubt about this, one need only examine the search results for three titles of digitized newspapers. A search for "giftware" in the *New York Times* from 1960–79 yields 3,765 results; in the *Washington Post* for the same time period, 984 results. The same search in the *Chicago Defender*, an African American newspaper, for 1959–75 yields just 18 results.

42   White's Quaint Shop, *Gifts*, 19.

43   Michael Hitchcock, introduction to *Souvenirs: The Material Culture of Tourism*, ed. M. Hitchcock and Ken Teague (Aldershot, UK: Ashgate, 2000), 4.

44   Nelson H. H. Graburn observes that "ethnically complex" handcrafts are often an "attempt of one group to portray members of another group using the creator group's symbolic and artistic traditions and attaching importance to those features that are important to the in-group rather than the out-group. This generally results in stereotypic portrayals that are satisfactory to the in-group but hilarious or insulting to the outsiders portrayed." Introduction to *Ethnic and Tourist Arts: Cultural Expressions from the Fourth World*, ed. N. H. H. Graburn (Berkeley: University of California Press, 1979), 29.

45   Tesori d'Italia Ltd., *1954–1955 Gift Catalog* ([New York]: n.p., 1954).

46   Province of Quebec, *Artisanal Handicrafts* ([Montreal?]: n.p., ca. 1960).

47   Shannon International, *Shopping and Mail Order Guide* (Dublin: Browne & Nolan, [1962]).

48   William Bornstein, "Wants Quality Souvenirs," *Souvenirs & Novelties* 18 (October–November 1965), 36.

49   Shopping International, *Shopping International Incorporated, American Trader, World Handicrafts* ([Hanover, NH?]: Shopping International, 1965), 2.

50   "Big Store in a Small Town," *Gifts & Decorative Accessories* 90.1 (January 1989), 126.

51   Michelle Nellett, "Individuality Counts Here," *Gifts & Decorative Accessories* 93.6 (June 1992), 120.

52   Maria Sagurton, "Recapturing America's Heritage," *Gifts & Decorative Accessories* 90.3 (March 1989), 50.

53   John Parris Franz, "Peachtree: Something for Everyone," *Gifts & Decorative Accessories* 90.1 (January 1989), 218.

54   Marie Lena Tupot and Doris Nixon, "Ceramics: Classically American," *Gifts & Decorative Accessories* 94.4 (April 1993), 56.

55   Tupot and Nixon, "Ceramics," 56.

56   Kammen, *In the Past Lane*, 157.

57   Phyllis Sweed, "Trends and Forecasts: A Casual Look at '94," *Gifts & Decorative Accessories* 94.12 (December 1993), 64.

58   Svetlana Boym, *The Future of Nostalgia* (New York: Basic Books, 2002), xv.

59   Phyllis Sweed, "So, It's Not a Melting Pot," *Gifts & Decorative Accessories* 94.3 (March 1993), 47.

60   Tupot and Nixon, "Ceramics," 56.

61   Nellett, "Individuality Counts Here," 120.

62   "Retailer's Guide to Candles and Candle Accessories," *Gifts & Decorative Accessories* 90.1 (January 1989), 64. Candle accessories included tea light holders, brass and wood candleholders, drip catchers, snuffers, and candle rings.

63   L. Knight, A. Levin, and C. Mendenhall, "Candles and Incense as Potential Sources of Indoor Air Pollution: Market Analysis and Literature Review," EPA/600/R-01/001 (Washington, DC: US Environmental Protection Agency, 2001), 1, https://cfpub.epa.gov/si/si_public_record_Report.cfm?dirEntryId=20899. This figure only increased in later years: in 2002 the Yankee Candle Company, then the industry leader, reported quarterly sales of $99 million. "Yankee Candle Sales Grow 17 Percent in Third Quarter," *Giftware Business* 60.11 (November 2002), 6.

64   "Four Companies Launch, Increase Consumer Advertising Campaigns," *Giftware Business* 60.10 (October 2002), 6; and Rachael Kelley, "Yankee Candle: A Brand Above," *Giftware Business* 60.7 (July 2002), 32.

65   Interestingly, about 96 percent of all candles are purchased by women. Knight et. al., "Candles and Incense as Potential Sources of Indoor Air Pollution," 16.

66   Melody Udell, "All Algow," *Giftware News* 36.2 (February 2011), 17.

67   Natalie Hope McDonald, "Scents and Sensibility," *Souvenirs, Gifts and Novelties* 53.1 (January 2014), 70.

68   Meredith Schwartz, "Sweet Smells of Success: Fragrance Houses Make the Calls on Upcoming Scent Trends," *Gifts & Decorative Accessories* 106.8 (August 2005), 38.

69   1803 Candle Company, "1803 Candle's Top 10 Scents," http://1803candles.com/blog/1803-candles-top-10-scents-for-april-2015/.

Chapter 8

1   T. Steele & Son, *What Shall I Buy for a Present: A Manual* ([Hartford: n.p., 1877]), 3, 5, 9.

2   Steele & Son, *What Shall I Buy*, 10, 12, 58.

3   Steele & Son, *What Shall I Buy*, 19.

4   "A Splendid Establishment," *Hartford* (CT) *Daily Courant*, December 13, 1875.

5   On the complex problems of the commodification of antiques, see Briann Greenfield, *Out of the Attic: Inventing Antiques in 20th-Century New England* (Amherst: University of Massachusetts Press, 2009), 28–35. See also Kent Grayson and Radan Martinec, "Consumer Perceptions of Iconicity and Indexicality and Their Influence on Assessments of Authentic Market Offerings," *Journal of Consumer Research* 31.2 (September 2004), 296–312.

6   Ellen Litwicki writes, "As Americans increasingly expressed affection through gift exchange, the market responded with a growing array of commodities that might be transformed into presents." "From the 'ornamental and evanescent' to 'good, useful things': Redesigning the Gift in Progressive America," *Journal of the Gilded Age and Progressive Era* 10.4 (October 2011), 476.

7   Mari Yoshihari, *Embracing the East: White Women and American Orientalism* (New York: Oxford University Press, 2003), 21, 23–4.

8   Kristin L. Hoganson, *Consumers' Imperium: The Global Production of American Domesticity, 1865–1920* (Chapel Hill: University of North Carolina Press, 2007), 33–4 and chap. 1 generally.

9   Yoshihari, *Embracing the East*, 26.

10   Fireside Studios, *Fireside Gifts* ([Adrian, MI?]: n.p., ca. 1931]), 35, 21, 38.

11   Edward W. Said, *Orientalism* (New York: Vintage, 1979), 108.

12   Fireside Studios, *Objets des Arts by Fireside* ([Adrian, MI?]: n.p., [1931]), 28, 42; Fireside Studios, *Fireside Gifts*, 36.

13   Russell Belk refers to this process as "positive contamination." See "Possessions and the Extended Self," *Journal of Consumer Research* 15.2 (September 1988), 139–68.

14   Fireside Studios, *Fireside Gifts*, 38–9.

15   G. C. Allen, "Japanese Industry: Its Organization and Development to 1937," in *The Industrialization of Japan and Manchukuo, 1930–1940*), ed. Elizabeth Boody Schumpeter (New York: Macmillan, 1940), 539–40.

16   James Rorty, *Our Master's Voice: Advertising* (New York: John Day, 1934), 216.

17   As Pierre Bourdieu has pointed out, "The whole relation to the work of art is changed when the painting, the statue, the Chinese vase or the piece of antique furniture belongs to the world of objects available for appropriation, thus taking its place in the series of the luxury goods which one possesses and enjoys *without needing to prove the delight they give and the taste they illustrate*, and which, even when not personally possessed, belong to the status attributes of one's group." *Distinction: A Social Critique of the Judgement of Taste*, trans. Richard Nice (Cambridge, MA: Harvard University Press, 1984 [1979]), 273, 278, my emphasis.

18   The *Oxford English Dictionary* contains no entry for "giftware." Only a handful of references to giftware appear in early twentieth-century American newspapers. One of the first occurs in an advertisement for "Law, The Gift-ware Man," in the December 23, 1907, issue of the *Springfield* (MA) *Republican*. Google's Ngram viewer shows "giftware" beginning to appear in meaningful numbers in the 1920s: https://books.google.com/ngrams/graph?content=giftware&year_start=1800&year_end=2000&corpus=15&smoothing=3&share=&direct_url=t1%3B%2Cgiftware%3B%2Cc0.

19   "Art and Gift Show Opens," *New York Times*, March 13, 1928.

20   In 1934 it became *Gift and Art Buyer*; it was renamed *Gifts & Decorative Accessories* in the mid-1960s.

21   Art-in-Trade was sponsored by Macy's as it sought to connect art and industry and thus become a cultural arbiter that could profit from "commercializing art while aestheticizing modern manufactured products." See "Art for Trade's Sake: The Fusion of American Commerce and Culture, 1927–1934," xroads.virginia .edu/~ma03/pricola/art/trade.html.

22   "Gift Shows Attract Many Buyers," *New York Times*, February 26, 1930.

23   "Vote Further Gift Shows," *New York Times*, September 5, 1928.

24   Don Herold, "We Can't Be Pikers," *Life* 98.2547 (August 28, 1931), 23.

25   W. W. Scott, "The Advertisement Reader Remembers Christmas," *Life* 92.2405 (December 7, 1928), 12.

26   *Life* 94.2457 (December 6, 1929), 27.

27   "Gift and Art Shows Open," *New York Times*, August 21, 1932.

28   "Trade Shows Opened," *New York Times*, January 8, 1933; "Gift Wares Buying Up Sharply," *New York Times*, August 24, 1933.

29   "Gift Ware Show Opens," *New York Times*, February 20, 1934; "Gift Show Opens Here," *New York Times*, August 21, 1934.

30   "Gift Show Opens; Buying Is Heavy," *New York Times*, February 26, 1935.

31   "Luxury Items Favored," *New York Times*, February 24, 1937.

32   "Rural Stores Buy Gifts," *New York Times*, August 25, 1937.

33   "Larger Stores Buying Giftwares," *New York Times*, February 24, 1938.

34   "American Gift Ware Boosted by the War," *New York Times*, August 4, 1940.

35   "American Gift Ware Boosted by the War."

36   "U.S. Goods Feature Gift Exhibit Here," *New York Times*, February 25, 1941.

37   Robert W. Kellogg Company, *Kellogg Selections for 1944* ([Springfield, MA?: n.p., 1944]), i, 1, 7, 11, 26.

38   Kellogg Company, *Kellogg Selections for 1944*, 10, 8, 24.

39   "Pottery Makers Go Modern," *BusinessWeek*, September 13, 1947.

40   "World Trade Fair Lists 5,000 Buyers," *New York Times*, August 7, 1950; "Sales Are Up 100% in Giftware Show," *New York Times*, February 10, 1951.

41   "U.S. Aids Ex-Enemy Economies," *BusinessWeek*, November 23, 1946, 15–16.

42   On marriage rates during the twentieth century, see Elaine Tyler May, *Homeward Bound: American Families in the Cold War Era* (New York: Basic Books, 1988), esp. 21, 79–91.

43   Norma Lanum, "You Needn't Get an Umbrella Rack," *Washington Post*, May 18, 1950.

44   May, *Homeward Bound*, 165. Spending on home furnishings rose by 240 percent; food expenditures increased by a more modest 33 percent, and clothing by 20 percent.

45   William H. Whyte Jr., *The Organization Man* (New York: Doubleday Anchor Books, 1957), 345, 347.

46   Madison House, *Gift Digest* ([Boston?]: n.p., 1953), 8, 9.

47   Bancroft's, *Bancroft's Out of This World Selections* ([Chicago]: n.p., ca. 1950s), 59, 3.

48   Whyte, *Organization Man*, 388.

49   Whyte, *Organization Man*, 177.

50   Game Room, *You Won't Believe This But . . .* ([Washington, DC: n.p., 1956?]).

51   *Gentry Gift Guide* ([New York]: n.p., ca. 1952); Alfred B. Zipser, "Displays Aim at Giver and Gourmet," *New York Times*, August 26, 1958.

52   Ann Peppart, "Dad's Game Room Solved Their Fuel Problem," *Better Homes and Gardens* 22 (November 1943), 8.

53   Game Room, *You Won't Believe This But . . .*

54   "Initialled Sales," *Business Week*, February 9, 1935, 23.

55   Anne Means, "Monograms in the Home," *American Home*, November 1933, 281.

56   Kellogg Company, *Kellogg Selections for 1944*, 8.

57   See, e.g., "Shop-hound: Tips on the Shop Market," *Vogue* 84.10 (November 15, 1934); Hanna Tauchau, "Marking Gift Linen," *American Home* 15 (December 1935), 34; and "200 Monogrammed Cigarettes," *New York Times*), December 13, 1936.

58   Madison House, *Gift Digest* ([Boston]: Madison House, 1953), 12.

59   Lillian Vernon, *An Eye for Winners: How I Built One of the Greatest Direct-Mail Businesses* (New York: HarperBusiness, 1996), 41.

60   Vernon, *An Eye for Winners*, 78.

61   "The Protocol of Monogramming," *House and Garden* 125 (June 1964), 38.

62   National Gift and Art Association, *44th Semi-Annual New York Gift Show* ([New York]: n.p., [1953]), 158.

63   National Gift and Art Association, *56th Semi-Annual New York Gift Show*, 89, 124.

64   Between 1939 and 1946, sales volumes increased an estimated 22–49 percent, representing thousands of retail shops and production concerns. Robert Shosteck, *Careers in Retail Business Ownership* (Washington, DC: B'nai B'rith Vocational Service Bureau, 1946), 325. National Gift and Art Association, *44th Semi-Annual New York Gift Show*, 332, 155, 171, 179.

65   "U.S. Ingenuity Helps Improve Foreign Crafts," *New York Times*, April 18, 1960.

66   Norcrest China Co., *Fine China and Gifts for 1959–60* ([Portland, OR]: Sweeney, Krist & Dimm, [1959]), 13.

67   Madison House, *Gift Digest*, 15; Bancroft's, *Bancroft's Out of This World Selections*, 14, 10; Giftime, *Shopping's Fun . . . Shopping's Done with These Exciting Gift Ideas* ([Philadelphia]: n.p., [1959]), 4; and Helen Gallagher-Foster House, *Fall/Winter 1964-65* ([Peoria?]: n.p., [1964]). The phrase "gayer oddities" appears in "Gift Shop Shows Many New Items," *New York Times*, September 7, 1951.

68   "Wants Quality Items," *Souvenirs & Novelties* 22 (June–July 1966), 6.

69   Eleanore Ziegler [Sloan], "Don't Ignore Customer's Intelligence," *Souvenirs & Novelties* 22 (June–July 1966), 18, 21.

Chapter 9

1   Steven Gelber refers to all objects that are collected as "collectibles," distinguishing between "primary" collectibles, manufactured specifically to be collected, and "secondary" ones—items made for another use. *Hobbies: Leisure and the Culture of Work in America* (New York: Columbia University Press, 1999), 63. I am using the term in Gelber's "primary" sense only.

2   On cabinets of curiosity, see Anthony Alan Shelton, "Cabinets of Transgression: Renaissance Collections and the Incorporation of the New World," in *The Cultures of Collecting*, ed. John Elsner and Roger Cardinal (Cambridge, MA: Harvard University Press, 1994), 177–203.

3   John Elsner and Roger Cardinal, introduction to Elsner and Cardinal, *The Cultures of Collecting*, 2.

4   Werner Muensterberger, *Collecting: An Unruly Passion: Psychological Perspectives* (Princeton, NJ: Princeton University Press, 1994), 3.

5   See, e.g., William Davies King, *Collections of Nothing* (Chicago: University of

Chicago Press, 2008); Philipp Blom, *To Have and to Hold: An Intimate History of Collectors and Collecting* (New York: Overlook Press, 2003); Russell Belk, "Possessions and the Extended Self," *Journal of Consumer Research* 15.2 (September 1988), 139–68; and Mihaly Csikszentmihalyi and Eugene Rochberg-Halton, *The Meaning of Things: Domestic Symbols of the Self* (Cambridge: Cambridge University Press, 1981). The literature on collecting is too broad to recap here, as it encompasses psychology, consumer culture, fine art, colonialism and imperialism, material culture, and history, among other fields.

6   Alvin F. Harlow, *Paper Chase: The Amenities of Stamp Collecting* (New York: Henry Holt, 1940), 25.

7   "Postage Stamps," *Boston Daily Advertiser*, September 14, 1860; "Collecting Postage Stamps," *Albany Evening Journal*, November 10, 1860. A circa 1890s circular issued by Henry Grossman of Asheville, North Carolina, announced he was "buying any kind of United States cancelled Postage Stamps of old and present issues, with the exception of the very commonest kind." Grossman bought by the thousand and considered "small lots" to be any numbering fewer than ten thousand stamps. "Henry Grossman, Another Early Florida Stamp Dealer," *Florida Postal History Society Journal* 20.1 (January 2013), 18.

8   This distinction has been long lasting and naturalized within the collecting world itself. At least one book, John Bedford's *The Collecting Man: A Concise Guide to Collecting as a Hobby and as Investment* (New York: David McKay, 1968), concerns itself with male collectors. He writes baselessly that "man is usually more interested than woman in history, in technical matters, in expertise, in the odd, and the curious; and when he sets out to collect he generally wants to make a real job of it," (7). For an example of the "Woman Collectors' Department," see *Philatelic West and Collectors' World* 35.1 (January 31, 1907), [30].

9   Ada Walker Camehl, *The Blue-China Book: Early American Scenes and History Pictured in the Pottery of the Time* (New York: Halcyon House, 1916), xiii. See also J. Samaine Lockwood, "Shopping for the Nation: Women's China Collecting in Late-Nineteenth-Century New England," *New England Quarterly* 81.1 (March 2008), 63–90.

10   Quoted in Elizabeth Stillinger, *The Antiquers: The Lives and Careers, the Deals, the Finds and the Collections of the Men and Women Who Were Responsible for the Changing Taste in American Antiques, 1850–1930* (New York: Alfred A. Knopf, 1980), 190, xiv.

11   "The Rage for Old Furniture," *Harper's Weekly*, November 16, 1878, 918. On the rise of antiquing during this time, see Gelber, *Hobbies*, esp. 129–33.

12   Irving W. Lyon, *Colonial Furniture of New England* (Boston: Houghton, Mifflin, 1891); Stillinger, *Antiquers*, xi.

13   On scrapbooks, see Ellen Gruber Garvey, *Writing with Scissors: American Scrapbooks from the Civil War to the Harlem Renaissance* (Oxford: Oxford University Press, 2012); Jessica Helfand, *Scrapbooks: An American History* (New Haven, CT: Yale University Press, 2008); and Susan Tucker, Katherine Ott, and Patricia Buckler, eds., *The Scrapbook in American Life* (Philadelphia: Temple University Press, 2006). Looking back on this era, philatelist Alvin Harlow remarked, "We are the most indefatigable nation of collectors in the world." He described the "fad which raged violently for years" for picture postcards following the Chicago World's Fair in 1893. Postcard collecting even had its own neologism, "philocarty," elevating it to the status of philately. Harlow claimed there was an entire book devoted to streetcar transfers, which "discuss[ed] with

a gravity worthy a scientific treatise such subjects as dating, types, condition, coloring, reversibility, the forming, arrangement, and indexing of the collection, and so on!" *Paper Chase*, 3.

14   "The Collector and the Poster," *Harper's Weekly*, February 9, 1895, 123.

15   George B. James Jr., comp., *Souvenir Spoons. Containing Descriptions and Illustrations of the Principal Designs Produced in the United States* (Boston: A. W. Fuller, 1891), i.

16   Edmund B. Sullivan, *American Political Badges and Medalets, 1789–1892*, 2nd edition (Lawrence, MA: Quarterman Publications, 1981; orig. pub. 1959).

17   Howland Wood, *The Commemorative Coinage of the United States*, ANS Numismatic Notes and Monographs 16 (New York: American Numismatic Society, 1922), 2.

18   Wood, *Commemorative Coinage*, 3, 4, 9.

19   See Marian Klamkin, *American Patriotic and Political China* (New York: Charles Scribner's Sons, 1973), esp. 60–9.

20   Mason Locke Weems, *The Life of Washington*, ed. Marcus Cunliffe (Cambridge, MA: Belknap Press of Harvard University Press, 2001/1962/1809), xv, xvxviii–xix, my emphasis. Weems assured Carey the biography would sell "with great rapidity for 25 or 37 Cents and it w^d not cost 10" (xv).

21   For a more complete inventory, see Robert H. McCauley, *Liverpool Transfer Designs on Anglo-American Pottery* (Portland, ME: Southworth-Anthoensen Press, 1942).

22   George L. Miller, "George M. Coates, Pottery Merchant of Philadelphia, 1817–1831," *Winterthur Portfolio* 19.1 (Spring 1984), 45.

23   Christina H. Nelson, "Transfer-Printed Creamware and Pearlware for the American Market," *Winterthur Portfolio* 15.2 (Summer 1980), 93–115. Account books of early nineteenth-century American pottery dealers show that cheap British chinaware was becoming increasingly fashionable. Philadelphia dealer George M. Coates, to name but one, sold over forty-nine thousand pieces to his rural customers (about 25 percent of his sales) between 1824 and 1830. See Miller, "George M. Coates," 45–6.

24   Phoebe Lloyd Jacobs, "John James Barralet and the Apotheosis of George Washington," *Winterthur Portfolio* 12 (1977), 32.

25   Barry Schwartz, "The Social Context of Commemoration: A Study in Collective Memory," *Social Forces* 61.2 (December 1982), 374.

26   Catherine E. Kelly notes that in the early republic objects from textiles to jewelry, signage, and crockery became "vehicles" to circulate George Washington's image. In many cases, a likeness of Washington would "enhance" an object, no matter how humble. *Republic of Taste: Art, Politics, and Everyday Life in Early America* (Philadelphia: University of Pennsylvania Press, 2016), 226.

27   The exceptions were gift books purchased annually, which were popular in the antebellum era and, as discussed, souvenir spoons. My thanks to Mark Barrow for the information on Bing & Grøndahl.

28   "The Plate That Took the Blue Ribbon!," *Hobbies: The Magazine for Collectors* 55.1 (March 1950), 86.

29   Herschell Gordon Lewis, "How 'Glad' Can Limited Tidings Be?," *Direct Magazine*, December 2003, 70.

30   Lauritz Coopersmith, "The Wall Street of the Plate Field: The Bradford Exchange," *Collectibles Illustrated* 2.4 (July/August 1983), 29.

31   "Collectible Industry Thriving in the U.S.," *Globe and Mail*, November 28, 1980.

32   Coopersmith, "The Wall Street of the Plate Field," 39.

33   Coopersmith, "The Wall Street of the Plate Field," 39.

34   Linda Ellis Fishbeck, "A Plate Portfolio: Breaking into Limited Editions," *Rarities* 1.1 (Fall 1980), 51.

35   "The Great Plate Explosion," *Collectibles Illustrated* 2.4 (July/August 1983), front cover, 25, 26.

36   "Collectible Industry Thriving in the U.S."

37   David Alexander, "The Rise and Fall of the Franklin Mint," *Coin Week*, December 28, 2015, www.coinweek.com/education/the-rise-and-fall-of-the-franklin -mint/.

38   Current assay value of collectible coins can be found at https://www.ngccoin .com/price-guide/coin-melt-values.aspx?MeltCategoryID=1 and can be compared with collectible value on sites such as eBay.

39   Alexander, "The Rise and Fall."

40   Alexander, "The Rise and Fall."

41   WK, "Product: Carson City Gold (1870–1893)," sell sheet from *The Franklin Mint Coin Redbook* (hereafter, *TFMCR*), ca. September 16, 2010. The *Redbook* is a three-ring binder containing internal sell sheets for sales agents (sales scripts and data) for over a hundred coins and coin sets. In the author's collection and purchased at the liquidation auction of the Franklin Mint Museum and Archive, October 20, 2016.

42   WK, "Product: 19th Century Morgan Silver Dollar" sell sheet, *TFMCR*, original emphasis.

43   WK, "Product: U.S. Mis-struck Error Coin Collection (Set of 4)" sell sheet, *TFMCR*.

44   WK, "Product: 2009 Silver Proof Coins—District of Columbia and U.S. Territories—6-coin Set—PF-69 Set" sell sheet, *TFMCR*.

45   WK, "Washington Presidential Dollar Error Coin 2007 (Missing Edge)" sell sheet, *TFMCR*, original emphasis.

46   WK, "Indian Head Gold Piece $20 'Saint-Gaudens' with Motto (1907–1933), *TFMCR*, original emphasis.

47   WK, "The First Five Years of the U.S. Silver Eagle Dollar—(1986–1990) Brilliant Uncirculated" sell sheet, *TFMCR*, original emphasis.

48   WK, "The Ultimate Nickel Collection—12 Coin Collection" sell sheet, *TFMCR*.

49   WK, "Presidential Dollar Completion Program (Pre-Pay)" sell sheet, *TFMCR*, original emphasis.

50   WK, "Presidential Dollar Completion Program (Pre-Pay)."

51   Indicating just how bogus these claims were, the company was sued by both Tiger Woods and the Princess Diana Memorial Fund because it did not have the rights to issue "authentic" and "official" collectibles featuring the celebrities. See Marius Meland, "Appeals Court Rejects Franklin Mint, Princess Diana Settlement," *Law360*, July 19, 2005, https://www.law360.com/articles/3730/appeals -court-rejects-franklin-mint-princess-diana-settlement.

52   This is a prime feature of humbug, which Max Black defines, in part, as "deceptive misrepresentation short of lying, especially by pretentious word or deed." "The Prevalence of Humbug," in *The Prevalence of Humbug and Other Essays* (Ithaca, NY: Cornell University Press, 1983), and at http://www.ditext.com/black /humbug.html.

53   WK, "Color Enhanced State Quarters Collector's Watch" sell sheet, *TFMCR*.

54   WK, "Washington Presidential Dollar Error Coin 2007 (Missing Edge)."

55   "Why Rarities," *Rarities* 1.1 (Fall 1980), 3, my emphasis.

56   "Give the Gift of Savvy," *Collectibles Illustrated* 2.3 (May/June 1983), 23.

57   Liane McAllister, "A Retailer's Guide to Collectors Clubs," *Gifts & Decorative Accessories* 93.10 (October 1, 1992), 1.

58   Malcolm Berko, "You've Been Snookered," Creators Syndicate, 2012, http://legacy.creators.com/lifestylefeatures/business-and-finance/taking-stock/-quot-you've-been-snookered-quot.html.

59   "Franklin Mint Review," BuySilver.org, © 2011, https://www.buysilver.org/online-dealers/franklin-mint-review/. See also Federal Trade Commission, "Investing in Collectible Coins," *FTC Consumer Alert*, 2011.

60   "Franklin Mint Review."

61   Wendy Cuthbert, "Little Profit in Plates: From Elvis to Puppies, Be Wary of Investment Potential," *Financial Post*, October 21, 1993, 27.

62   *Charal v. Andes*, 81 F.R.D. 99 (E.D. Pa. 1979). I would like to thank Jay Stiefel for this reference.

63   Walter Benjamin's seminal piece from 1936 is relevant here: "The Work of Art in the Age of Its Technological Reproducibility [First Version]," trans. Michael W. Jennings, *Grey Room* 39 (Spring 2010), 11–37.

Chapter 10

1   Bonita LaMarche, "Staffordshire Figurines from the Jerome Irving Smith Collection," *Bulletin of the Detroit Institute of Arts* 65.4 (1990), 41, 43. See also P. D. Gordon Pugh, *Staffordshire Portrait Figures and Allied Subjects of the Victorian Era* (Woodbridge, Suffolk: Antique Collectors' Club, 1970, rev. and rpt. 1987); and Reginald G. Haggar, *Staffordshire Chimney Ornaments* (London: Phoenix House, 1955).

2   Simon Shaw, *History of the Staffordshire Potteries; and the Rise and Progress of the Manufacture of Pottery and Porcelain* (1829; rpt. London: Scott, Greenwood, 1900), xix.

3   World Collectors Net, "Victoriana, Victorian Staffordshire and Victorian Collectables," posted October 4, 2013, http://www.worldcollectorsnet.com/articles/victoriana-victorian-staffordshire-victorian-collectables/.

4   Thomas Balston, *Staffordshire Portrait Figures of the Victorian Age* (London: Faber & Faber, 1958), 15. Experts have written, wryly, that flatbacks "seem now to have a childlike charm, perhaps because they were actually made by children." "A–Z of Ceramics," Victoria and Albert Museum, https://www.vam.ac.uk/articles/a-z-of-ceramics.

5   See Haggar, *Staffordshire Chimney Ornaments*, who points to a house sale inventory from 1796 listing "six large and elegant China Chimney Ornaments" yet describes the pieces as having only "*intrinsic* value": i.e., no real monetary worth (14). See also Louise Stevenson, "Virtue Displayed: The Tie-Ins of *Uncle Tom's Cabin*," http://utc.iath.virginia.edu/interpret/exhibits/stevenson/stevenson.html. For additional statistics, see "Imports and Exports of the United States," *Niles' Weekly Register*, May 30, 1835, 220.

6   LaMarche, "Staffordshire Figurines," 45.

7   The British cottagers who collected these figurines were characterized as "fervently patriotic, sentimental, and nonconformist in their religious beliefs, and great lovers of animals. They scarcely read at all." LaMarche, "Staffordshire Figurines," 47.

8   LaMarche, "Staffordshire Figurines," 46; Haggar, *Staffordshire Chimney Ornaments*, 15. See also Samuel S. Scriven, *Employment of Children and Young*

Persons in the District of the North Staffordshire Potteries (1843); and Children's Employment Commission (1862), *First Report of the Commissioners* (London: George Edward Eyre and William Spottiswoode, 1863), 1–4 for statistics and vii–xlix for descriptions of harrowing and hazardous working conditions.

9   *Encyclopedia of Antique Restoration and Maintenance* (New York: Clarkson N. Potter, 1974), 26, 27.

10   Because of its porous surface and propensity to chip and scratch, plaster is a very fragile and ephemeral material. But it was cheap and abundant—a hundred thousand tons of plaster were shipped to the United States in 1832 alone. "Art. IX—Remarks on the Mineralogy and Geology of the Peninsula of Nova Scotia . . . ," *American Monthly Review* 1.5 (May 1832), 402.

11   James Fenimore Cooper, *Excursions in Italy*, vol. 1 (London: Richard Bentley, 1838), 177.

12   Calvert Vaux, *Villas and Cottages* (New York: Harper & Bros., 1857), 84.

13   Quoted in Steven Gelber, *Hobbies: Leisure and the Culture of Work in America* (New York: Columbia University Press, 1999), 137. See also Elizabeth Stillinger, *The Antiquers: The Lives and Careers, the Deals, the Finds and the Collections of the Men and Women Who Were Responsible for the Changing Taste in American Antiques, 1850–1930* (New York: Alfred A. Knopf, 1980), 256.

14   Thomas E. Hudgeons III and Nancy Smith, eds., *The Official 1982 Price Guide to Hummel Figures & Plates* (Orlando, FL: House of Collectibles, 1981), 7.

15   Susan Stewart observes, "The miniature always tends toward tableau rather than toward narrative, toward silence and spatial boundaries rather than toward expository closure. Whereas speech unfolds in time, the miniature unfolds in space. The observer is offered a transcendent and simultaneous view of the miniature, yet is trapped outside the possibility of a lived reality of the miniature. Hence the nostalgic desire to present the lower classes, peasant life, or the cultural other within a timeless and uncontaminated miniature form." *On Longing: Narratives of the Miniature, the Gigantic, the Souvenir, the Collection* (Durham, NC: Duke University Press, 1996), 66.

16   Ken Armke, *Hummel: An Illustrated Handbook and Price Guide* (Radnor, PA: Wallace-Homestead Book Company, 1995), 207–8. Collectors have to make a special pilgrimage to the Hummel Museum in Texas "to see the rare figurine *Silent Night with Black Child*"—so rare, its picture is not even included in key price guides (177).

17   Schmid purchased Hummels from Goebel beginning in 1937. See also Heidi Ann Von Recklinghausen, *The Official M. I. Hummel Price Guide, 2nd Edition, Figurines & Plates* (Iola, WI: Krause Publications, 2013), 144.

18   Marie Lynch, ed., *The Original "Hummel" Figures in Story and Picture* (Boston: Schmid Brothers, 1955), 3.

19   Lynch, *Original "Hummel" Figures*, 3.

20   Lynch, *Original "Hummel" Figures*, [4].

21   Hudgeons and Smith, *Official 1982 Price Guide*, 7.

22   Hudgeons and Smith, *Official 1982 Price Guide*, 13.

23   Armke, *Hummel*, 24–5.

24   Liane McAllister, "Collectibles/Gift Mix Boosts Sales for the '90s," *Gifts & Decorative Accessories* 91.9 (September 1990), 54.

25   N. R. Kleinfield, "Among Hummel Fans, Details Mean So Much," *New York Times*, September 20, 1990.

26 Some scholars argue that in the world of fine art and antiques collecting, "competition in social markets may magnify small differences in perceived quality among classes of objects into very large differences in equilibrium prices." Collectors are constantly looking for verification from experts, and emulating collecting "leaders" when they consider what to buy. Gary Becker and Kevin Murphy (with William Landes), *Social Economics: Market Behavior in a Social Environment* (Cambridge, MA: Belknap Press of Harvard University Press, 2000), 79.

27 For more on neoteny and animals, see Ralph H. Lutts, "The Trouble with Bambi: Walt Disney's *Bambi* and the American Vision of Nature," *Forest & Conservation History* 36.4 (October 1992), 160–71.

28 George Monaghan, "Precious Moments: Tiny Figures Become Giants as Popular Collectibles," *Star-Tribune* (Minneapolis–St.Paul, MN), December 30, 1986.

29 Dallie Miessner, *The Precious Moments Story: Collectors' Edition* (Huntington, NY: Portfolio Press, 1986).

30 Eric Morgenthaler, "Host of Characters: Precious Moments Is Brilliant at Answering Its Collectors' Prayers," *Wall Street Journal*, September 14, 1993.

31 Morgenthaler, "Host of Characters."

32 Monaghan, "Precious Moments."

33 Monaghan, "Precious Moments." As with other forms of kitsch, "latent problems are projected into objects whose value is purely fictitious, because it makes things easier." Aleksa Celebonovic, "Notes on Traditional Kitsch," in *Kitsch: The World of Bad Taste*, ed. Gillo Dorfles (New York: Bell Publishing, 1969), 289. Celebonovic describes kitsch as "objects which have no real use or meaning" and that only offer "an image of well-being," allowing "the master of the house to abandon themselves to a game imbued with a puerile and immature imagination" (280).

34 Too little has been written about race and collecting, though see Elvin Montgomery, "Recognizing Value in African American Heritage Objects," *Journal of African American History* 89.2 (Spring 2004), 177–82. Most scholarship leaves the impression that African Americans tend to collect "Black Americana" (i.e., overtly racist materials as a way to reclaim them) or "tribal art." Neither conclusion has been borne out by hard data. See, e.g., Gloria Canada, "Living with Black Americana," *Antiques and Collecting*, September 1990, 28–31; Stacey Menzel Baker, Carol M. Motley, and Geraldine R. Henderson, "From Despicable to Collectible: The Evolution of Collective Memories for and the Value of Black Advertising Memorabilia," *Journal of Advertising* 33.3 (Autumn 2004), 37–50; Kenneth Goings, *Mammy and Uncle Mose: Black Collectibles and American Stereotyping* (Bloomington: Indiana University Press, 1994); Eric Robertson, "African Art and African-American Identity," *African Arts* 27.2 (April 1994), 1, 6, 8, 10; and Paula Rubel and Abraham Rosman, "The Collecting Passion in America," *Zeitschrift für Ethnologie* 126.2 (2001), 313–30.

35 Benj Gallander and Ben Stadelmann, "Can a Wizard's Spell Save Kitsch Merchandiser Enesco?," *Globe and Mail*, November 25, 2000.

36 Miessner, *Precious Moments Story*, 230, 241.

37 McAllister, "Collectibles/Gift Mix Boosts Sales for the '90s," 54.

38 Liane McAllister, "A Retailer's Guide to Collectors Clubs," *Gifts & Decorative Accessories* 93.10 (October 1992), 1.

39 Mary Ann Fergus, "Invasion of the Beanie Babies," *Pantagraph*, June 23, 1996, C1.

40 Giles Austin, "The Man behind the Beanie Baby—Profile—Ty Warner," *Times*

(London), November 11, 1999. Warner boasted, "We filled three 747s from Korea and China with tens of thousands of Beanie Babies to get them here for Easter."

41   Dave Saltonstall, "Beanie Boom Goin' Bust? Cuddly Toys Losing Steam," *New York Daily News*, June 14, 1998.

42   Becker and Murphy note that rationing is a form of advertising that helps make items seem popular and desirable (and therefore hard to come by), while also creating publicity that "raises future demand for these goods, or raises demand for other goods that [the producers] sell." *Social Economics*, 140.

43   "Interview: Beanie Babies Spokesperson Pat Brady Discusses the Newest Beanie Baby," NBC News, *Today*, July 9, 1998. Transcript accessed via Factiva.com, lightly edited for clarity.

44   Austin, "The Man behind the Beanie Baby."

45   Anne VanderMey, "Lessons from the Great Beanie Babies Crash," *Fortune*, March 11, 2015, http://fortune.com/2015/03/11/beanie-babies-failure-lessons/. See also Zac Bissonnette, *The Great Beanie Baby Bubble: The Amazing Story of How America Lost Its Mind over a Plush Toy—and the Eccentric Genius behind It* (New York: Portfolio, 2015).

46   Karen Thomas, "Beanie Baby to Honor Diana: Purple Bear Available Soon," *USA Today*, December 5, 1997.

47   "Ask Dr. Beanie," *Pantagraph*, August 3, 1997.

48   Hattie Kaufman, "Are Beanie Babies Retired for Good?" *CBS This Morning*, September 2, 1999. Transcript accessed via Factiva.com.

49   "Beanie Heists Continue in Syracuse," *Associated Press Newswires*, October 1, 1998; Chris Clair, "Thousands in Beanies Stolen in Rolling Meadows," *Chicago Daily Herald*, October 2, 1999.

50   "Couple Told to Split Beanie Babies," AP Online, November 5, 1999.

51   Austin, "The Man behind the Beanie Baby."

52   Saltonstall, "Beanie Boom Goin' Bust?" See also William S. McTernan, "It's a Beanies Bear Market: Dump What You Can, Experts Advise," *Austin American-Statesman*, October 8, 1998. Becker and Murphy (with Edward Glaeser) note that copies—whether look-alikes or outright counterfeits—"can greatly dilute the exclusivity of trademarked goods" and thus can easily harm the market. *Social Economics*, 98.

53   Kauffman, "Are Beanie Babies Retired for Good?"

54   Kate N. Grossman, "Ty Inc. Putting Beanies' Fate to a Vote," Associated Press, December 25, 1999.

55   Mary Ethridge, "Past Their Prime, Beanie Babies Celebrate Their 10th Birthday," *Akron Beacon Journal*, March 16, 2003.

56   "Beanie Baby Collection to Benefit Children Overseas," *Record* (Stockton, CA), May 30, 2011.

57   This conforms to Michael Thompson's "rubbish theory." At some point, Beanie Babies, like Stevengraphs, might shift in category from rubbish to transient to durable. But this is only likely if the majority of Beanie Babies vanish. *Rubbish Theory: The Creation and Destruction of Value* (Oxford: Oxford University Press, 1979).

58   Meredith Schwartz, "Is the Collectibles Industry Proving to Be a Limited Edition?" *Gifts & Decorative Accessories* 101.6 (June 2000), 123.

59   Craig Wilson, "Happiness Is a Hummel; Figurines Fetch Big Bucks at Fair," *USA Today*, July 24, 1989.

60   Morgenthaler, "Host of Characters: Precious Moments."

61   "Madison County Prosecutor Says He Will Fight Clemency Requests," Associated Press, June 10, 2002.
62   Liane McAllister, "Is There a Future in Collectibles?" *Gifts & Decorative Accessories* 99.6 (June 1998), 95.
63   McAllister, "Is There a Future."
64   McAllister, "Collectibles/Gift Mix Boosts Sales for the '90s," 54.
65   Cook DuPage, "Enesco to Lay off 120 in 2 Locations," *Chicago Daily Herald*, May 4, 2001.
66   Julie Jargon, "Ending Deal to Sell Precious Moments; Fragile Sales of Figures Prompt Firm to Terminate Distribution License Early," *Crain's Chicago Business* 28.22 (May 30, 2005), 12.

Chapter 11

1   "An Essay on Novelty," *Monthly Miscellany* 1.4 (April 1774), 183.
2   "Novelty," *New-York Gazette*, April 4, 1821; "Agricultural Novelty," *Southern Chronicle* (Camden, SC), November 19, 1823; "A Novelty," *Southern Chronicle* (Camden, SC), February 12, 1823; "The following account . . . ," *Hallowell* (ME) *Gazette*, December 29, 1819; "The Sea Serpent," *Lancaster* (PA) *Journal*, September 5, 1817.
3   Peter Benes, *For a Short Time Only: Itinerants and the Resurgence of Popular Culture in Early America* (Amherst: University of Massachusetts Press, 2016); Brett Mizelle, "'I Have Brought my Pig to a Fine Market': Animals, Their Exhibitors, and Market Culture in the Early Republic," in *Cultural Change and the Market Revolution in America, 1789–1860*, ed. Scott C. Martin (Lanham, MD: Rowman & Littlefield, 2005), 181–216. See also, e.g., "American Museum," *National Advocate* (New York), July 3, 1822; and "Anniversary of American Independence," *National Advocate* (New York), July 4, 1822.
4   "LAST NIGHT. At Mr. Bulet's Assembly Rooms," *American Commercial Daily Advertiser* (Baltimore, MD), June 25, 1816.
5   "Novelty," *Eastport* (ME) *Sentinel*, February 21, 1824.
6   Daniel Wickberg, *The Senses of Humor: Self and Laughter in Modern America* (Ithaca, NY: Cornell University Press, 2015), 122. Chap. 4 specifically discusses "The Commodity Form of the Joke."
7   By the late 1860s there were tens of thousands of traveling sales agents. See Timothy B. Spears, *100 Years on the Road: The Traveling Salesman in American Culture* (New Haven, CT: Yale University Press, 1995), 53.
8   "Aesthetic of abundance" comes from Miles Orvell, *The Real Thing: Imitation and Authenticity in American Culture, 1800–1940* (Chapel Hill: University of North Carolina Press, 1989), 42. The aesthetic might have its origins in "fancy," which created more intense experiences by combining dissimilar things. See David Jaffee, *A New Nation of Goods: The Material Culture of Early America* (Philadelphia: University of Pennsylvania Press, 2011), 240.
9   Fargo Novelty Company, *Illustrated Catalogue Mail Order* ([Frenchtown, NJ?]: n.p., [1908]), [28].
10 Eureka Trick and Novelty Co., *Illustrated Manual of Tricks, Novelties, Musical Instruments, Scientific Toys, &c., &c.* (New York: Eureka Trick and Novelty Co., 1877), 27.
11 The trade in novelty goods, in fact, accounted in large measure for an uptick in commerce between Japan and the United States. Gary Cross and Gregory Smits, "Japan, the U.S., and the Globalization of Children's Consumer Culture,"

Journal of Social History 38.4 (Summer 2005), 876. Early American wholesalers likely forged connections with merchants in the southern Pacific Rim: novelty producer and distributor Johnson Smith & Co. began in Australia before moving to Chicago. A New Zealand newspaper advertisement from 1895 promoted a "Great Exhibition of Japanese Novelties . . . Specially Suitable for Presents. Direct Shipment" (*Otago* [New Zealand] *Daily Times*, November 25, 1895). In 1916 in New York City alone there were twelve different *categories* of novelty suppliers and well over 660 novelty shops and stores. *R. L. Polk & Co.'s 1916 Trow's New York City Directory* (New York: R. L. Polk & Co., 1916), 2219–20. On novelty, planned obsolescence, and Americanness, see Robert Lekachman, "The Cult of Novelty," *Challenge* 8.7 (April 1960), 7–11; and Edmund W. J. Faison, "The Neglected Variety Drive: A Useful Concept for Consumer Behavior," *Journal of Consumer Research* 4.3 (December 1977), 172–5.

12   Kipp Brothers Collection (M0850). Manuscript and Visual Collections Department, William Henry Smith Memorial Library, Indiana Historical Society, Indianapolis. Figures taken from loose ledger page removed from Kipp Brothers minute book, 1891–1944, showing accounts payable and receivable for 1902.

13   Jackson Lears, *Fables of Abundance: A Cultural History of Advertising in America* (New York: Basic Books, 1994), 24, 49; H. C. Wilkinson & Co., *H. C. Wilkinson & Co.'s Illustrated Catalogue* ([New York]: n.p., ca. 1895), 26.

14   A. Coulter & Co., *Wholesale Price List*, inside front cover, [29].

15   Bennet & Co., *Bennet & Co.'s Wholesale Catalogue of 'Xmas Goods and Novelties* ([Montreal?]: n.p., 1883]).

16   Fargo Novelty Co., *Illustrated Catalogue*, 30.

17   T. Ombrello, "Resurrection Plant," http://faculty.ucc.edu/biology-ombrello/pow /resurrection_plant.htm. A compelling video of the plant coming back to life can be found here: http://vimeo.com/25485145.

18   Fargo Novelty Co., *Illustrated Catalogue*, 32.

19   Universal Distributors, *Illustrated Catalogue of Novelties* ([Stamford, CT?]: n.p., ca. 1915), [27].

20   Frederick J. Augustyn Jr., *Dictionary of Toys and Games in American Popular Culture* (New York: Routledge, 2015), n.p., entry for "Magic Rocks." Todd Coopee, "Sea-Monkeys," Toytales, posted May 12, 2015, https://toytales.ca/sea -monkeys/. Space limitations preclude giving Magic Rocks and Sea Monkeys the analysis they deserve.

21   Reproduction of advertisement at "Chia Pet Evolution" blog entry at the Chia Power website, http://www.chiativity.org/chia_pet/. Both Murro and the What-Is-It? looked to earlier theatrical traditions likely familiar to consumers at the time. For more on the learned pig, see Mizelle, "'I Have Brought My Pig to a Fine Market.'"

22   "Grass Sprouted on the Pottery Head of 'Paddy O'Hair,'" *Pittsburgh Press*, January 16, 1983.

23   Johnson Smith & Co., *Novelties* (Detroit: Johnson Smith & Co., 1947), 341.

24   Susan Stewart, *On Longing: Narratives of the Miniature, the Gigantic, the Souvenir, the Collection* (Durham, NC: Duke University Press, 1993), 80.

25   Zubeck Novelty Co., *Illustrated Catalogue and Price List of Jokers' Articles, Prizes* ([New Jersey?]: n.p., ca. 1915), [9–10].

26   Crest Trading Co., *The Crest "Fun for All" Catalogue* ([New York]: n.p., ca. 1915), 16, 21.

27   William V. Rauscher, *S. S. Adams: High Priest of Pranks and Merchant of Magic* (Oxford, CT: 1878 Press, 2002), 2.

28 Each catalog page is what graphic design expert Edward Tufte calls a "visual confection"—"an assembly of visual events . . . brought together and juxtaposed." "By means of a multiplicity of image-events," he continues, "confections illustrate an argument, show and enforce visual comparisons, combine the real and the imagined, and tell us yet another story." *Visual Explanations: Images and Quantities, Evidence and Narrative* (Cheshire, CT: Graphics Press, 1997), 121.

29 Eureka Novelty Company, "Vol. 332 Minute and Invoice Book, 1899 Nov.–1900 Nov. 9." Ohio History Connection, Columbus. The engravings were in both wood and electrotype, indicating the company purchased copies of some illustrations and commissioned original work for others. Eureka outsourced an additional $30 worth of catalogs to another printer.

30 Fargo Novelty Co., *Illustrated Catalogue*, back cover.

31 Johnson Smith is still in business and can be found at http://www.johnsonsmith .com/.

32 Quoted in Stanley Elkin, *Pieces of Soap* (New York: Simon & Schuster, 1992), 217.

33 Keith L. Eggener, "'An Amusing Lack of Logic': Surrealism and Popular Entertainment," *American Art* 7.4 (Autumn 1993), 38.

34 Quoted in Eggener, "'An Amusing Lack of Logic,'" 31.

35 For more on the cultural aspects of the pastoral ideal, see Alan Trachtenberg, *The Incorporation of America: Culture and Society in the Gilded Age* (New York: Hill & Wang, 1982; 25th anniversary ed. with a new preface, 2007); Leo Marx, *The Machine in the Garden: Technology and the Pastoral Ideal in America*, 2nd ed. (New York: Oxford University Press, 2000); and Siegfried Giedion, *Mechanization Takes Command: A Contribution to Anonymous History* (1948; Minneapolis: University of Minnesota Press, 2013).

36 Jose Rosales, "Of Surrealism & Marxism," *Blindfield Journal*, https:// blindfieldjournal.com/2016/12/01/of-surrealism-marxism/.

37 "Comic Christmas Cards, Joke-Trick Items in Demand," *Billboard* 52.49 (December 7, 1940), 56.

38 Quoted in Mary Ann Caws, ed. *Surrealism: Themes and Movements* (London: Phaidon, 2004), 84.

39 Aaron Jaffe, "Modernist Novelty," *Affirmations: Of the Modern* 1.1 (Autumn 2013), 125.

40 Jaffe, "Modernist Novelty," 107.

41 Arthur Power Dudden, "American Humor," *American Quarterly* 37.1 (Spring 1985), 8, 9. Dudden writes that much American humor is not only laced with "vulgarity and violence" but tinged with "the skeptical, the sardonic, the mocking," and even "the deliberately cruel."

42 For more on popular responses to market dupes, see Edward J. Balleisen, *Fraud: An American History from Barnum to Madoff* (Princeton, NJ: Princeton University Press, 2017); Corey Goettsch, "'The World Is but One Vast Mock Auction': Fraud and Capitalism in Nineteenth-Century America," in *Capitalism by Gaslight: Illuminating the Economy of 19th-Century America*, ed. Brian P. Luskey and Wendy A. Woloson (Philadelphia: University of Pennsylvania Press, 2015); and Wendy A. Woloson, "Wishful Thinking: Retail Premiums in Mid-Nineteenth-Century America," *Enterprise & Society* 13.4 (2012), 790–831.

43 Henry Wysham Lanier, *A Century of Banking in New York, 1822–1922* (New York: Gilliss Press, 1922), 60.

44 From George W. Peck, *Peck's Bad Boy and His Pa* (Chicago: Belford, 1883), 14, quoted in Alfred Habegger, "Nineteenth-Century American Humor: Easygoing Males, Anxious Ladies, and Penelope Lapham," *PMLA* 91.5 (October 1976), 884.

45   See also [Sharpshooter], "The Bull's Eye," *Commercial West* 13.13 (March 28, 1908), 9–10.

46   Maurice Zolotow, "The Jumping Snakes of S. S. Adams," *Saturday Evening Post*, June 1, 1946, 26.

47   Jaffe, "Modernist Novelty," 107.

48   Eureka Trick and Novelty Co., *Illustrated Manual*.

49   Fargo Novelty Co., *Illustrated Catalogue*.

50   Royal Novelty Co., *Illustrated Catalogue* ([Norwalk, CT?]: n.p., ca. 1910).

51   Gellman Bros., *Annual Buyer's Guide Catalog for 1937* (Minneapolis: Gellman Bros., 1937).

52   Royal Novelty Co., *Illustrated Catalogue*.

53   According to Dennis Hall, "Culture is not so much a set of things . . . as a process, a set of practices. Primarily, culture is concerned with the production and exchange of meanings—the 'giving and taking of meaning' between members of a society or group." "Gag Gifts: Borders of Intimacy in American Popular Culture," *Journal of American and Comparative Cultures* 24.3/4 (2001), 172, 173.

54   Even humor scholars have deemed women's humor distinct from "ordinary" humor—an aberration against which "normal" and "mainstream" (i.e., male) humor is defined. Traditionally, women are the victims, objects, and punch lines of jokes, rather than their creators or deliverers. See, e.g., Mary Crawford and Diane Gressley, "Creativity, Caring, and Context: Women's and Men's Accounts of Humor Preferences and Practices," *Psychology of Women Quarterly* 15 (1991), 217–31.

55   In 1902 H. W. Boynton observed the very different relationships that men and women have to humor and joking: "I think men are often unfair when after such [pranking] experiments, painful enough . . . , they accuse the woman of not seeing the joke. She does see it, but it does not appeal to her as the funniest thing in the world. She has heard other jokes, and is ignorant of the necessity for all this side-holding and slapping on the back. She therefore finishes her tea in quietude of spirit long before the last reminiscent detonations have ceased to echo in the masculine throat." "Books New and Old: American Humor," *Atlantic Monthly* 90 (September 1902), 418.

56   Fargo Novelty Co., *Illustrated Catalogue*; Universal Distributors, *Illustrated Catalogue of Novelties*.

57   *Secret Service* no. 691 (April 19, 1912), 29–30.

58   Zubeck Novelty Co., *Illustrated Catalogue*; C. J. Felsman, *Novelties, Jokes, Tricks, Puzzles, Magic from All Over the World and Every Where Else* (Chicago: n.p., ca. 1915).

59   Hall, "Gag Gifts," 172.

60   I. Sheldon Posen, "Pranks and Practical Jokes at Children's Summer Camps," *Southern Folklore Quarterly* 38 (1974), 302. Posen notes that many pranks in children's summer camps are of a sexual or scatological nature, and some of them, like boys stealing a girl's bra and hoisting it up a flagpole, have "a more aggressive function." "The prank was then," he continues, "a manner of conquest—however vicarious—of a sexual object, which demanded for the boys the respect of their peers and to an extent, derision for their victim" (306).

61   Gershon Legman, *Rationale of the Dirty Joke: An Analysis of Sexual Humor*, vol. 1 (New York: Simon & Schuster, 2006), 9.

62   US Patent #871,252, issued to George E. Ames, "Joke Box," filed Nov. 14, 1906, granted November 17, 1907.

63   Claimed to be "Perfectly Harmless," by the Royal Novelty Company in a 1917 advertisement in *Popular Mechanics.*

64   US Patent #887,759, issued to Julius Bing, "Trick Cigar," filed Sept. 1, 1906, granted May 19, 1908.

65   Bengor Products Co., *1936 Wholesale List of Tricks-Jokes-Novelties* ([New York]: n.p., [1936]).

66   Many novelties were devised by the German toy industry, which had been thriving for centuries. The bulk of imports from other countries—especially papier-mâché, celluloid, and tin goods from Japan—were often knock-offs of German examples. Companies like Erfurt-based J. C. Schmidt and N. L. Chrestensen specialized in ornamental plants, party supplies, games, novelties, and other weird stuff. They exported many of their wares to American mass merchandisers of cheap goods and novelty shops. See, e.g., J. C. Schmidt, *Preisbuch über Cotillon-Ball- und Scherzartikel, Saaldekorationen, Sommerfestartikel usw. Saison 1911/12* (rpt. Hildesheim, Zürich, and New York: Olms Presse, 1999); and N. L. Chrestensen, *My Dumb Traveler* (Erfurt: n.p., 1910).

67   Quoted in Mark Newgarden, *Cheap Laffs: The Art of the Novelty Item* (New York: Harry N. Abrams, 2004), 34.

68   "Banquet of New York Paper Dealers," *Paper Trade Journal*, December 12, 1912, 8.

69   Gellman Bros., *Annual Buyer's Guide Catalog for 1937*, 282, 283.

70   Zolotow, "The Jumping Snakes of S. S. Adams," 27.

71   Adams invented and/or brought to market over six hundred novelties and some of the most popular, including the Bingo Shooting Device, the Dribble Glass, and the Joy Buzzer. By the mid-twentieth century he had patented thirty-seven of them. Many, like the Joy Buzzer, were pirated nevertheless. On Adams's early life, see Rauscher, *S. S. Adams*, esp. 1–10. Rauscher claimed the active ingredient in Cachoo Powder was Dianisidine, which, according to www.reference .md, is a "highly toxic compound which can cause skin irritation and sensitization." When it was discontinued in the 1940s because of its toxicity, retailers soon began making knock-offs. "Sneezing Powder," *American Druggist and Pharmaceutical Record* 52 (January–June 1908), 119. The *Washington Post* reported "an outbreak of cachoos" in Wilmington in 1907, an unknown perpetrator having distributed the powder in a theater and a restaurant and "about the principal streets." "Starts Audience Sneezing," *Washington Post*, March 25, 1907. In 1910 a group of boys did the same in Los Angeles: "Joke of Flies," *Los Angeles Times*, May 12, 1910. The "gold dust" quote comes from Gardner Soule, "Fun's Henry Ford Is Still Inventing," *Popular Science*, January 1955, 125. Various things have been used to make itching powder, including maple seeds, rose hips, fiberglass (!), and *Mucuna pruriens*, which is no longer available on the American market. Itching powder was also eventually banished. Michael R. Albert, "Vignette in Contact Dermatology: Novelty Shop 'Itching Powder,'" *Australasian Journal of Dermatology* 39 (1998), 188. Because "racketeers" used stink bombs to "terrorize movie theaters and shopkeepers," they, too, were taken off the market. Irving D. Tressler, "Our Native Industries: I—Rubber Doughnuts—Dribble Glasses," *Life* 102.2599 (February 1935), 19.

72   On "bad goods," see Rebecca L. Walkowitz and Douglas Mao, *Bad Modernisms* (Durham, NC: Duke University Press, 2006).

73   E. J. Tangerman, "Adams' 'Ribs' Aren't Missing," *American Machinist* 90.18 (August 15, 1946), 107.

74   "Britain's Secret WWII Weapons Revealed," BBC News, October 26, 1999, http://news.bbc.co.uk/2/hi/uk_news/486391.stm.

75   *How to Be a Spy: The World War II SOE Training Manual*, introduction by Dennis Rigden (Suffolk, Eng.: Printed by St. Edmundsbury Press, 2004), 25–6.

76   "US Military Malodorant Missiles Kick Up a Stink," *New Scientist*, June 2, 2012, http://www.newscientist.com/article/mg21428676.800-us-military-malodorant-missiles-kick-up-a-stink.html.

77   Stephanie Pain, "Stench Warfare," *New Scientist* 171 (July 7, 2001), 42.

78   Johnson Smith & Co., *Supplementary Catalogue of Surprising Novelties, Puzzles, Tricks, Joke Goods, Useful Articles, Etc.* (Racine, WI: n.p., ca. 1920s), 12.

79   Soule, "Fun's Henry Ford," 125.

80   Johnson Smith & Co., *Our Most Popular Novelties. Extra! Many New Items never before shown!* (Detroit: Johnson Smith & Co., 1941).

81   "Sizzling Hot!" *Billboard* 43.49 (December 5, 1931), 93.

82   Tangerman, "Adams' 'Ribs' Aren't Missing," 107, 108. Inflation value calculated from http://www.saving.org/inflation/inflation.php?amount=1&year=1932; for prices, see http://www.thepeoplehistory.com.

83   Joel Sayre, "From Gags to Riches," *Scribner's Commentator* 9.5 (March 1941), 75–7. A practical joker all his life, Adams worked at one time for a newspaper, was a successful competitive trapshooter, and eventually became a picture frame salesman who pranked his customers with exploding cigars and other practical jokes.

84   Zolotow, "The Jumping Snakes," 26. See also Sayre, "From Gags to Riches."

85   Tressler, "Our Native Industries," 19.

86   Quoted in Elkin, *Pieces of Soap*, 218.

87   American Law Institute, *Proceedings of the 71st Annual Meeting* (Philadelphia: American Law Institute, 1994), 120–1.

88   People no doubt also made their own dirty toys; physicians' anatomical manakins were also sometimes used for prurient purposes. On the nineteenth-century trade in racy novelties, see Donna Denis, *Licentious Gotham: Erotic Publishing and Its Prosecution in Nineteenth-Century New York* (Cambridge, MA: Harvard University Press, 2009), 170; and Paul J. Erickson, "Economies of Print in the Nineteenth-Century City," in Luskey and Woloson, *Capitalism by Gaslight*.

89   See Brian P. Luskey, *On the Make: Clerks and the Quest for Capital in Nineteenth-Century America* (New York: New York University Press, 2010); and Patricia Cline Cohen, Timothy J. Gilfoyle, and Helen Lefkowitz Horowitz, *The Flash Press: Sporting Male Weeklies in 1840s New York* (Chicago: University of Chicago Press, 2008).

90   *Grand Fancy Catalogue of the Sporting Man's Bazaar for 1870* ([United States?: n.p., 1870]). Library Company of Philadelphia. "Hand one of these to Some Friend" was handwritten at the top of this flyer, indicating the sub rosa way information about these goods was circulated.

91   For examples, see the collections at the Kinsey Institute, Indiana University, https://kinseyinstitute.org/collections/index.php. In *Intimate Matters: A History of Sexuality in America*, John D'Emilio and Estelle B. Freedman write that one officer complained directly to Abraham Lincoln that pornography was "quite commonly kept and exhibited by soldiers and even officers" and that men could

purchase 12" x 15" dirty pictures for twelve cents each (Chicago: University of Chicago Press, 1997), 132–3.

92   C. W. Philo, *Philo's Army Purchasing Agency* ([Brooklyn?: n.p., ca. 1861–5]). Library Company of Philadelphia.

93   David S. Sparks, ed., *Inside Lincoln's Army* (New York: Yoseloff, 1964), 255–6, 257.

94   The Comstock Act prohibited sending "any obscene book, pamphlet, paper, writing advertisement, circular, print, picture, drawing or other representation, figure, or image on or of paper or other material, or any cast, instrument, or other article of an immoral nature" through the mail. George P. Sanger, ed., *Statutes at Large and Proclamations of the United States of America, from March 1871 to March 1873*, vol. 17, chap. 258, "An Act for the Suppression of Trade in, and Circulation of, obscene Literature and Articles of immoral Use," March 3, 1873 (Boston: Little, Brown, 1873), 598.

95   See, e.g., entries throughout Johnson Smith & Co., *Our Most Popular Novelties Extra!*; and Ardee Manufacturing Company, *Manufacturers, Importers and Jobbers of Toys Novelties and Other Mail Order Merchandise* ([Stamford, CT: Ardee Manufacturing Company, ca. 1903]), loose printed insert. "Peek-a-Boos—With, Without," *Billboard* 49.48 (November 27, 1937), 110.

96   "A Real Hot Number," *Billboard* 43.36 (September 5, 1931), 99.

97   "New! Sensational!" *Billboard* 47.30 (July 29, 1935), 101.

98   Dowst and Company, maker of pot metal novelties in Cracker Jack boxes and also the famous Tootsietoy line of toys, conceived of these toilets as dollhouse furniture, but Fishlove often repurposed them for his gag boxes.

99   On gag boxes, see Lisa Hix, "How Your Grandpa Got His LOLs," *Collectors Weekly*, August 24, 2012 (http://www.collectorsweekly.com/articles/how-your-grandpa-got-his-lols/), which features the collection of Mardi and Stan Timm.

100   Game Room, *You Won't Believe This But . . .* ([Washington, DC: n.p., 1956?]).

101   Johnson Smith & Co. *2,000 Novelties* ([Detroit]: Johnson Smith & Co, 1958).

102   Bengor Products Co., *1936 Wholesale List*.

103   Examples of these novelties can be found in any number of catalogs over time. Description of the inflatable legs appears in Johnson Smith & Co., *2,000 Novelties*.

104   Bengor's Silver Skin condoms were alleged to be poor quality. *U.S. v. 39 Gross of Rubber Prophylactics*, District of Maryland Court, Baltimore, MD, case no. 28732, "Adulteration and misbranding of rubber prophylactics," seizure date December 9, 1937, issue date November 1938, http://archive.nlm.nih.gov/fdanj/handle/123456789/60537.

105   La France Novelties, "Agents and Distributors . . . OUR HOT CARDS . . . Best Yet," *Billboard* 54.43 (October 24, 1942), 47.

106   For more, see Jim Linderman, *Smut by Mail: Vintage Graphics from the Golden Age of Obscenity*, ebook (2011), http://vintagesleaze.blogspot.com/2011/02/smut-by-mail-new-vintage-sleaze-dull.html#.U6madsflLyc.

107   *Stag FUN Package* ([Philadelphia]: n.p., ca. 1950).

108   For example, advertisements in the back pages of *Spicy Stories* 5.8 (August 1935). Similar periodicals include *Bedtime Stores*, the *Gay Parisienne*, and *French Night Life*.

109   For more on Wesley Morse, see Topps Company, *Bazooka Joe and His Gang*, introduction by Nancy Morse (New York: Harry N. Abrams, 2013).

110   Zolotow, "The Jumping Snakes," 26.

111   "Whoopee Cushions Got Their First Airing Here," *Toronto Star*, March 31, 2008.

112   Royal Novelty Co., *Illustrated Catalogue*.

113   Lisa Hix, "The Inside Scoop on Fake Barf," *Collectors Weekly*, August 23, 2011, http://www.collectorsweekly.com/articles/the-inside-scoop-on-the-fake-barf-industry/; quote from Newgarden, *Cheap Laffs*, 28.

114   Mikhail Bakhtin, *Rabelais and His World*, trans. Hélène Iswolsky (original Russian, 1965; Bloomington: Indiana University Press, 1984).

115   Simon Dickie, "Hilarity and Pitilessness in the Mid-Eighteenth Century: English Jestbook Humor," *Eighteenth-Century Studies* 37.1 (Fall 2003), 4.

116   J. A. Leo Lemay, *The Life of Benjamin Franklin*, vol. 2, *Printer and Publisher, 1730–1747* (Philadelphia: University of Pennsylvania Press, 2006), esp. 210–13.

117   Letter written by Benjamin Franklin and addressed to the Royal Academy at Brussels, 1781, Early Americas Digital Archive, http://mith.umd.edu/eada/html/display.php?docs=franklin_bagatelle2.xml.

118   According to Thomas P. Lowry, the catalog was found among the personal effects of Union soldier Pvt. Edmon Shriver, Company F, 42nd Ohio, and is in a private collection. *The Stories Soldiers Wouldn't Tell: Sex in the Civil War* (Mechanicsburg, PA: Stackpole Books, 1994), 55.

119   Palmer Gift Shop, *Jokes, Puzzles, Party Favors. Catalogue Number 4* ([Chicago]: Palmer Gift Shop, 1935), 2; Bengor Products Co., *1936 Wholesale Price List*.

120   *The Bazar Book of Decorum: The Care of the Person, Manners, Etiquette, and Ceremonials* (New York: Harper & Bros., 1870), 49, 63, 76–7. On nineteenth-century refinement, see John Kasson, *Rudeness and Civility: Manners in Nineteenth-Century Urban America* (New York: Hill & Wang, 1990), esp. chap. 6.

121   Most of these examples come from the ca. 1908 Fargo Novelty Co.'s *Illustrated Catalog*.

122   Gershon Legman, *No Laughing Matter: An Analysis of Sexual Humor*, vol. 2 (Bloomington: Indiana University Press, 1975), 812, 813.

123   Zubeck Novelty Co., *Illustrated Catalogue*, 4.

124   Crest Trading Co., *The Crest "Fun for All" Catalog* ([New York]: n.p., ca. 1915), 20.

125   Bengor Products, *1936 Wholesale List*.

126   Johnson Smith & Co., *2,000 Novelties*.

127   Zolotow, "The Jumping Snakes," 26.

128   Jack Hitt, "Sea-Monkey Fortune," *New York Times*, April 15, 2016. The inventor of Sea Monkeys, Harold von Braunhut, also invented X-Ray Spex, among other toys and novelties. Many of today's Sea Monkeys are Chinese knock-offs.

129   "Chia Pets," https://www.chia.com/chia-pets/about-chia-pets/.

130   "Chia Pets Often Euthanized after Novelty Wears Off," *Derf Magazine*, http://www.derfmagazine.com/news/business/chiapet-euthanized.

131   Quoted in Newgarden, *Cheap Laffs*, 126.

Epilogue

1   "Happy with crappy" comes from James Fallows, *China Airborne* (New York: Pantheon, 2012), quoted in Ian Johnson, "China's Lost Decade," *New York Review of Books*, September 27, 2012, 82.

2   On the history of recycling, see Susan Strasser, *Waste and Want: A Social History of Trash* (New York: Henry Holt, 1999). For a more theoretical take on trash, see Michael Thompson, *Rubbish Theory: The Creation and Destruction of Value* (Oxford: Oxford University Press, 1979).

3   Doyle Rice, "World's Largest Collection of Ocean Garbage Is Twice the Size of Texas," *USA Today*, March 22, 2018, https://www.usatoday.com/story/tech /science/2018/03/22/great-pacific-garbage-patch-grows/446405002/. See also the Ocean Cleanup Foundation, https://www.theoceancleanup.com/. For air pollution, see David G. Streets et al., "Modeling Study of Air Pollution Due to the Manufacture of Export Goods in China's Pearl River Delta," *Environmental Science and Technology* 40.7 (April 2006), 2099–107; and Rob Hengeveld, *Wasted World: How Our Consumption Challenges the Planet* (Chicago: University of Chicago Press, 2012).

4   "Hong Kong, Thanks to China, Tops Toy Market," *Australian Financial Review*, January 18, 1989.

5   Peter Hessler, "China's Instant Cities," *National Geographic* 211.6 (June 2007), 88–117. On outsourcing and factory zones, see Naomi Klein, *No Logo*, 10th anniversary ed. (New York: Picador, 2009); Ellen Ruppel Shell, *Cheap: The High Cost of Discount Culture* (New York: Penguin, 2009); and Elizabeth Cline, *Overdressed: The Shockingly High Cost of Cheap Fashion* (New York: Portfolio/Penguin, 2012).

6   Hessler, "China's Instant Cities."

7   Hessler, "China's Instant Cities."

8   Steven Husted and Shuichiro Nishioka, "China's Fare Share? The Growth of Chinese Exports in World Trade," *Review of World Economies* 149.3 (2013), 565–6. In 2010, 16.7 percent of all merchandise imports to the United States came from China (56–9), Jianqing Ruan and Xiaobo Zhang, "Low-Quality Crisis and Quality Improvement: The Case of Industrial Clusters in Zhejian Province," in *Industrial Districts in History and the Developing World*, ed. Tomoko Hashino and Keijiro Otsuka, (Singapore: Springer Science+Business Media, 2016), 170. Peter K. Schott, "The Relative Sophistication of Chinese Exports," *Economic Policy* 23.1 (January 2008), 21, 35.

9   Schott, "The Relative Sophistication of Chinese Exports," 10.

10   A small sample of literature on anti-chain-store issues includes Alfred G. Buehler, "Anti-Chain-Store Taxation," *Journal of Business of the University of Chicago* 4.4 (October 1931), 346–69; Hugh A. Fulton, "Anti-Chain Store Legislation," *Michigan Law Review* 30.2 (December 1931), 274–79; Edward W. Simms, "Chain Stores and the Courts," *Virginia Law Review* 17.4 (February 1931), 313–24; Esther M. Love, *Operating Results of Limited Price Variety Stores in 1948: Chains and Independents* (Boston: Harvard Graduate School of Business Administration, 1949); and F. J. Harper, "'A New Battle on Evolution': The Anti-Chain Store Trade-At-Home Agitation of 1929–1930," *Journal of American Studies* 16.3 (December 1982), 407–26.

11   Harper, "'A New Battle,'" 410n7.

12   "German Porcelain Maker Halts Hummel Production," *Deutsche Welle*, June 19, 2008.

13   Bess Lovejoy, "How SkyMall Captured a Moment of Technological and American History," *Smithsonian*, January 27, 2015, https://www.smithsonianmag.com /travel/how-skymall-captured-moment-technological-and-american-history -180954043/?no-ist.

14   Datamonitor, *Dollar Tree Stores Company Profile*, August 24, 2007, 7. On the merger of Dollar Tree and Family Dollar, see Shawn Tully, "How the Dollar Store War Was Won," *Fortune*, April 23, 2015, www.fortune.com. For general profiles of each company, see Jim Piller and John S. Strong, "The High Price of Dollar Stores: Dollar Tree and Dollar General Battle for Family Dollar," http://www.babson.edu

/executive-education/thought-leadership/strategy-innovation/Pages/the-high-price-of-dollar-stores.aspx. See also Barbara Kahn et. al., "Consumer and Managerial Goals in Assortment Choice and Design," *Marketing Letters* 25.3 (September 2014), 293.

15   MarketLine, *Company Profile Dollar Tree, Inc.*, January 8, 2016, 4, https://www.marketline.com, reference code E7F38462-EF1E-4309-904C-496EA358DBA1; Datamonitor, *Dollar Tree Stores Company Profile*, 4.

16   Datamonitor, *Family Dollar Stores Company Profile*, July 21, 2008, 4; MarketLine, *Company Profile Family Dollar Stores, Inc.*, July 1, 2015, 3, https://www.marketline.com, reference code 9EC011A2-C7DA-4A45-B933-E74BC6E64EBB.

17   In 2016 Dollar Tree reported that approximately 40 percent of its merchandise was imported, mostly from China due to low labor costs. MarketLine, *Company Profile Dollar Tree, Inc.*, 10.

18   Datamonitor, *Family Dollar Stores Company Profile*, 6; MarketLine, *Company Profile Family Dollar Stores, Inc.*, 6; MarketLine, *Company Profile Family Dollar Stores, Inc.*, 6; Datamonitor, *Dollar General Corporation Company Profile*, January 31, 2011, 7.

19   Beth Macy, *Factory Man: How One Furniture Maker Battled Offshoring, Stayed Local—and Helped Save an American Town* (New York: Little, Brown, 2014); James Fallows and Deborah Fallows, *Our Towns: A 100,000-Mile Journey into the Heart of America* (New York: Pantheon, 2018).

20   Lori G. Kletzer, "Trade Related Job Loss and Wage Insurance: A Synthetic Review," *Review of International Economics* 12.5 (2001), 724–48; Barbara Ehrenreich, *Nickel and Dimed: On (Not) Getting by in America* (New York: Henry Holt, 2001).

21   On Walmart, see, e.g., Greg Spotts, *Wal-Mart: The High Cost of Low Price* (New York: Disinformation Company, 2005); Anthony Bianco, *The Bully of Bentonville: How the High Cost of Wal-Mart's Everyday Low Prices Is Hurting America* (New York: Currency, 2006); Stacy Mitchell, *Big-Box Swindle: The True Cost of Mega-Retailers and the Fight for America's Independent Businesses* (Boston: Beacon, 2006).

22   Datamonitor, *Family Dollar Stores SWOT Analysis*, 2008–15; MarketLine, *Dollar Tree SWOT Analysis*, 2017.

23   It would be quixotic to attempt to provide a comprehensive account of studies of material culture, consumption, and identity. Key works include Karl Marx, *Capital*, vol. 1, *A Critique of Political Economy* (1867; New York: Penguin Classics, 1992); Thorstein Veblen, *The Theory of the Leisure Class: An Economic Study of Institutions* (1899; New York: Modern Library, 1934); Roland Barthes, *Mythologies*, trans. Annette Lavers (New York: Hill & Wang, 1972); Mary Douglas and Baron Isherwood, *The World of Goods: Toward an Anthropology of Consumption* (New York: Basic Books, 1979); Mihaly Csikszentmihalhyi and Eugene Rochberg-Halton, *The Meaning of Things: Domestic Symbols of the Self* (Cambridge: Cambridge University Press, 1981); Pierre Bourdieu, *Distinction: A Social Critique of the Judgment of Taste* (original French, 1979; London: Routledge Classics, 2010); Arjun Appadurai, ed., *The Social Life of Things: Commodities in Cultural Perspective* (Cambridge: Cambridge University Press, 1986); Colin Campbell, *The Romantic Ethic and the Spirit of Modern Consumerism* (Oxford: Blackwell, 1987); Russell W. Belk, "Possessions and the Extended Self," *Journal of Consumer Research* 15.2 (September 1988), 139–68; Grant D. McCracken, *Culture and Consumption II: Markets, Meaning, and Brand Manage-*

ment (Bloomington: Indiana University Press, 2005); and Russell W. Belk, "The Extended Self in a Digital World," *Journal of Consumer Research* 40.3 (October 2013), 477–500.

24   Ben Guarino, "Ikea to Settle for $50 Million after Its Dressers Tipped Over, Killing Three Young Boys," *Washington Post*, December 22, 2016, https://www .washingtonpost.com/news/morning-mix/wp/2016/12/22/ikea-to-settle-for -50-million-after-its-dressers-tipped-over-killing-three-young-boys/?noredirect =on&utm_term=.78e0b4b93593; Oliver Roeder, "The Weird Economics of Ikea," *FiveThirtyEight*, October 21, 2016, https://fivethirtyeight.com/features/the-weird -economics-of-ikea/.

25   Bessie Nestoras, "High Style: Sophisticated, Design-Driven American Art Glass," *Gifts & Decorative Accessories* 106.6 (June 2005), 46.

26   Wanda Jankowski, "Trends in Crafts Retailing," *Gifts & Decorative Accessories* 97.3 (March 1996), 40; Michelle Nellett, "Individuality Counts Here," *Gifts & Decorative Accessories* 93.6 (June 1992), 120.

27   Nestoras, "High Style."

28   Jankowski, "Trends in Crafts Retailing."

29   Promotional Products Association International, *PPAI 2015 Annual Distributor Sales Summary* ([Irving, TX?]: Promotional Products International, 2016), 1, www .ppai.org/members/research/.

30   Promotional Products Association International website, www.ppai.org/about /association-history; Jane Von Bergen, "Norman Cohn: At 83, the 'Sultan of Swag' Is Still Go-Go on the Logo," *Philadelphia Inquirer*, October 30, 2016.

31   Promotional Products Association International (PPAI), *Mapping Out the Modern Consumer*, 2016, https://advocate.ppai.org/Documents/PPAI%202017 %20Consumer%20Study%20Report.pdf, 7–9. People receiving free stuff reported feeling happy (71 percent), interested (52 percent), thankful (46 percent), appreciated (43 percent), and special (28 percent).

32   Henry Bunting, *Specialty Advertising: A New Way to Build Business* (Chicago: Novelty News Press, 1914), 29.

33   Alexandra Lewin, Lauren Lindstrom, and Marion Nestle, "Commentary: Food Industry Promises to Address Childhood Obesity: Preliminary Evaluation," *Journal of Public Health Policy* 27.4 (2006), 331, 338.

34   Including Disney mini dolls and cameras and a wide variety of *Star Trek* merchandise, such as dinner plates, movie passes, flash-drive wristbands/bracelets, t-shirts, and badges. Other items: cereal-to-go cups; beverage cups and sippy cups; alarm clocks; cash cards; children's books and audiobooks; toy cookware sets; costumes based on brand spokescharacters; Lego toys; NASCAR toy race cars and other model cars; Shrek squirters and highlighters; pencil toppers; t-shirt iron-ons; stickers; tattoos; fortunes (printed on fruit snacks); music downloads; ringtones; horoscopes; discounted DVDs and movie-related merchandise; children's magazines; t-shirts; soccer jerseys; soccer balls; sweat bands; towels; frisbees; basketballs and hoop stands; water bottles; key chains; whistles and noisemakers; backpacks; trading cards; flash cards; flash drives; magnets; stuffed animals; Halloween trick-or-treat bags; coloring and activity books; fanzines (in connection with movie promotions); school supplies; numerous small toys and games—many connected to current movies; and myriad downloadable items. Federal Trade Commission, *A Review of Food Marketing to Children and Adolescents*, 79. On the relative effectiveness of children's premiums, see Tali Te'Eni-Harari, "Sales Promotion, Premiums, and Young People in the 21st Century," *Journal of Promotion Management* 14 (2008), 17–30. For an extensive

study on advergaming, see Elizabeth S. Moore, *It's Child's Play: Advergaming and the Online Marketing of Food to Children* (Menlo Park, CA: Henry J. Kaiser Family Foundation, July 2006).

35   Rob Neyer, "The Question? To Bobble, or Not to Bobble," July 3, 2015, Fox Sports, http://www.foxsports.com/mlb/just-a-bit-outside/story/bobblehead -giveaways-numbers-brewers-tigers-dodgers-policy-070315.

36   John V. Petrof, "Relationship Marketing: The Wheel Reinvented?," *Business Horizons* 40.6 (November–December 1997), 29.

37   David Cheal, "'Showing Them You Love Them': Gift Giving and the Dialectic of Intimacy," *Sociological Review* 35.1 (February 1987), 159, 166.

38   In 2001 Pfizer spent approximately $86 million on promotional products. R. Stephen Parker and Charles E. Pettijohn, "Direct-to-Consumer Advertising and Pharmaceutical Promotions: The Impact on Pharmaceutical Sales and Physicians," *Journal of Business Ethics* 48.3 (December 2003), 284.

39   For a full definition of "relationship marketing," see Josie Fisher, "Business Marketing and the Ethics of Gift Giving," *Industrial Marketing Management* 36 (2007), 99–108. The *Journal of the Academy of Marketing Science* published a special issue on "relationship marketing" in 1995.

40   Lou Harry and Sam Stall, *As Seen on TV: 50 Amazing Products and the Commercials That Made Them Famous* (Philadelphia: Quirk Books, 2002), 18, 32, 40, 59; German Electric Belt Agency, *The German Electric Belts and Appliances* ([Brooklyn?]: n.p., ca. 1887), 11.

41   Amazon user reviews by "Scott Fraser," September 7, 2011; "H. Baril," January 26, 2013; "Anonymous Customer," June 5, 2017; "Thomas Peterson," January 6, 2014; and "Amazon Customer," October 3, 2016, https://www.amazon.com /Loftus-Looks-Real-Fake-Barf/product-reviews/B0006GKGXE/ref=cm_cr_arp _d_paging_btm_2?ie=UTF8&reviewerType=all_reviews&pageNumber=2.

INDEX